ADVANCES IN SPACE RESEARCH

The Official Journal of the Committee on Space Research (COSPAR)
A Scientific Committee of the International Council of Scientific Unions (ICSU)

VOLUME 20, NUMBER 10

LIFE SCIENCES: LIFE SUPPORT SYSTEM STUDIES-I

This volume of Advances in Space Research is
dedicated to the Memory of
Dr Willy Z. Sadeh

Symposium F4.6

Program Committee

P. Loretan, U.S.A.
J. L. Garland, U.S.A.
R. M. Wheeler, U.S.A.
A. Trotman, U.S.A.
J. I. Gitelson, Russia
M. André, France

Symposium F4.8

Program Committee

T. W. Tibbitts, U.S.A.
N. S. Pechurkin, Russia
R. Strayer, U.S.A.

Symposium F4.2

Program Committee

S. S. Nielsen, U.S.A.
P. Chagvardieff, France
F. B. Salisbury, U.S.A.
R. M. Wheeler, U.S.A.
V. Rygalov, Russia
Y. Tako, Japan
D. Verger, France

Symposium F4.9

Program Committee

C. A. Mitchell, U.S.A.
W. Z. Sadeh, U.S.A.
R. D. MacElroy, U.S.A.
D. Auslander, U.S.A.

LIFE SCIENCES: LIFE SUPPORT SYSTEM STUDIES-I

Proceedings of the F4.6, F4.8, F4.2 and F4.9 Symposia of COSPAR Scientific Commission F which were held during the Thirty-first COSPAR Scientific Assembly, Birmingham, U.K., 14–21 July 1996

Edited by

R. M. WHEELER

Biomedical Operations Office, Mail Code JJ-G, NASA, John F. Kennedy Space Center, FL 32899, U.S.A.

J. L. GARLAND

Dynamac Corporation, Mail Code DYN-3, NASA, John F. Kennedy Space Center, FL 32899, U.S.A.

T. W. TIBBITTS

Department of Horticulture, University of Wisconsin, 1575 Linden Drive. Madison WI 53706-1590, U.S.A.

S. S. NIELSEN

Department of Food Science, Purdue University, 1160 Smith Hall, West Lafayette IN 47907-1160, U.S.A.

and

C. A. MITCHELL

NASA Specialized Center of Research and Trainning (NSCORT) in Bioregenerative Life Support, Purdue University, 1165 Horticulture Building, West Lafayette IN 47907-1165, U.S.A.

Published for

THE COMMITTEE ON SPACE RESEARCH

PERGAMON

U.K. Elsevier Science Ltd, The Boulevard, Langford Lane, Kidlington, Oxford OX5 1GB, U.K.

U.S.A. Elsevier Science Inc., 660 White Plains Road, Tarrytown, New York 10591-5153, U.S.A.

JAPAN Elsevier Science Japan, Tsunashima Building Annex, 3-20-12 Yushima, Bunkyo-ku, Tokyo 113, Japan

---

First edition 1997

ISBN: 9780080433073

In order to make this volume available as economically and as rapidly as possible the author's typescript has been reproduced in its original form. This method unfortunately has its typographical limitations but it is hoped that they in no way distract the reader.

Printed and bound in Great Britain by
CPI Antony Rowe, Chippenham and Eastbourne

CONTENTS

PRODUCTION, PROCESSING AND WASTE RECYCLING IN A CELSS

PRODUCTION, PROCESSING AND WASTE RECYCLING IN A CELSS

Proceedings of the F4.6 Symposium of COSPAR Scientific Commission F which was held during the Thirty-first COSPAR Scientific Assembly, Birmingham, U.K., 14–21 July 1996

Edited by

R. M. WHEELER

Biomedical Operations Office, Mail Code JJ-G, NASA, John F. Kennedy Space Center, FL 32899, U.S.A.

and

J. L. GARLAND

Dynamac Corporation, Mail Code DYN-3, NASA, John F. Kennedy Space Center, FL 32899, U.S.A.

Pergamon

Adv. Space Res. Vol. 20, No. 10, p. 1799, 1997
Published by Elsevier Science Ltd on behalf of COSPAR
Printed in Great Britain
0273-1177/97 $17.00 + 0.00

PII: S0273-1177(97)00843-0

PREFACE

COSPAR Session F4.6, "Production, Processing, and Waste Recycling in a CELSS", covered a range of topics in bioregenerative life support research. Presentations included results from studies using plants for atmospheric regeneration, along with studies of plant responses to conditions in closed habitats, similar to what might occur in space. Presentations also included unique findings from waste treatment and recycling studies, where plant production systems were linked to, and sustained by inputs from waste streams. An additional paper reported on exploratory studies to use extraction techniques to produce soybean oil for CELSS applications. The session was organized by Dr. Phil Loretan from Tuskegee University, Alabama, USA, and co-chaired by Dr. Audry Trotman, also of Tuskegee University. The contributions of the following manuscript referees are gratefully acknowledged: D.L. Bubenheim, G.D. Goins, J.Y. Lu, C.L. Mackowiak, O. Monje, D. Mortley, M.T. Patterson, G.W. Stutte, A. Trotman, T.W. Tibbitts, and K. Wignarajah.

Pergamon

Adv. Space Res. Vol. 20, No. 10, pp. 1801–1804, 1997

Printed in Great Britain
0273-1177/97 $17.00 + 0.00

PII: S0273-1177(97)00843-0

DIRECT UTILIZATION OF HUMAN LIQUID WASTES BY PLANTS IN A CLOSED ECOSYSTEM

G. M. Lisovsky, J. I. Gitelson, M. P. Shilenko, I. V. Gribovskaya and I. N. Trubachev

Institute of Biophysics-Russian Academy of Sciences, Siberian Branch, Krasnoyarsk 660036, Russia

ABSTRACT

Model experiments in phytotrons have shown that urea is able to cover 70% of the demand in nitrogen of the conveyer cultivated wheat. At the same time wheat plants can directly utilize human liquid wastes. In this article by human liquid wastes the authors mean human urine only. In a long-term experiment on "man-higher plants" system with two crewmen, plants covered 63 m^2, with wheat planted to - 39.6 m^2. For 103 days, complete human urine (total amount - 210.7 l) wassupplied into the nutrient solution for wheat. In a month and a half NaCl supply into the nutrient solution stabilized at 0.9-1.65 g/l. This salination had no marked effect on wheat production. The experiment revealed the realistic feasibility to directly involve liquid wastes into the biological turnover of the life support system. The closure of the system, in terms of water, increased by 15.7% and the supply of nutrients for wheat plants into the system was decreased.

Closedness of biological turnover of matter in a man-made "man - higher plants" ecological system might involve, among other processes, direct utilization of human liquid wastes by plants. The amount of urine comprises 15-20% of the total amount of water cycling within the system including water as part of food, household, hygiene and potable water necessary for man. What is more, it they contains most nitrogen-bearing compounds emitted by man, almost all of the NaCl and some other substances involved in the biological turnover.

Human liquid wastes can be utilized either by preliminary physical-chemical treatment (evaporating or freezing out the water, finally oxidizing the organic matter, isolating the mineral components required for plants, etc.) and further involvement of the obtained products or by direct application into the nutrient solution for plants.

The challenge of direct utilization is that plants have no need of Na^+ and Cl^-, and also the organic forms of nitrogen emitted by man cannot fully meet the demand of plants forthis element. Besides, hygienic and/or psychological reasons make it desirable to avoid direct use of liquid wastes in the nutrient solutions that would have direct contact with edible part of plants (tubers, roots, bulbs).

Feasibility of direct utilization of liquid wastes by plants in a closed "man - higher plants" ecosystem has been experimentally studied on wheat - grain culture as a model plant with the edible part in the form of seeds spatially dissociated with the nutrient medium. The wheat covered 60-65% of the area under higher plants. The studies have been carried out in "Bios-3"experimental facility described in detail elsewhere (Lisovsky, 1979; Gitelson *et al.*, 1989).

Preliminary experiments on conveyer cultivation of plants of wheat have been conducted in "Bios-3" without closing the system. The major components of human liquid wastes - urea and sodium chloride were added daily to the nutrient medium for the wheat plants of different age - from sprouts to seed formation - was daily added. Older plants were transferred onto pure water obtained from the transpiration moisture condensate without addition of mineral components and urine (Lisovsky, 1979). It has been experimentally found that with nitrate nitrogen completely substituted in the nutrient medium for urea the plants' condition deteriorated and the yield decreased. With the part of urea nitrogen in the total nitrogen nutrition of wheat not more than 70-75% and NaCl concentration no greater than 2 g/l the condition of plants did not deteriorate and their productivity did not decrease.

With sodium chloride introduced daily into the wheat nutrient medium at the ratio of 10-12 g of NaCl (daily consumption of the table salt by one man) per 0.6-0.7 kg of the dry biomass synthesized by wheat (providing for the daily demand of man in oxygen), the NaCl concentration in the nutrient medium gradually increased to stabilize at about 2 g/l. This is direct evidence of the equilibrium established between NaCl supply and its removal from the medium as part of the wheat biomass. This also demonstrated that this degree of salinity in the solution that did not bring about a decrease in plant productivity. NaCl content in the inedible part of the dry biomass of wheat increased to 1.5-1.9% (dry biomas), at this.

Having succeeded with applying urea and NaCl to the nutrient medium, we conducted experiments to introduce human urine with equal amounts of NaCl and urea into the medium. These experiments yielded negative results. On the 10th-12th day after the beginning of urine application the younger plants started to yellow, the stems decelerated their growth in height and the roots - in length. However, in the following experiments with plant growth the schedule for feeding urine into the medium was altered with additions made 2-3 times each day and plants were grown on the altered nutrient solution during the entire vegetation period. The daily amount of urine from one crewman (1000-1300 ml) is introduced into 200-220 l of the nutrient medium required for 16-18 m^2 of the area under wheat capable of synthesizing daily 600-700 g of dry biomass in "Bios-3" experimental facility conditions. Two long-term experiments (150 and 190 days) conducted with conveyerculture of wheat in the phytotrons without closure demonstrated that the application of human urine into the nutrient solution did not make the wheat yield decrease of This mode of growing increased the capacity of the maturing plants to passively absorb NaCl from the solution.

The biochemical composition of seeds of wheat grown on the purely mineral Knoppe medium and on the medium with addition of human urine was practically identical (Table). Roots of maturing plants have been found to build up large amounts of sodium chloride (1.61% of dry weight), smaller amounts have been found in the straw (0.54%) and trace amounts (0.003%) - in the seeds.

These results provided new prospects to conduct in "Bios-3" biofacility a long-term experiment on "man - higher plants" system with regeneration of atmosphere, water and plant food. For 5 months the system accommodated two crewmen and higher plants, of which wheat covered 40 m^2 and other cultures (sedge-nut, carrots, beet-roots, radish, etc.) - 20 m^2 of the sowing area.

From the first days of experiment the nutrient solution for wheat was added all sanitary, hygienic and household water from the crew, from the middle of the second month of the experiment, - human urine was added. The entire amount of liquid wastes of two crewmen - 1.5 to 2.5 liters daily - was introduced into the nutrient solution for wheat. In 103 days, from the start of utilization to the end of experiment, the urine amounted to 210.7 liters.

Table. Biochemical composition of wheat seeds grown on Knoppe nutrient medium (I) and on the medium with Knoppe nutrient medium and human urine (II).

Substance	Content, % of dry biomass	
	I	II
Indispensable amino acids	3.83	3.70
Total amino acids	15.40	15.03
Water soluble sugars	4.1	4.2
Starch	63.7	64.3
Cellulose	2.6	2.7
Riboflavin (B_2), mg/%	0.40	0.35
Thiamin (B_1), mg/%	0.49	0.45
Uric acid	absent	absent
Creatinine	absent	absent

The application of urine into the nutrient medium for wheat made the medium gradually build up sodium chloride, both elements of which are not indispensable for plants (Fig. 1). Sodium chloride concentration in the medium increased concurrently with the content in the wheat biomass, mainly due to passive absorption by the maturing plants. As a result, the supply of sodium chloride into the nutrient medium from the crewmen and the increasing removal with the biomass reached an equilibrium. Thus, during the first 4 week period from the beginning of the experiment, sodium chloride concentration in the nutrient medium increased by 512 mg/l, during the second 4 week period, - by 603 mg/l, during the third 4 week period by 136 mg/l only, and during the last period by 68 mg/l, thus by the last period; so, supply of sodium chloride into the medium with urine and its removal with the wheat biomass were practically reaching the equilibrium. Sodium chloride content in the wheat biomass (on absolutely dry biomass basis) increased in this time from trace amounts (tenths and hundredths of per cent) during the control period to 3.50-4.05% in the roots, 1.12 - 1.37% in the straw and only 0.01% in the seeds by the end of experiment. This "salination" of the nutrient medium and the wheat biomass had no considerable effect on wheat productivity.

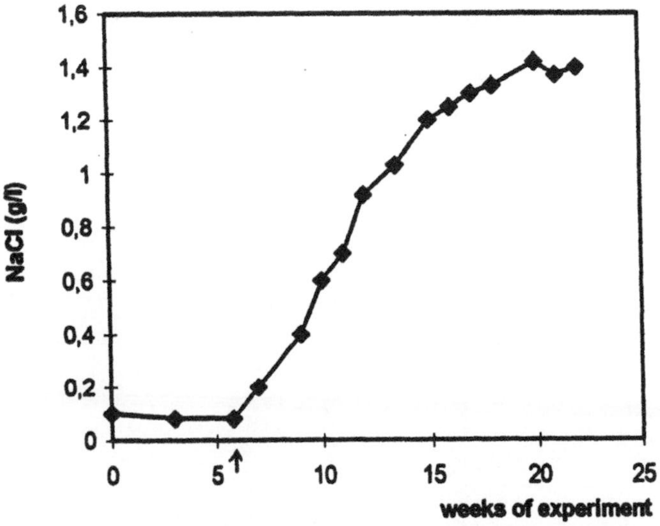

Fig. 1. Sodium chloride accumulation in the nutrient medium for wheat. ↑ is the time the human liquid wastes began to be introduced.

Thus, the experiment on direct utilization of human urine by wheat plants in a closed system has demonstrated the following:

a) feasibility to reduce the daily demand of the system for water from external sources by 2.046 l; essentially - by 15.7%; - enhancing the closedness of the system

b) feasibility to utilizemineral components of urine which make possible reduce consumption of salts, acids and alkali required for nutrient media for the wheat. In terms of individual elements this reduction was: for nitrogen - 70%, for potassium and sulfur - 20-26%, for phosphorus and calcium - 10-12%;

c) to avoid special physical-chemical facilities and technologies to treat liquid human wastes in a bioregenerative life support system.

The many months-long experiment on direct utilization of liquid wastes in a closed "man - higher plants" system has, on the whole, demonstrated this method to be acceptable for biological regeneration in human habitats in CES.

REFERENCES

Gitelson I.I., I.A. Terskov, B.G. Kovrov, G.M. Lisovsky, Yu.N. Okladnikov *et al*, Long-Term Experiments on Man Stay in Biological Life-Support System. Adv, Spase Res., 9, (8)65 (1989).

Lisovsky G.M. (Edit) *Closed System: Man - Higher Plants.* Publ. H. "Nauka", Novosibirsk (in Russian), (1979).

Pergamon

Adv. Space Res. Vol. 20, No. 10, pp. 1805–1813, 1997

Printed in Great Britain
0273-1177/97 $17.00 + 0.00

PII: S0273-1177(97)00845-4

INTEGRATING BIOLOGICAL TREATMENT OF CROP RESIDUE INTO A HYDROPONIC SWEETPOTATO CULTURE

A. A. Trotman, P. P. David, C. K. Bonsi, W. A. Hill, D. G. Mortley and P. A. Loretan

Tuskegee University NASA Center for CELSS, Tuskegee University, Tuskegee, AL 36088, U.S.A.

ABSTRACT

Residual biomass from hydroponic culture of sweetpotato [*Ipomoea batatas* (L.) Lam.] was degraded using natural bacterial soil isolates. Sweetpotato was grown for 120 days in hydroponic culture with a nutrient solution comprised of a ratio of 80% modified half Hoagland solution to 20% filtered effluent from an aerobic starch hydrolysis bioreactor. The phytotoxicity of the effluent was assayed with 'Waldmann's Green' lettuce (*Lactuca sativa* L.) and the ratio selected after a 60-day bioassay using sweetpotato plants propagated vegetatively from cuttings. Controlled environment chamber experiments were conducted to investigate the impact of filtrate from biological treatment of crop residue on growth and storage root production with plants grown in a modified half Hoagland solution. Incorporation of bioreactor effluent, reduced storage root yield of 'Georgia Jet' sweetpotato but the decrease was not statistically significant when compared with yield for plants cultured in a modified half Hoagland solution without filtrate. However, yield of 'TU-82-155' sweetpotato was significantly reduced when grown in a modified half Hoagland solution into which filtered effluent had been incorporated. Total biomass was significantly reduced for both sweetpotato cultivars when grown in bioreactor effluent. The leaf area and dry matter accumulation were significantly ($P < 0.05$) reduced for both cultivars when grown in solution culture containing 20% filtered effluent.

INTRODUCTION

Ipomoea batatas (L.) Lam. (sweetpotato) is one of several crops selected by the U. S. National Aeronautics and Space Administration (NASA) with potential for use in a CELSS (Hill et al., 1984). The sweetpotato has been extensively studied in controlled environments and has great potential for supplying carbohydrates, vitamins, antioxidants and a retinue of menu items to meet the demands of long duration human exploration of space (Hill *et al.*, 1992; Bonsi *et al.*, 1992). The outcome from degradation of inedible crop biomass is conversion of organic wastes into inorganic salts, reduced organic compounds, microbial biomass and harmless by-products of microbial metabolism such as H_2O, CO_2, CH_4 (Bayer and Lamed, 1992). The reuse of nutrients is of considerable importance in a bioregenerative approach to crop production (Finger and Alazraki, 1995). The development of a nutrition management strategy that incorporates nutrients recycled from senesced and harvested biomass using a biological approach to sustainability in a controlled environment agriculture scheme is desirable (Strayer and Cook, 1995). The controlled environment experiments reported in this paper examined the effects of incorporating filtered (0.2 µm) bioreactor effluent into the crop nutrient medium for the hydroponic growth and storage root production of sweetpotato. This research integrated waste resource recycling and crop production and provided preliminary data for establishing guidelines and alternatives for realizing a sustainable life support system required for long duration manned space endeavors.

METHODS

Biodegradation of Sweetpotato Biomass

Harvest biomass from growth chamber sweetpotato experiments was ground and sieved (75 µm) before use in biodegradation. The biodegradation of sweetpotato biomass was achieved under continuous stirring using a 45-L Wheaton Turbo-Lift Bioreactor (346450, Fisher Scientific, Atlanta, GA), incubated at room temperature (30 ± 2°C). Oxygen content was monitored and sterile air (0.22 µm Whatman filter, Millipore Co., Bedford, MA) added continuously with an aquarium pump to maintain aerobic conditions in the bioreactor. Sterile (121°C; 15 min.; 1.0×10^5 Pa— 3 cycles), ground (75 µm) harvest biomass was added to the bioreactor at a fresh weight of 100 g (weighed after autoclaving). A basal salts medium (28 L) containing the following salts per liter: 0.5 g $MgCl_2$; 0.5 g K_2HPO_4; 0.5 g $NaNO_3$; 0.5 g KCl; 0.004 g $FeSO_4$, was prepared as previously described (Trotman *et al.*, 1996). The basal culture medium was sterilized by autoclaving (121°C; 20 min.; 1.0 x10^5 Pa) in 4- and 8-L batches and then transferred to the bioreactor. The bioreactor loading rate was effectively 20 percent (dry weight/volume). The bioreactor was inoculated with 1.5 L of a 26-hour-old culture of natural bacteria,—WDSt 3A, previously isolated from sweetpotato fields (Trotman *et al.*, 1996). At 14-day intervals, 20 L of the liquid fraction of the bioreactor was siphoned off using a peristaltic pump (Vera Varistaltic Pump Plus, # 72-305-000, Fisher Scientific, Atlanta, GA) and three sterile in-line filters (Whatman 2 --> 0.45 µm --> 0.22 µm). Care was taken and aseptic practices observed to ensure that the content of the bioreactor was not contaminated. After each harvest, the bioreactor was re-seeded with 2.0 L of a 24-h WDSt 3A culture and 18 L of basal salts medium. A new batch of sterile, sweetpotato harvest foliage (100 g) was added to ensure filtrate quality in succeeding bioreactor effluent harvests. The experiment was sampled (20 mL aliquots) at inoculation and at 14-day intervals for the duration of the study. Measurements of pH, TDS, NO_3, total organic carbon and total viable counts were completed on samples of effluent. The filtered bioreactor effluent was diluted with modified half-strength Hoagland (MHH) solution at a 1 part filtrate to 4 parts MHH. The diluted (20% filtrate) effluent was used as a crop nutrient medium in the controlled environment growth studies with sweetpotato.

Phytotoxicity Assay (Lettuce & Sweetpotato)

Phytotoxicity tests were conducted to determine if filtrate from microbial degradation of sweetpotato harvest biomass was phytotoxic and at what levels when used as a growth medium. For both the lettuce (*Lactuca sativa* L.) and sweetpotato [*Ipomoea batatas* (L.) Lam.] cuttings studies, filtered effluent (filtrate) was obtained from laboratory scale 4-L aerobic bioreactors in which rates of biodegradation of sweetpotato biomass using two bacterial inocula (WDSt 3A and WLSt 7A—bacterial soil isolate with high amylolytic potential) were compared. At harvest, total plant biomass was dried (75°C) in an oven (IsoTemp, 650G, Fisher Scientific, Atlanta, GA) and weights recorded after 48 h. The percent decrease in total dry biomass/nutrient treatment from the phytotoxicity studies was used in selecting the filtrate ratio of filtered bioreactor effluent that would be incorporated in MHH solution for subsequent 120-day growth chamber studies with sweetpotato. The control solution in each crop growth experiment was a modified (N:K = 1: 2.4) one-half strength Hoagland—MHH (Mortley *et al.*, 1993).

In a growth chamber study, lettuce plants were grown from 'Waldmann's Green' seeds for 30 days (April 27 - May 26, 1995) at temperatures of 18°C (light)/ 12°C (dark) and a 12-h photoperiod with 400 µmol m^{-2} s^{-1} light. Plants were grown in 50-mL polypropylene bottles with varying ratios of MHH solution. Four seeds were laid on filter paper funnels placed in each bottle with 30 mL of the nutrient medium treatment. Filtrate:MHH-ratio treatments were: (1) 100% MHH solution plus 0% filtrate (100 mL); (2) 80% MHH solution plus 20% WDSt 3A filtrate (100 mL); (3) 50% MHH solution plus 50% WDSt 3A filtrate (100 mL); (4) 20% MHH solution plus 80% WDSt 3A filtrate (100 mL); (5) 0% MHH solution plus 100% WDSt 3A filtrate (100 mL); (6) 80% MHH solution plus 20% WLSt 7A filtrate (100 mL); (7) 50% MHH solution plus 50% WLSt 7A filtrate (100 mL); (8) 20% MHH solution plus 80% WLSt 7A filtrate (100 mL); (9) 0% MHH solution plus 100% WLSt 7A filtrate (100 mL). A completely randomized design with three replications was used in the lettuce study.

The phytotoxicity assay with sweetpotato was conducted using vine cuttings grown in jars, over a 60-day period, (May 27 - July 06, 1995) in a growth room (Environmental Growth Chambers) study with fluorescent lighting maintained at 700 µmol m^{-2} s^{-1}, a 16-h photoperiod and light/dark temperatures of 28°C/22°C. Ratios of filtrate and modified half Hoagland solution were prepared as in the lettuce study and used as growth medium for seven-day old sweetpotato cuttings, 'TU-82-155'. Single, rooted cuttings (15 cm) with leaves removed were placed in 150 mL glass jars containing 100 mL of nutrient medium treatment. Each container was sterilized with 10% Clorox prior to start of the experiment. All vials were covered with aluminum foil to preclude entry of light into the rooting zone. Vials were closed with filter paper through which the cutting protruded. A completely randomized design with two replications was used in the sweetpotato cutting study. Solution was replenished to 100 mL, with appropriate treatment, when levels fell below the 90-mL mark.

Sweetpotato Growth Studies

The studies were conducted with 'TU-82-155' and 'Georgia Jet' sweetpotato and the effects of treatment application were examined independently in two experiments. In each experiment, the sweetpotato crop was grown for 120 days in two growth chambers (Conviron CMP 3244). The environmental conditions were light/dark temperatures of 28/22°C, a light level of 700 µmoL m^{-2} s^{-1}, and a 14-h photoperiod (0600-2200). Cumulative light for each experiment was 35.3 $\pm$ 0.4 mol m^{-2} d^{-1} supplied with incandescent and fluorescent lamps (50:50 input wattage). A randomized block design was used in each experiment for treatment allocation to channels. Plants were grown hydroponically in 0.2 m^2 growth channels. Each channel was supplied with nutrients from a 45-L reservoir using a recirculating nutrient film technique (Hill *et al.*, 1989).

In Experiment 1, 'TU-82-155' sweetpotato was grown in two different nutrient solution treatments. The sweetpotato cultivar 'TU-82-155' was maintained in two growth channels in each of two chambers. Four vine cuttings (15 cm), bare of leaves, were placed in each channel and grown for 120 days (August 12 — December 29, 1995). The two crop nutrient treatments were: 20% WDSt 3A filtrate+80% MHH or "20% Filtrate" and Control or "100% MHH". The treatments were each supplied in 15.2 L volumes to one channel per growth chamber. The MHH solution and the "20% Filtrate" nutrient solution were replenished when the volume dropped below 10% of 15.2 L. Replenishment was with either water or treatment solution based on the measured electrical conductivity (EC) readings. The set-point for EC for the Filtrate:MHH and the MHH treatment was 2300 μS and replenishment was made when EC dropped below 2000 or increased to 2500 $\mu S.cm^{-1}$. When EC measurements were below the set-point, the treatment solution was used for volume adjustment. But for EC values higher than the set-point, water was used for volume adjustment. A 10X stock of MHH was used for regaining EC set-point values for the MHH treatment. The pH was adjusted to maintain a range of 6.0 - 6.5 ± 0.2 units.

Solution volume, pH, EC, and salinity were checked each Monday, Wednesday, Friday for the duration of each experiment. The water use by plants was monitored throughout each experiment, as the recorded volume of water or solution used to replenish each reservoir at Monday-Wednesday-Friday level adjustment to original volume. At harvest, the foliar biomass for each plant was removed from each channel and maximum plant height measured. Total fresh foliar biomass per plant was measured and four random measures made of the stem diameter. The leaf laminae were removed from each plant and total laminae weight recorded prior to leaf area measurements. Leaf area for each lamina on each plant was measured with a model LI 3000 leaf area meter (LI-COR Inc., Lincoln, NE). The root biomass was removed from each channel and storage roots (diameter greater than or equal to 1.0 cm) were counted, and the length/diameter ratio and total fresh storage root biomass data recorded. The thickness of the fibrous root mat and fresh weight were recorded. Harvested material was dried at 75°C for 48 h and weighed. Mean dry matter accumulation was calculated for each treatment as the total crop biomass/square meter.day. Data were analyzed by analysis of variance on each experiment separately. Pairwise differences were compared to Fisher's LSD values at $P < 0.05$. Chemical characteristics of the 20% Filtrate:MHH and MHH solutions are presented in Table 1.

Table 1. Summary of selected chemical parameters in nutrient solution composition in Experiments 1 and 2

Parameter	20% Filtrate:80%MHH	MHH
pH	7.19	5.95
Salinity, mmho/cm	0.95	0.45
Electrical Conductivity, μS/cm	1506.70	1010
NO_3^-, (mM)	351.60	182.80
$H_2PO_4^-$, (mM)	199.80	44.82
Cl^-, (μM)	204.60	462.45
SO_4^{2-}, (μM)	86.20	65.85

In Experiment 2, the sweetpotato cultivar 'Georgia Jet' was grown, for 120 days (February 12 to June 11, 1996), in the same growth chambers as Experiment 1, under the same environmental conditions. Using this sweetpotato cultivar, that is also adapted to hydroponic culture, we compared plant growth and yield responses when filtered effluent and supplemental inorganic nutrients were added to the nutrient solution. In addition to the two treatments used in Experiment 1, a 20% Filtrate:MHH solution supplemented with an inorganic nutrient stock (490 μmol N, 450 μmol P and K at 1.2 mmol) was used as a crop nutrient medium. The treatments were designated: 100% MHH, 20% Filtrate:80% MHH and 20% Filtrate:80% MHH+N-P-K. The replenishment protocol was the same as in Experiment 1 for all the treatments of Experiment 2. In the 20% Filtrate:80% MHH+N-P-K treatment, the 20% Filtrate:MHH nutrient solution was used to bring the EC to 2000-2500, and supplemental N-P-K (stock containing ten times the concentrations of N-P-K in MHH), to increase the EC range of 2600-3000 µS.cm-1.

Microbial Population Enumerations

Microbial populations were determined by serial dilution plating of the plant nutrient solution and bioreactor contents at 7- and 14-day intervals, respectively. Serial dilutions were made of each sample by adding 1 mL of plant nutrient solution or bioreactor effluent to 9.0 mL of sterile water and plating appropriate dilutions (100 µL) in duplicate onto both Tryptic Soy Agar and Potato Dextrose Agar plates (Difco, Detroit, MI). Inoculated and uninoculated plates were incubated to evaluate possible contamination. Colonies were counted after 4 days with final counts made at 21 days of incubation at 28°C.

RESULTS AND DISCUSSION

Nutrient Solution Characteristics

A significant ($P < 0.05$) treatment effect was observed on mean temporal E.C. levels based on measurements made over the duration of the experiments. The use of the E.C. level as the determinant for nutrient replenishment when filtrate is incorporated into the nutrient was inadequate for optimum plant growth as observed in the reduced total biomass obtained in this research. The total microbial populations were not influenced by nutrient solution treatment. However, higher numbers of fungi were recovered from the MHH than for filtrate amended solution(Figure 1). The pH of the MHH nutrient solution dropped to acidic ranges (3.1 - 5.0) between adjustments and may account for the higher population of fungi recovered in nutrient samples. For nutrient solution with filtrate, pH values increased between periods of adjustment and ranged from 6.8 to 7.4 pH units.

Phytotoxicity Assays (Lettuce & Sweetpotato)

Lettuce plants grew at all filtrate:MHH ratios tested. Visual comparison of leaf color between control and treated plants showed non-specific chlorosis in effluent-treated plants. The difference in leaf number and total plant biomass was not significant between control (MHH) and 20% WDSt 3A treated-plants (Figure 2). The plants grown in 100% WDSt 3A had fewer leaves than plants of all other treatments and the specific leaf weight (total leaf dry weight/number of leaves) was the highest.

Sweetpotato cuttings grew in filtrate concentrations higher than 20% but were chlorotic and fibrous root development was retarded. At a 50% or greater filtrate ratio the cuttings were severely yellow and total biomass at 60 days was low (Table 2).

Table 2 . Mean dry weight of sweetpotato vines planted as 15 cm cuttings, harvested after growing for 60 days in nutrient solution with selected ratios of filtered effluent

Treatment	Storage Root Initiation	Total Dry Weight g/plant
WDSt 3A:MHH		
20:80	NONE	2.005±0.025
50:50	NONE	1.430±0.020
80:20	NONE	0.090±0.020
100:0	NONE	1.25±0.150
100% MHH	Good Root Initiation	2.505±0.045
WLSt 7A:MHH		
20:80	Good Root Initiation	1.845±0.015
50:50	NONE	0.317±0.263
80:20	NONE	0.025±0.005
100:0	NONE	0.000

[1] Total Dry weight (root+stem+leaves) data are mean±standard error measured for 2 plants.

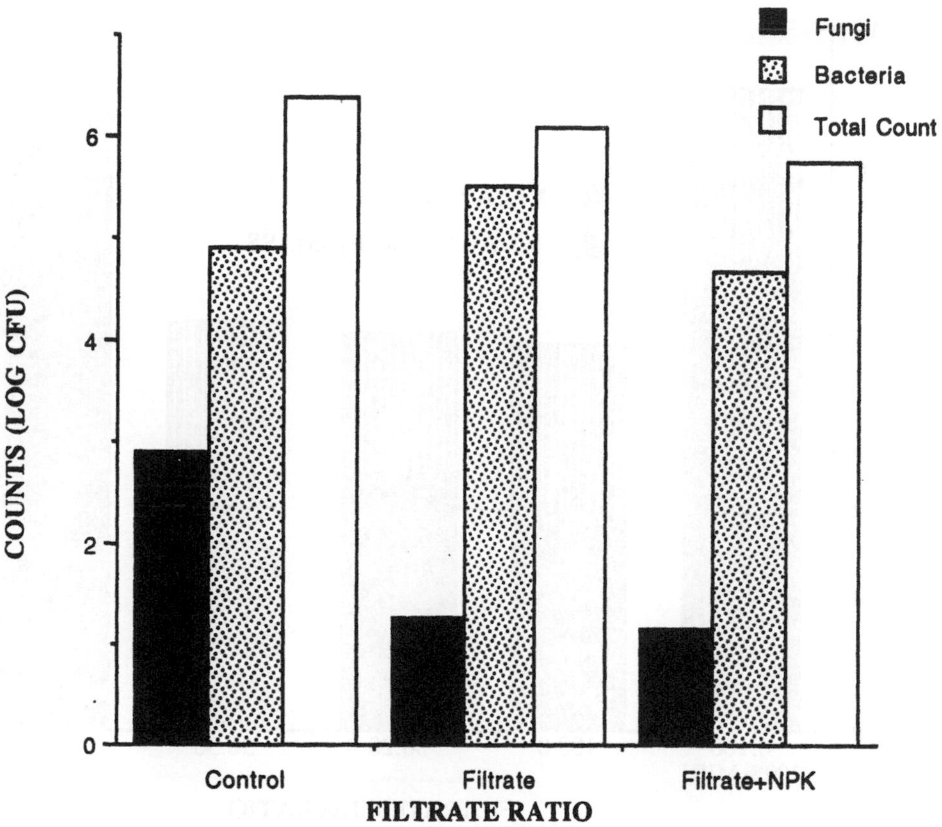

Fig. 1. Mean counts of microorganisms in solution samples collected throughout the 120 day duration of the crop.

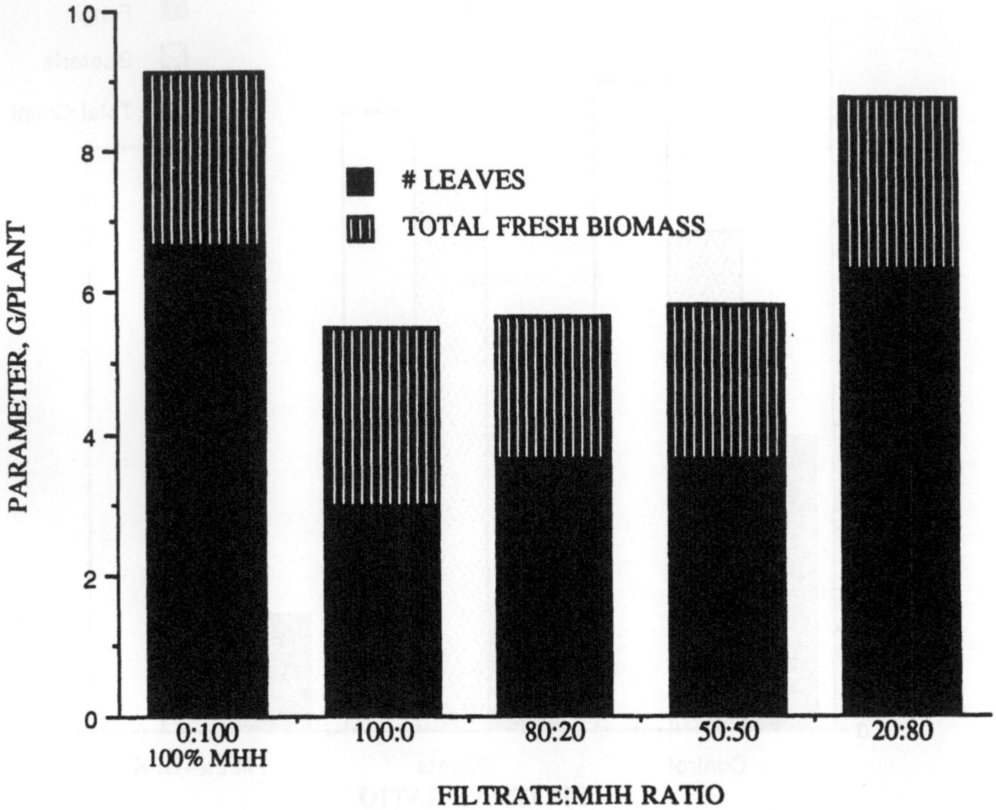

Fig. 2. Effect of growing lettuce in various ratios of MHH:filtrate from biodegradation with WDSt 3A.

Table 3. Mean dry matter accumulation in two sweetpotato cultivars as influenced by crop nutrient medium. Each value is the mean of eight plants, at 120 days after planting.

	Crop Nutrient Medium					
	MHH		20%Filtrate/MHH		20% Filtrate/MHH +N-P-K[1]	
Parameter	TU	G J	TU	G J	TU	G J[2]
Dry Matter Accumulation, g $m^{-2}d^{-1}$, LSD $_{(0.05)}$ TU / GJ = 0.6/3.55	7.2	12.69	1.4	5.27	NA	2.05
Leaf Area, $cm^2 Plant^{-1}$ LSD $_{(0.05)}$ TU / GJ = 1192.7 / 3483.8	8026.3	13291.90	1581.10	3346.40	NA	2.05
Mean Leaf Area LSD $_{(0.05)}$ TU / GJ = 5.2/8.5	21.1	46.0	15.1	16.3	NA	27.90
Leaf DW LSD $_{(0.05)}$ TU / GJ = 3.2/11.29	23.60	37.82	7.10	9.97	NA	9.20
Shoot:Root Ratio LSD $_{(0.05)}$ TU / GJ = 0.6/1.26	1.86	1.23	1.86	0.70	NA	3.69
Storage Root Number LSD $_{(0.05)}$ TU / GJ = 2.18/1.91	5	3	1	0	NA	5
Storage Root DW[3] LSD $_{(0.05)}$ TU / GJ = 12.3/51.9	83.60	110.50	4.10	66.71	NA	0.00
Fibrous Mat DW[4]	11.31±0.1	11.31±0.1	6.89.±0.2	10.90±2.6	NA	7.85±1.5
Fibrous Mat Thickness, cm LSD $_{(0.05)}$ TU / GJ = 1.1/0.6	4.9	5.5	2.5	2.3	NA	4.6
Pencil Root DW LSD $_{(0.05)}$ TU / GJ = 14.4/37.8	64.00	84.20	32.80	35.80	NA	27.5
Stem DW LSD $_{(0.05)}$ TU / GJ = 6.83/41.23	45.02	117.56	11.39	34.88	NA	28.76
Total DW LSD $_{(0.05)}$ TU/GJ = 21.1184.38	173.48	302.3	34.81	13.2	NA	49.2
EBI[5] LSD $_{(0.05)}$ TU/GJ = 0.11/Not Significant	0.48	55.43	0.09	17.90	NA	0
Total Inedible Biomass DW LSD $_{(0.05)}$ TU/GJ = 8.54/41.01	65.46	153.96	23.58	53.52	NA	41.42

[1]NA=20% Filtrate/MHH+N-P-K: Treatment was not included in Experiment 1.
[2]TU = 'TU-82-155' ; GJ = 'Georgia Jet'
[3]DW = dry weight in grams/plant; each plant grew in an area of 0.2 square meters.
[4]Fibrous Mat DW= Mean±S.E. (n=4)
[5]EBI = Edible biomass index (storage root DW/total DW) x 100%.

Sweetpotato Growth in Filtered Effluent

Both, 'TU-82-155' and 'Georgia Jet' sweetpotato used in this study showed Filtrate:MHH-induced chlorosis on the leaf lamina beginning approximately 7 days after incorporation of filtrate into the crop nutrient solution. The plants generally recovered over the duration of the experiment, but there remained visual differences in green coloration until harvest. The leaves in MHH supplied plants were dark green in color throughout the study. At 14 days after planting, leaf number and size were reduced and this continued throughout the study, for filtrate-treated plants. In 'Georgia Jet' sweetpotato, leaves on filtrate-treated plants were thicker. Leaf senescence occurred earlier in Filtrate:MHH-treated plants, ca. days into the experiment (in both cultivars). Senescence in control plants was observed at 42 and 49 days for 'TU-82-155' and 'Georgia Jet' sweetpotato, respectively.

When 'TU-82-155' sweetpotato was grown in hydroponic culture with 20% Filtrate:MHH, storage root yield was significantly decreased (Table 3). Also, total crop biomass was significantly reduced ($P < 0.05$) for 'TU-82-155' sweetpotato grown in 20% Filtrate:MHH solution. The calculated individual leaf area for 'TU-82-155' indicated that incorporating bioreactor effluent into the nutrient solution significantly reduced leaf size (Table 3). Similar reductions occurred in dry mass accumulation. The mean dry matter accumulation for plants supplied 20% Filtrate:80% MHH solution was reduced 5.8 g $m^{-2}.d^{-1}$ compared with 'TU-82-155' grown in MHH solution (Table 3). The shoot/root ratio increased significantly ($P < 0.05$) with values of 1.86 and 0.73 for plants grown in 20% Filtrate:80% MHH and 100% MHH, respectively. The increased shoot-to-root dry matter ratio indicated that there was an increase in dry matter allocation to the shoot for effluent-treated plants. The reduction in root number and biomass and in leaf number and area (Table 3) represents an unfavorable situation since leaf area has a large impact on light interception and transpiration, and root biomass is an important component for nutrient and water uptake. Reduced root development could have adversely affected uptake of nutrients as well as water and would explain the observed reduction in total plant biomass (Table 3).

Mean storage root yield, although reduced, was not significantly different for 'Georgia Jet' sweetpotato grown in 20% Filtrate:80% MHH compared to the control (Table 3). While storage root weight was reduced, filtered bioreactor effluent appeared to cause an increase in the number of 'Georgia Jet' storage roots formed (Table 3). The leaf area per plant and the thickness of the fibrous root mat of 'Georgia Jet' sweetpotato measured at 120 days after planting were significantly higher for the control than the 20% Filtrate:MHH-treatment (Table 3). The significant ($P < 0.05$) decrease in the average daily rate of dry matter accumulated in the 20% Filtrate:MHH-treated plants (Table 3) resulted in a significant decrease in total dry matter/plant at harvest, when compared to the control (Table 3).

The [20% Filtrate:80% MHH]+N-P-K treatment was deleterious to sweetpotato growth and resulted in an 84% decrease in mean dry matter accumulation between treated and control plants (Table 3). The use of an N-P-K stock solution containing ten times the nitrogen, phosphorus and potassium concentrations of MHH and with E.C. values as the determining factors for nutrient replenishment resulted in a distortion of the plant response to the filtered effluent in the nutrient solution. Although the plants in this supplemental N-P-K treatment were larger in the first four weeks of growth, there was increased senescence and a rapid decline in leaf size as Experiment 2 progressed. At harvest, some N-P-K-treated plants showed interveinal chlorosis and 25% of the plants had split stems. There was differentiation for storage roots but only pencil roots (diameter ≤ 10 mm) were obtained.

SUMMARY

Incorporating effluent from biological (bacterial soil isolate) treatment of inedible sweetpotato biomass into the crop nutrient medium though not causing death of 'TU-82-155' or 'Georgia Jet' sweetpotato, resulted in a decrease in total plant biomass. Growth parameters measured at harvest for both cultivars indicated that plants were significantly influenced by incorporation of filtrate in the nutrient solution. Increased shoot/root ratio suggest that plants did not have the optimum root biomass required for nutrient uptake to maintain growth/development. Reduced leaf area indicated that the incorporation of filtrate impacted the area available for receptors of photosynthetically active radiation. The results obtained in this study can be largely attributed to the use of single culture biodegradation and the plant growth effects may be due to remnants of toxic by-products in the effluent used for crop nutrient medium. Also, the build-up of toxic levels of co-ions such as sodium, chloride and sulfate may have contributed to the reduced yield obtained.

For 'Georgia Jet' sweetpotato, leaf area was significantly reduced when 20% Filtrate:MHH was incorporated into the solution; however, storage root biomass was not affected. Supplementing the 20% Filtrate:MHH with a N-P-K stock and using EC set point of 2600 - 3000 μS cm-1 for replenishment further reduced ($P < 0.05$) the rate of biomass accumulation. Plants of both sweetpotato cultivars showed non-specific leaf chlorosis and total biomass was the lowest for plants receiving 20% Filtrate:MHH + N-P-K. These plants also showed higher levels of leaf senescence.

The results from this research suggest that using bioreactor effluent for crop production is likely to influence dry matter allocation. The degree of biomass reduction effected by effluent treatment may be influenced by the degree of decomposition of the crop residue in the bioreactor and the potential for residual toxic organics. This study highlights the need for more information on appropriate methods for incorporation of bioreactor effluent into hydroponic crop production.

In the development of guidelines for integrating biological treatment of crop residue into crop production, the influence of the composition of the effluent on growth parameters should be carefully considered. Detailed research examining a range of factors (electrical conductivity, pH, polyphenolic composition, C:N levels, enzymes associated with phenolic compound metabolism) are required before accurate protocols can be developed to predict crop performance in effluent-treated culture.

ACKNOWLEDGMENT

The research was funded by grants from the National Aeronautics and Space Administration (NAG-2940) and the United States Department of Agriculture (ALX-SP-1D). Technical support by Tuskegee University G.W. Carver Agricultural Experiment Station personnel—Jill Hill, Jacquelyn Carlisle, Jennifer Seminara, Doris Douglas and Kendra Stanciel—was extremely beneficial and appreciated. Contribution No. 265 of the George Washington Carver Agricultural Experiment Station, Tuskegee University.

REFERENCES

Bayer, E. A. and R. Lamed, The cellulose paradox: pollutant *par excellance* and/or a reclaimable natural resource? *Biodegrad.*, 3, 171 (1992).

Bonsi, C. K., W. A. Hill, D. G. Mortley, P. A. Loretan, C. E. Morris and E. R. Carlisle, Growing sweetpotatoes for space missions using NFT, In: *Sweetpotato Technology for the 21st Century*, edited by W. A. Hill, C. K. Bonsi, and P. A. Loretan, p. 110, Tuskegee University, Tuskegee, AL, (1992).

Finger, B. W., and M. P. Alazraki. Development and integration of a breadboard-scale aerobic bioreactor to regenerate nutrients from inedible crop residues, pp. 1-6, SAE Technical Paper Series, Society of Automotive Engineers, Paper 951498, Warrendale, PA (1995).

Hill, W. A. Selection of root and tuber crops for space missions, In: *The Sweet Potato for Space Missions*, edited by W. A. Hill, P. A. Loretan, and C. K. Bonsi, pp. 3-12, Tuskegee University, Tuskegee, AL, (1984).

Hill, W. A., D. G. Mortley, C. L. MacKowiak, P. A. Loretan, T. W. Tibbitts, R. M. Wheeler, C. K. Bonsi, and C. E. Morris, Growing root, tuber and nut crops hydroponically for CELSS, *Adv. Space Res.* 12, 125 (1992).

Hill, W. A., P. A. Loretan, C. K. Bonsi, C. E. Morris, J. Y. Lu and C. R. Ogbuehi, Utilization of sweetpotatoes in Controlled Ecological Life Support Systems (CELSS), *Adv. Space Res.* 9, 29 (1989).

Mortley, D. G., C. K. Bonsi, W. A. Hill, P. A. Loretan, and C. E. Morris. Irradiance and nitrogen to potassium ratio influences sweetpotato yield in nutrient film technique, *Crop Sci.*, 33, 782 (1993).

Strayer, R. F. and K. Cook. Recycling plant nutrients at NASA's KSC-CELSS breadboard project: Biological performance of the breadboard-scale aerobic bioreactor during two runs, pp 1-6, SAE Technical Paper Series, Soc. Auto. Eng. Tech. Paper 951708, Warrendale PA (1995).

Trotman, A. A., A. M. Almazan, A. D. Alexander, P. A. Loretan, X. Zhou and J. Y. Lu, Biological degradation and composition of inedible sweetpotato biomass, *Adv. Space Res.* 18, 267 (1996).

Pergamon

Adv. Space Res. Vol. 20, No. 10, pp. 1815–1820, 1997

Printed in Great Britain
0273-1177/97 $17.00 + 0.00

PII: S0273–1177(97)00846–6

USE OF BIOLOGICALLY RECLAIMED MINERALS FOR CONTINUOUS HYDROPONIC POTATO PRODUCTION IN A CELSS

C. L. Mackowiak*, R. M. Wheeler**, G. W. Stutte*, N. C. Yorio* and J. C. Sager**

**Dynamac Corporation, Mail Code Dyn-3, Kennedy Space Center, FL 32899, U.S.A.*
***NASA Biomedical Operations and Research, Mail Code MD-RES, Kennedy Space Center, FL 32899, U.S.A.*

ABSTRACT

Plant-derived nutrients were successfully recycled in a Controlled Ecological Life Support System (CELSS) using biological methods. The majority of the essential nutrients were recovered by microbiologically treating the plant biomass in an aerobic bioreactor. Liquid effluent containing the nutrients was then returned to the biomass production component via a recirculating hydroponic system. Potato (*Solanum tuberosum* L.) cv. Norland plants were grown on those nutrients in either a batch production mode (same age plants on a nutrient solution) or a staggered production mode (4 different ages of plants on a nutrient solution). The study continued over a period of 418 days, within NASA Breadboard Project's Biomass Production Chamber at the Kennedy Space Center. During this period, four consecutive batch cycles (104-day harvests) and 13 consecutive staggered cycles (26-day harvests) were completed using reclaimed minerals and compared to plants grown with standard nutrient solutions. All nutrient solutions were continually recirculated during the entire 418 day study. In general, tuber yields with reclaimed minerals were within 10% of control solutions. Contaminants, such as sodium and recalcitrant organics tended to increase over time in solutions containing reclaimed minerals, however tuber composition was comparable to tubers grown in the control solutions.

INTRODUCTION

Ash (inorganic nutrients) can comprise over 10% of inedible plant dry mass when the plants are grown in recirculating hydroponics (McKeehen *et al.*, 1996; Wheeler *et al.*, 1994). A majority of these nutrients are easily recovered by solubilizing the biomass in water (leaching) or treating the biomass in an aerobic bioreactor (Garland *et al.*, 1993; Finger and Strayer, 1994). In a single potato production study, recycling nutrients from bioreactor processing resulted in crop yields equal to the control treatment composed of reagent-grade salts (Mackowiak *et al.*, 1996).

When the nutrients are solubilized quickly, as in the case of leaching, a large amount of organic compounds are also solubilized, which directly or indirectly impair plant growth (Garland *et al.*, 1993; Mackowiak *et al.*, 1996). The majority of soluble organics are labile, resulting in rapid degradation in an aerobic bioreactor. Some of the organics are oxidized to recalcitrant phenolic compounds which give the nutrient solution a (tea-colored) appearance (Kumada, 1965). The concentration of these organic compounds can be monitored using spectrophotometric methods (Schnitzer, M., 1978; Mackowiak *et al.*,

1994). In addition to monitoring the spectral composition of the aging nutrient solutions, elemental analyses of the solution and the plant biomass from successive generations provided data on the potential for build-up of elemental contaminants.

MATERIALS AND METHODS

The study was conducted in NASA's Biomass Production Chamber (BPC), located at Kennedy Space Center (Wheeler *et al.*, 1996). The BPC is one component of the Breadboard Project's test of bioregenerative systems for advanced life support (Prince and Knott, 1989). Potato (*Solanum tuberosum* L.) cv. Norland were started from in vitro nodal explants and thinned to four plants per culture tray at 10 days-after-planting (DAP). The upper chamber compartment (10 m^2) had plants grown in a batch culture system (all plants grown to maturity and then immediately replanted), whereas the lower chamber compartment had plants grown in staggered procession (with plantings made at 26-day intervals resulting in four different age groups). Each chamber was comprised of two levels for growing plants. Each chamber had nutrient recycling tested on one level to compare with a modified half-strength Hoagland's control, thus providing four different treatments (two culture systems and two nutrient formulations) and each treatment had 16 culture trays (5 m^2 growing area).

Details on bioreactor engineering requirements and effluent production methods using the Breadboard-scale aerobic bioreactor (B-SAB) have been described by Finger and Alazraki, (1995) and Strayer and Cook, (1995). Details on nutrient solution recipes and replenishment methods can be found in Mackowiak *et al.* (1996). The starting nutrient solution was a modified half-strength Hoagland's solution (Mackowiak *et al.*, 1996). A concentrated replenishment solution, used to replace nutrients removed by the plants, is presented in column 2 of Table 1. Since our goal was to recycle 50% (by mass) of the crop's nutrient requirement, we had to add back some nutrients as reagent grade chemicals, mainly Ca, P, Mg, and Fe (Table 1). Micronutrient composition (using ICP analysis) varied quite a bit over time in the effluent, so in most cases over 80% of the micronutrients were amended with reagent-grade chemicals to assure adequate amounts in the replenishment solution. The effluent was analyzed for elemental content whenever changes in the bioreactor processing occurred (Table 1). The analytical information let us know what nutrients needed to be amended with reagent-grade chemicals. Weekly samples from the nutrient delivery systems also were analyzed for elemental composition, using colorimetric methods for N and P and inductively coupled plasma spectrometer (ICP) for the other nutrients. Bioreactor effluent and nutrient solution were analyzed periodically for total organic carbon (TOC) using a high temperature furnace analyzer.

All plants received the same environmental conditions, with lighting provided by high-pressure sodium (HPS) lamps as a 12-h light / 12-h dark photoperiod and PPF levels averaged 814 ± 15 $\mu mol\ m^{-2}\ s^{-1}$. Temperature was set at 20/16 °C (light/dark) and relative humidity controlled at 70 ± 10%. Carbon dioxide was added during the light phase to maintain 1200 $\mu mol\ mol^{-1}$ (0.12 kPa). Plants were harvested at 104 DAP (except for the first harvest at 105 DAP). It took 105 days to get the staggered planting system to steady-state (four plant ages on a single system). Tuber, root, and shoot biomass were freeze-dried for dry mass determinations. Tissue samples were analyzed with a DC-arc spectroscopy multielement analysis system.

RESULTS AND DISCUSSION

At 418 days the entire BPC was harvested and all the dry mass values that had been collected during the course of the study tallied to obtain overall biomass production rates. Using recycled minerals for potato production resulted in biomass yields as good as yields from using only reagent-grade chemicals (Figure 1). The staggered planting system resulted in greater biomass than the batch planting, which may have been a result of improved light interception by the plants grown on that system (Stutte *et al.*, this issue). The average total biomass production rates from this study were within 10% of single batch values from previous BPC potato studies, while tuber production was within 18% (Wheeler *et al.*, 1996). The somewhat lower average production values in this study may be related to a buildup of biogenic compounds in the nutrient solutions over time (Stutte *et al.*, 1995). These compounds tend to hasten tuberization and provide an extremely strong induction signal, resulting in reduced biomass (Engels and Marschner, 1986; Wheeler *et al.*, 1995). Nonetheless, it does not appear that plants grown on recycled nutrients were affected more than those on the control solutions (Table 2).

Table 1. Composition of effluent from bioreactor harvests over the course of the study.

Nutrient	Inorganic Stock	Effluent Composition*						
		0 DOE**	41 DOE	73 DOE	105 DOE	217 DOE	350 DOE	C.V.***
	(mM)	(mM)	(mM)	(mM)	(mM)	(mM)	(mM)	(%)
NO_3-N	62	56	73	138	56	49	53	48
PO_4-P	9	2.4	2.2	2.6	2.4	3.7	2.1	23
K	48	51	40	50	52	49	43	10
Ca	9	1.8	3.4	2.6	2.1	4.0	2.9	29
Mg	10	5.0	4.6	3.8	3.8	4.2	8.5	36
	(μM)	(μM)	(μM)	(μM)	(μM)	(μM)	(μM)	
Fe	134	0	0	0	0	89	0	
Mn	74	0	18	4	3.2	4.0	15	
Zn	9.6	0	54	0	2.9	4.6	0.8	
Cu	10.4	0	0	0	25	0	1.3	
B	95	0	0	0	4.8	0	6.4	
Mo	0.1	0	0	0	0	0	0	

*Data normalized to a bioreactor loading rate of 20 g L^{-1}. If required (41, 73 DOE), effluent was diluted with chamber condensate water to provide 50% of the nutrient delivery system's potassium requirements.
**DOE = Day of experiment.
*** C.V. = coefficient of variation of reactor harvests.

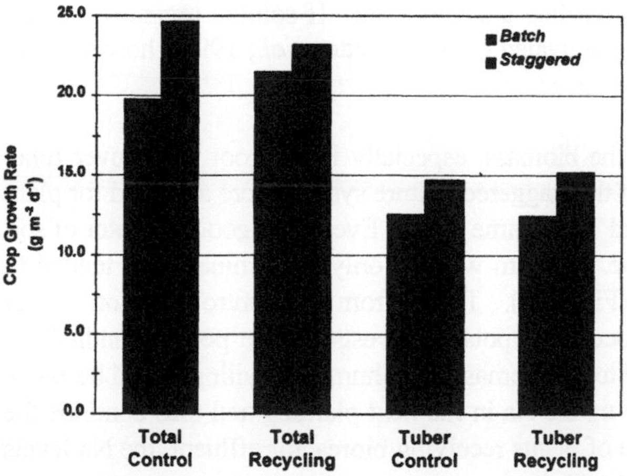

Fig 1. Culture system and nutrient recycling effects on potato tuber and total dry mass production rates. Production rates were calculated from the biomass summations for the entire 418 day study.

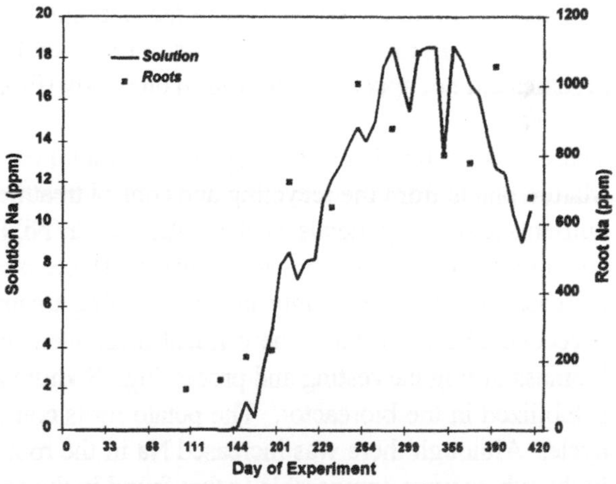

Fig. 2. Na concentration of nutrient solution and root tissue in the recycling treatment of the staggered culture system. There were no changes in the control treatment.

The coefficient of variation for macronutrients suggest that the effluent's potassium content was the most uniform over successive bioreactor harvests (Table 1). In most cases there were single instances where a nutrient value would be quite different from surrounding values, as was the case with NO_3-N at 73 DOE and PO_4-P at 217 DOE (Table 1). In the case of Ca and Mg, there was variation throughout the study. Robinson *et al.* (1993) reported that P, Ca, and Mg release by the fungal degradation of wheat varied by fungal species. It may be in our own case that as the bioreactor microbial population adjusted, release of these elements was affected. In the case of micronutrients, the low concentrations made accurate analysis difficult. A study which covered several U.S. analytical labs found a wide variation in the levels of Ca, Fe, Cu, and B reported for the same plant tissue (Sterret et. al., 1987). Minor adjustments were made in the quantities of reagent-grade nutrients needed (approximately every 8 weeks), which was based on effluent inorganic analysis. This kept the recycled nutrient replenishment solution comparable to the control replenishment.

Weekly spectral absorbency scans of hydroponic solutions from the four treatments were taken in the range of 220 - 420 nm to track organic loading of the nutrient solutions over time. This range was selected because light absorption of humic substances becomes greater with decreasing wavelengths (Kumada, 1965; Baes and Bloom, 1990). Absorbency values were unchanged in the control treatment but values increased in the nutrient recycling treatment of the staggered culture system by a factor of 4.5. Similar trends were seen with the batch culture systems, until activated carbon was added to both the control and recycling treatments at the beginning of the last 104 d cycle, which was done to test the removal of some potato exudates (Wheeler *et al.*, 1995). The two day carbon treatment reduced absorbency levels approximately 10%. Nutrient solution total organic carbon (TOC) was also measured. The bioreactor effluents varied from 150 - 230 ppm, and the nutrient recycling treatments increased over time (from 22 to 150 ppm), but the control treatments had only a small increase (5 to 15 ppm). The recalcitrant organics accumulating in treatments with recycled nutrients may be humic substances, viz., humic acid and fulvic acid. Humic substances have measured half-lives of 200 - 2000 years (Martin and Haider, 1986), and make up over 70% of the organic matter in most soils. At pH 6, humic substances can form water-soluble complexes with many organics and inorganics, as in the case of fulvic acid, or adsorb to organics and inorganics, as in the case of humic acid (Schnitzer, 1986). Due to their nature, humic substances provide some buffering capabilities. Acid use efficiency (biomass per amount of acid added) was greater in the effluent treatments than the controls, where the staggered planting effluent > batch effluent > staggered planting control > batch control (1.67, 1.25, 1.13, and 1.02 g biomass $mmol^{-1}$ HNO_3, respectively). Although we did not see a plateau in TOC or in spectral absorbency of organics, it may have eventually reached a plateau as organic degradation in the nutrient solutions reached a steady-state. If soluble organics become unacceptable in the solution, it is easily removed by using activated carbon (Stutte *et al.*, 1995), however, the organics did not appear deleterious in our study (Figure 1).

There was no build-up of heavy metal contaminants in the biomass, especially in the root tissue over time. Mature plants from the recycling and control treatments of the staggered culture system were analyzed for plant nutrients and heavy metals, such as Al, Co, Cr, Ni, Pb, and Ti, to name a few. Even Si, a good indicator of soil contamination from shoes, etc., remained stable over time. Sodium was the only contaminant that tended to increase over time, but only in the recycling treatment (Figure 2). Tissue from the control solution never exceeded 280 ppm Na. The effluent originated from successive potato harvests, where people handled the biomass during harvesting and processing. Sodium added to the biomass from human handling would be easily solubilized in the bioreactor. The potato roots could take up the Na in the next planting and thus continue the cycle. Although there was increased Na in the root tissue of plants receiving bioreactor effluent, the Na levels in the tubers were comparable to that found in the control (< 90 ppm).

CONCLUSIONS

Recycling 50% of the nutritional mass requirements (based on K needs) for potato was successful in recirculating hydroponics with either batch or staggered production, as compared to using only reagent-grade salts. The K concentrations in bioreactor effluent were most consistent over successive bioreactor harvests, whereas Ca and Mg were much more variable. The changes in recovery may be related to changes in microbial community composition of the bioreactor over time and/or variation caused by nutrient analytic equipment. Recalcitrant soluble organics increased over time in the nutrient recycling treatments but not in the control solutions. Although continued build-up may be a problem in longer studies, it did not hinder production in our 418 day study and may even improve nutrient solution buffering capacity, resulting in a lower acid requirement for pH control. It was notable that heavy metals did not build up in plant tissues over successive harvests during this study, however Na did increase in nutrient solution and plant roots of the recycling treatments. The Na probably came from human handling of the harvested biomass prior to its being recycled and thus could be alleviated by automating the harvesting process.

REFERENCES

Baes, A.U. and P.R. Bloom, Fulvic acid ultraviolet-visible spectra: influence of solvent and pH, Soil Sci. Soc. Am. J. 54:1248-1254 (1990).

Engels, C.H. and H. Marschner, Allocation of photosynthate to individual tubers of *Solanum tuberosum* L. 3: Relationship between growth rate of individual tubers, tuber weight and stolon growth prior to tuber initiation, J. Expt. Bot., 185:1813-1822 (1986).

Finger, B.W. and M.P. Alazraki, Development and integration of a Breadboard-scale aerobic bioreactor to regenerate nutrients from inedible crop residues, Society of Automotive Engineers Tech. Paper 951498 (1995).

Finger, B.W. and R.F. Strayer, Development of an intermediate-scale aerobic bioreactor to regenerate nutrients from inedible crop residues, Society of Automotive Engineers Tech. Paper 941501 (1994).

Garland, J.L., C.L. Mackowiak, and J.C. Sager, Hydroponic crop production using recycled nutrients from inedible crop residues, Society of Automotive Engineers Tech. Paper # 932173 (1993).

Kumada, K., Studies on the colour of humic acids, Soil Science and Plant Nutrition, 11(4):11-16 (1965).

Mackowiak, C.L., J.L., Garland, and G.W. Stutte, Growth regulator effects of water soluble materials from crop residues for use in plant hydroponic culture, Proc. 21st Annual Plant Growth Regulator Society of America (1994).

Mackowiak, C.L., J.L. Garland, R.F. Strayer, B.W. Finger, and R.M. Wheeler, Comparison of aerobically-treated and untreated crop residue as a source of recycled nutrients in a recirculating hydroponic system, Adv. Space Res., 18(1/2):281-287 (1996).

Martin, J.P. and K. Haider, Influence of mineral colloids on turnover rates of soil organic carbon, pp. 283-304, In P.M. Huang and M. Schnitzer (eds). Interactions of Soil Minerals with Natural Organics and Microbes. Soil Sci. Soc. Am. Special Publication #17. Madison, WI (1986).

McKeehen, J.D., D.J. Smart, C.L., Mackowiak, R.M. Wheeler, and S.S. Nielsen, Effect of CO_2 levels on nutrient content of lettuce and radish, Adv. Space Res., 18(4/5):85-92 (1996).

Prince, R.P. and W.M. Knott, CELSS Breadboard Project at the Kennedy Space Center, In: D.W. Ming and D.L. Henninger (eds) Lunar Base Agriculture: Soils for Plant Growth, Amer. Soc. Agron., Madison WI, USA (1989).

Robinson, C.H., J. Dighton, J. C. Frankland, and P.A. Coward, Nutrient and carbon dioxide release by interacting species of straw-decomposing fungi, Plant and Soil, 151:139-142 (1993).

Schnitzer, M., Humic substances: Chemistry and reactions, pp. 1-64, In M. Schnitzer and S.U. Khan (ed.) Soil Organic Matter, Elsevier Science Publishing Co., New York (1978).

Schnitzer, M., Binding of humic substances by soil mineral colloids, pp. 77-102, In P.M. Huang and M. Schnitzer (eds). Interactions of Soil Minerals with Natural Organics and Microbes. Soil Sci. Soc. Am. Special Publication #17. Madison, WI (1986).

Sterrett, S.B., C.B. Smith, M.P. Mascianica, and K.T. Demchak, Comparison of analytical results from plant analysis laboratories, Commun. In Soil Sci. Plant Anal,. 18(3):287-299 (1987).

Strayer, R.F. and K. Cook, Recycling plant nutrients at NASA's KSC-CELSS Breadboard Project: Biological performance of the Breadboard-scale aerobic bioreactor during two runs, Society of Automotive Engineers Tech. Paper 951708 (1995).

Stutte, G.W., C.L. Mackowiak, N.C. Yorio, and R.M. Wheeler, Characterization of possible tuber-inducing factor in a continuous use hydroponic system, Proc. Internat. Congress on Plant Growth Substances. Twenty-first annual meeting. Plant Growth Regulator Society of America (1995).

Wheeler, R.M., C.L. Mackowiak, J.C. Sager, W.M. Knott, and W.L. Berry, Proximate nutritional composition of CELSS crops grown at different CO_2 partial pressures, Adv. Space Res., 14(11):171-176 (1994).

Wheeler, R.M., G.W. Stutte, C.L. Mackowiak, N.C. Yorio, and L.M. Ruffe, Accumulation of possible potato tuber-inducing factor in continuous-use recirculating NFT system, HortSci. 30(4):790 (1995).

Wheeler, R.M., C.L. Mackowiak, G.W. Stutte, J.C. Sager, N.C. Yorio, L.M. Ruffe, R.E. Fortson, T.W. Dreschel, W.M. Knott, and K.A. Corey, NASA's biomass production chamber: A testbed for bioregenerative life support studies, Adv. Space Res. 18(4/5):215-224 (1996).

Pergamon

Adv. Space Res. Vol. 20, No. 10, pp. 1821–1826, 1997

Printed in Great Britain
0273-1177/97 $17.00 + 0.00

PII: S0273-1177(97)00847-8

INTEGRATION OF WASTE PROCESSING AND BIOMASS PRODUCTION SYSTEMS AS PART OF THE KSC BREADBOARD PROJECT

J. L. Garland, C. L. Mackowiak, R. F. Strayer and B. W. Finger

Dynamac Corporation, Mail Code DYN-3, Kennedy Space Center, FL 32899, U.S.A.

ABSTRACT

After initial emphasis on large-scale baseline crop tests, the Kennedy Space Center (KSC) Breadboard project has begun to evaluate long-term operation of the biomass production system with increasing material closure. Our goal is to define the minimum biological processing necessary to make waste streams compatible with plant growth in hydroponic systems, thereby recycling nutrients into plant biomass and recovering water via atmospheric condensate. Initial small and intermediate-scale studies focused on the recycling of nutrients contained in inedible plant biomass. Studies conducted between 1989-1992 indicated that the majority of nutrients could be rapidly solubilized in water, but the direct use of this crop "leachate" was deleterious to plant growth due to the presence of soluble organic compounds. Subsequent studies at both the intermediate scale and in the large-scale Biomass Production Chamber (BPC) have indicated that aerobic microbiological processing of crop residue prior to incorporation into recirculating hydroponic solutions eliminated any phytotoxic effect, even when the majority of the plant nutrient demand was provided from recycled biomass during long term studies (i.e. up to 418 days) Current and future studies are focused on optimizing biological processing of both plant and human waste streams.

STATEMENT OF PURPOSE

The purpose of this paper is to present the research approach, accomplishments, and goals related to the integration of waste processing and biomass production as part of the Kennedy Space Center (KSC) Breadboard Project.

KSC BREADBOARD PROJECT

Phase I Results

The objective of the KSC Breadboard project is to test the feasibility of using biological systems to provide life support for humans during long term space missions. The initial phase of the project focused on quantifying the production of life support requirements (atmospheric regeneration, food production, and water purification via transpiration) by candidate ALS crops (wheat, potato, soybean, and lettuce) under controlled environmental conditions in a large-scale (20 m^2) system known as the Biomass Production Chamber (BPC). Data from these baseline plant growth tests (Wheeler et al. 1993a, Wheeler et al. 1993b, Wheeler et al. 1994, Wheeler et al. 1996) provide the basis for the second phase of the Breadboard project; integration of waste recycling into system operation.

The baseline crop studies quantified the production rates of inedible plant biomass. Harvest index (edible biomass/total biomass) of most of ALS candidate crops that were tested was below 50%, while that of potato was near 80% (Wheeler et al. 1996). If one assumes that a person will require approximately 750 g dry mass of a vegetarian diet per day, and that the average harvest index is 55% for a mixed diet, inedible biomass production would range from 300-600 g dm $person^{-1}$ day^{-1} in a system in which plants provide 50-100% of diet. The value is 10-20 times greater than estimates of dry fecal mass, which range from 20-32 g dm $person^{-1}$ day^{-1} (Parker and Gallagher 1988, Shubert et al. 1984).

Another relevant finding from these studies was that, in terms of human requirements, production of water via transpiration was 5-7 times greater than the rate of atmospheric regeneration (i.e., CO_2 removal and O_2 production) and 10-20 times the rate of food production. Water purification is particularly relevant in a regenerative system since liquid wastes (urine and washwater) will be the dominant waste streams. Estimates of urine production range from 1.3-2.1 L $person^{-1}$ day^{-1}, while graywater production (i.e., liquid waste from the shower, clotheswasher, and dishwasher) is estimated in the range of 25 L $person^{-1}$ day^{-1} (Wydeven and Golub 1990). Combined, the mass of liquid wastes will be 50 times greater than inedible biomass and 1000 times greater than fecal dry matter. A biomass production system scaled to meet food requirements for one human would produce approximately 40 L of atmospheric condensate day^{-1}. Our approach at the KSC Breadboard project is to test the feasibility of using plant transpiration as the major water purification component within the system.

Results from plant growth studies indicated that the costs of storage or resupply to provide plant nutrient requirements would be significant. Mackowiak et al. (1996a) estimated that the mass of reagent-grade salts required to support plant growth would be equivalent to 30% of the mass of human food requirements. The high nutrient demand of the system appears to be partially a result of significantly higher accumulation of nutrients in hydroponically-grown plants relative to field-grown crops (McKeehen et al. 1996). Decreasing nutrient levels in the hydroponic solutions may reduce excessive nutrient uptake, but may also increase the potential risk of nutrient deficiency. Even if plant uptake can be minimized, inorganic nutrient requirements will remain a significant mass flux in the system.

Phase II Approach and Results

Waste processing in a bioregenerative system can be separated into three major components: 1) water recycling, 2) CO_2 regeneration from organic matter, and 3) inorganic nutrient recycling. Water recycling, as indicated above, is the primary processing requirement in terms of mass, and is dominated by the need to recycle graywater. Our approach to liquid waste (both graywater and urine) processing will be described in the research plan section below. The oxidation of waste streams is necessary to resupply CO_2 for plant growth. Humans are important oxidizers of organic matter (i.e., food) in the system, and in a system partially open with respect to food, human respiration represents a net influx of CO_2 into the system. Since food storage is likely for first generation advanced life support systems, complete oxidation of wastes generated within the system is unnecessary. Given the microbiological and psychological issues related to recycling human waste, and its low overall mass, we have focused our efforts on recycling inedible plant biomass. As stated above, recycling the nutrients contained in the plant biomass represents a significant potential savings. The feasibility of biological processing methods have been evaluated initially because of their lower energy requirements, lower potential for volatilization of inorganic nutrients, and lower operation temperatures and pressures (i.e., near ambient) relative to physical-chemical approaches.

Initial studies found that the majority of the nutrients in inedible plant material are rapidly solubilized in water at room temperature and pressure (Garland and Mackowiak 1990, Garland 1992). Direct use of these crop leachates in hydroponic systems inhibited plant growth due to either direct (i.e., phytotoxic compounds) and/or indirect (high biological oxygen demand in the root zone) effects caused by the presence of dissolved organic carbon compounds (Garland and Mackowiak 1990, Garland et al. 1993, Mackowiak et al. 1996b). Biological treatment of the crop residues decreased the dissolved organic content and eliminated plant growth inhibition (Garland and Mackowiak 1990, Garland et al. 1993). Simple, continuously stirred tank reactors (CSTR) were designed for processing the inedible biomass (Finger and Strayer 1994, Finger and Alazraki 1995). Effluents from these reactors have been used to supply approximately 50% of the plant nutrient demand in studies with wheat, potato, and lettuce without any deleterious effects on plant growth, even after 418 days of continuous operation (Mackowiak et al. 1996a, Mackowiak et al. this volume a).

While biological processing appears to be an effective method for recycling the majority of nutrients within inedible plant biomass, its ability to oxidize organic material is limited. The degree of organic matter oxidation in the bioreactors is dependent on the retention time of the reactor, ranging from 20% at a 1-day retention time to nearly 80% at 48-day retention time (Strayer et al., this volume a). The labile soluble organic compounds are rapidly degraded, while longer periods are necessary to degrade the more recalcitrant polymeric material such as cellulose and hemicellulose. Physical-chemical methods such as incineration would allow for complete oxidation at relatively short (i.e., a few hours) retention times (Bubenhiem et al. 1993), and may be useful for treating residual material from bioreactors following nutrient extraction.

Future Studies

The KSC Breadboard project plan contains three major goals regarding the integration of resource recovery and biomass production in the next 1-2 years; 1) optimize a hybrid biological/physical-chemical approach for processing inedible plant material, 2) evaluate the effects of recycling human graywater through plant production systems (with or without bioreactor pretreatment), and 3) quantify NaCl partitioning in ALS candidate crops.

Past studies indi ıte that biological processing is a reliable, low-energy method for recycling the majority of inorganic nutrients contained in inedible plant biomass (Finger and Strayer 1994, Finger and Alazraki 1995, Mackowiak et al. 1996a). Since the majority of nutrients are recovered from rapid abiotic solubilization within the bioreactor, the microbiological processing is only necessary to reduce the relatively labile soluble organic material. Therefore, we are testing effluents from bioreactors with retention times as low as 1 day. More efficient reactor designs (i.e., smaller volume and mass) such as fixed-f. .n systems can be utilized if solids are not processed. In this approach, inedible biomass would be separated initially into soluble and insoluble fractions. The soluble fraction could be rapidly processed in small bioreactors with a rapid turnover to recycle the majority of nutrients. Residual solids could be oxidized with physical-chemical systems to regenerate the majority of the CO_2. This approach is potentially advantageous since it utilizes the strengths of biological and physical-chemical treatment technologies by largely separating the processes of nutrient recycling and CO_2 regeneration. Biological processing regenerates nutrients in a form directly compatible with plant growth. Physical-chemical treatment would rapidly oxidize solid residues with lower inorganic nutrient content, thereby reducing the potential for volatilization of certain elements (i.e., nitrogen to NOx. compounds).

As stated above, the major mass of waste to be processed within an Advanced Life Support system is human graywater. Two potential harmful side effects of recycling graywater are 1) soap toxicity in plant growth systems, and 2) microbiological contamination. Many estimates of soap concentrations in graywater streams in space-based facilities are less than 200 ppm (Wydeven and Golub 1990, Friedman et al. 1992). Preliminary plant growth studies indicate that acute soap toxicity occurs at levels greater than 250 ppm (Bubenhiem, pers. comm.). Given that the volume of graywater (25 L $person^{-1}$ day^{-1}) is less than the probable volume of nutrient solution, direct recycling of the graywater into the hydroponic systems is not likely to result in soap toxicity. Accumulation of soap is unlikely since the soap compound can be rapidly degraded by microorganisms, and bacterial densities are nearly 10^{11} cells/ g dry wt. root in the plant production systems (Garland 1994). Given the high microbial densities in the plant growth system, proliferation of deleterious microorganisms washed from humans, and concomitant health risks in working with or harvesting plants systems, is a concern.

Our plan is to sequentially test the effects of directly recycling human graywater in small, intermediate, and large-scale plant growth experiments. Different loading rates of soap will be examined to determine maximum capacity (on a per m^2 growing area) of the plant production systems to process graywater. These studies will involve monitoring of 1) plant growth to determine acute and chronic plant growth effects, 2) soap concentrations to quantify rates of degradation, 3) microbial density and composition to quantify the potential survival of human-associated organisms, and 4) chemical and microbiological quality of atmospheric condensate to determine potential impacts on water recycling.

Even if direct incorporation of the graywater into plant growth systems does not have any deleterious effects, biological pretreatment may be a logical approach from a water use perspective. Approximately 30 L of water $person^{-1}$ day^{-1} will be needed to process inedible biomass at the loading rates utilized in the CSTR (20 g/L). Use of graywater as the diluent in waste processing systems may be more efficient than contaminating relatively clean water sources such as atmospheric condensate. Therefore, we will evaluate the effects of graywater incorporation on the performance of the aerobic bioreactors as part of our overall analysis of graywater recycling.

Urine recycling is desirable because it represents a significant source of water and nitrogen within the system. Water balances were discussed in the introduction. Based on an average urea content of 7000 mg/L (Putnam 1972), urine production would contain nearly 900 mmol of N $person^{-1}$ day^{-1}. Average N use for ALS candidate crops such as wheat and potato equals approximately 35 mmol m^{-2} day^{-1} (Mackowiak et al. 1996a). Assuming a crop growth area requirement of 40 m^2 per person, the estimated N flux in urine is over half of the plant N requirement in the system. The simplest approach for recycling the nitrogen and water within urine would be direct incorporation into the plant growth system. The major problems with direct recycling are potential phytotoxic effects of 1) ammonium as N source (urea can be readily converted to ammonium by microorganisms), and 2) NaCl accumulation. Na levels as lowe as 50 mM are toxic to some plants (Marschner 1995). Physical-chemical methods for removal of NaCl from the urine prior to recycling, and microbiological conversion of ammonium to nitrate (nitrification) could eliminate potential phytotoxicity. The KSC Breadboard project has begun development and testing of nitrification reactors (Strayer et al. this volume b), as well as evaluation of plant growth on mixed nitrogen (ammonium and nitrate) sources (Mackowiak et al. this volume b) to address the issue of ammonium in recycled waste streams. We are also conducting plant growth experiments to quantify the partitioning of NaCl in the tissue of candidate ALS in order to determine to what extent removal of NaCl from recycled urine is necessary. If sufficient NaCl could be partitioned into edible structures (such as potato tubers), NaCl inputs in the diet could be minimized to prevent NaCl accumulation in the systems.

CONCLUSIONS

1) Aerobic biological processing is an effective method for recovering the majority of inorganic nutrients from inedible biomass, and the recovered nutrients can be directly recycled to plant growth systems without any deleterious effects.

2) A hybrid biological/PC approach for processing inedible plant biomass should be tested as a means for both recycling nutrients and regenerating CO_2.

3) Direct recycling of human graywater in plant growth systems, or incorporation into bioreactors used to recycle inedible biomass, warrants immediate emphasis.

4) Urine recycling is necessary to recover N required for plant growth, but new approaches are needed to reduce NaCl accumulation in the system and concomitant plant toxicity.

REFERENCES

Bubenheim, D.L., K. Wignarajah, and T. Wydeven, Incineration in a controlled ecological life support system: a method for resource recovery from inedible biomass, *SAE Tech. Paper* 932249 (1993)

Bubenheim, D., K. Wignarajah, W. Berry, and T. Wydeven, Phytotoxic effects of graywater due to surfactants, *Amer. Soc. Hort Sci.* (in press)

Finger, B.W. and M.P. Alazraki, Development and integration of a breadboard-scale aerobic bioreactor to regenerate nutrients from inedible crop residues. *SAE Tech. Paper* 951498 (1995)

Finger, B.W. and R.F. Strayer, Development of an inermediate-scale aerobic bioreactor to regenerate nutrients from inedible crop residues. *SAE Tech. Paper* 941501 (1994)

Friedman, M.A., T.E. Styczynski, S.H. Schwartzkopf, B.E. Tiemiat, and M.C. Tieimiat, Gray water recycling with unique vapor compression distillation (VCD) design. *SAE Tech. Paper* 921318 (1992)

Garland, J.L., Characterization of the water soluble component of inedible residue from candidate CELSS crops. *NASA Tech. Mem.* 107557 (1992)

Garland, J.L., The structure and function of microbial communities in recirculating hydroponic systems. *Adv. Space Res.* 14:383-386 (1994)

Garland, J.L. and C.L. Mackowiak, Utilization of the water soluble fraction of wheat straw as a plant nutrient source. *NASA Tech. Mem.* 107544 (1990)

Garland, J.L., C.L. Mackowiak, and J.C. Sager, Hydroponic crop production using recycled nutrients from inedible crop residues. SAE Tech. Paper 932173 (1993)

Mackowiak, C.L., J.L. Garland, and J.C. Sager, Recycling crop residues for use in recirculating hydroponic crop production, *Acta Horticulturae* 440:19-24 (1996a)

Mackowiak, C.L., J.L. Garland, R.F. Strayer, B.W. Finger, and R.M. Wheeler, Comparison of aerobically-treated and untreated crop residue as a source of recycled nutrients in a recirculating hydroponic system. *Adv. Space Res.* 18:281-287 (1996b)

Mackowiak, C.L., R.M. Wheeler, G.W. Stutte, N.C. Yorio, and J.C. Sager, Use of biologically reclaimed minerals for continuous hydroponic potato porduction in a CELSS. *Adv. Space Res.* (this volume a)

Mackowiak, C.L., G.W, Sutte, J.L. Garland, B.W. Finer, and L.M. Ruffe, Hydroponic potato production on nutrients derived from anaerobically-processed potato plant residues. *Adv. Space Res.* (this volume b)

Marschner, H., Mineral nutrition of higher plants, 2nd Ed., Academic Press, New York (1995)

McKeehan, J.D., C.A. Mitchell, R.M. Wheeler, B. Bugbee, and S.S. Nielson, Excess nutrients in hydroponic solutions alter nutrient content of rice, wheat, and potato, Adv. Space Res. 18:73-83 (1996)

Parker, D.N. and S.K. Gallagher, Distribution of human waste samples in relation to sizing waste processing in space, Paper LBS-88-107, Symposium on lunar base and space activities in the 21st Century, Houston, Tx. (1988)

Putnam, D.F., Composition and concenrtrative properties of human urine. NASA CR-1802 (1971)

Shubert, F.H., R.A. Wydeven, and P.D. Quatrone, Advanced regenerative environmental control and life support systems: air and water regeneration. *Adv. Space Res.* 4:279-288 (1984)

Strayer, R.F., B.W. Finer, and M.K. Alazraki, Effects of bioreactor retention time on aerobic microbial decomposition of CELSS crop residues. *Adv Space Res.* (this volume a)

Strayer, R.F., B.W. Finger, and M.P. Alazraki, Evaluation of an anaerobic digestion system for processing CELSS crop residues for resource recovery. *Adv. Space Res.* (this volume b)

Wheeler, R.M., W.L. Berry, C.L. Mackowiak, K.A. Corey, J.C. Sager, M.M. Heeb, and W.M. Knott, A data base of crop nutrient use, water use, and carbon dioxide exchnge in a 20 square meter growth chamber: I. Wheat as a case study, *J. Plant Nutr.* 16, 1881-1915 (1993a).

Wheeler, R.M., K.A. Corey, J.C. Sager, and W.M. Knott, Gas exchange characteristics of wheat stands grown in a closed, controlled environment, *Crop Sci.* 33, 161-168 (1993b)

Wheeler, R.M., C.L. Mackowiak, J.C. Sager, N.C. Yorio, W.M. Knott, and W.L. Berry, Growth and gas exchange of lettuce stands in a closed environment, *J. Amer. Soc. Hort. Sci.* 119, 610-615 (1994)

Wheeler, R.M., C.L. Mackowiak, G.W. Stutte, J.C. Sager, N.C. Yorio, L.M. Ruffe, R.E. Fortson, T.W. Dreschell, W.M. Knott, and K.A. Corey, NASA's Biomass production chamber: a testbed for bioregenerative life support studies. *Adv. Space Res.* 18,215-224 (1996)

Wydeven, T., and M.A. Golub. 1990. Generation rates and chemical composition of waste streams in a typical crewed space habitat. NASA Tech. Mem. 102799

Pergamon

Adv. Space Res. Vol. 20, No. 10, pp. 1827–1832, 1997

Printed in Great Britain
0273-1177/97 $17.00 + 0.00

PII: S0273–1177(97)00848–X

WASTE BIOREGENERATION IN LIFE SUPPORT CES: DEVELOPMENT OF SOIL ORGANIC SUBSTRATE

N. S. Manukovsky, V. S. Kovalev, V. Ye. Rygalov and I. G. Zolotukhin

Institute of Biophysics (Russian Academy of Sciences, Siberian Branch), Krasnoyarsk 660036, Russia

ABSTRACT

An experimental model of matter turnover in the biotic cycle: plants (plant biomass) → mushrooms (residual substrate + mushroom fruit bodies) → worms (biohumus) → microorganisms (soillike substrate) → plants is presented. The initial mass of soillike substrate was produced from wheat plants grown in a hydroponic system. Three cycles of matter turnover in the biotic cycle were carried out. Grain productivity on soillike substrate was 21.87 g/m^2 day^1. The results obtained were used for designing a CES containing man, plants, soillike substrate, bioregeneration module and aquaculture. It was shown, that the circulating dry mass of the CES is 756 kg. The main part (88%) of the circulating mass accumulates in the soillike substrate and bioregeneration module.

INTRODUCTION

It is known that to develop a CES two problems are to be solved:

- co-ordination of matter turnover;
- maintenance of full value diet.

Under natural conditions co-ordination of matter turnover occurs in the soil. Simulation of this process in man-made ecosystems implies solving the following problems (Shepelev *et al.*, 1983):

- maintaining reliable linkage between constructive and destructive branches of the ecosystem with minimum impacts on control and energy expenditures;
- creating a buffer of carbon and other biogeneous elements to compensate for possible imbalance due to changes in the system structure and accidents;
- forming a versatile filter to adsorb and degrade volatile microimpurities and microbial and other particulates to reduce the toxicity of the system atmosphere;
- conditioning the atmosphere to maintain the composition and the specific "bouquet" of volatile compounds within the boundaries favorable for humans.

The idea to simulate biological turnover employing a soil substrate formed on the basis of natural soil from the organic matter produced inside a CES has been frequently discussed - Alexander *et al.* (1989); Ming (1989); Saugier *et al.* (1990); Smernoff *et al.* (1993). Nevertheless, in these publications no data are presented on intensity of matter turnover and values of masses of CES components.
The application of natural soils in space stations presents the following problems (Daunicht, 1990):

- low productivity of natural soils as compared with hydroponic systems;
- the need to use the immense mass of soil resulting from low productivity;
- vast transport expenditures;
- poor control of soil microflora.

In this connection the transport of natural soils should be decreased or excluded. Our approach is based on forming a soillike substrate from plant biomass without using a natural soil. Prestocked food and a nutrient medium for plant growing in a hydroponic system could be the primary sources for its forming. For example, carbon of the prestocked food in consequence of human metabolism is transfered to carbon dioxide and hereupon is absorbed by growing plants. The inedible plant biomass is converted in a biohumus by means of oyster mushrooms and worms. The biohumus could be used as a soil substrate or as organic part of soil substrate. Such a substrate is "endogenous"(ESS) because it is produced inside a CES. The transition of the entire CES to the steady state shall be determined by the time the ESS build up the steady mass. In particular, this time is determined by the decomposition constant of ESS. For example, proceeding from the experimentally found decomposition constant value of 0.001 day^{-1} the time for the soil substrate to build up the estimated steady-state biomass would take 1000 days (Manukovsky *et al.*, 1996).

Using ESS implies the increase of autonomy of a CES and the decrease or exclusion of the prestocked food. That should not be a reason to decrease a value of diet. In this connection an aquaculture could be considered as a perspective source of food protein. An aquaculture is supposed to be a part of a biological-physical-chemical life support system (Gitelson *et al.,* 1995). Nevertheless, it is unknown at what degree the wastes of aquaculture will be involved in the matter turnover of a CES by means of physical-chemical methods. That is why no data are presented in the publication on the intensity of matter turnover and values of stationary masses of the CES components. The recovery of these wastes seems to be more effective by means of biological methods. In this connection it is worth-while to consider an aquaculture as a part of a CES containing bioregeneration module and soil substrate.

The aims of this work are:

- to study the prefeasibility of ESS production;
- to perform a preliminary design of a CES containing ESS, bioregeneration module and aquaculture.

EXPERIMENTAL

In the course of forming and testing an ESS we carried out the following operations:

- growing plants (P);
- growing mushrooms (M);
- growing worms and forming biohumus (W).

These operations were conducted in the following succession (Fig. 1.):

- Operation - P_0 (10-11 weeks). Wheat (*Triticum aestuvi L*, variety 232) was grown and harvested in a hydroponic nutrient delivery system. The plant biomass containing a wheat straw and grains was directed to a mushroom growing system.
- Operation - M_0 (10-11 weeks). Edible oyster mushroom (Pleurotus florida) was grown on a substrate prepared from the biomass of wheat (*Triticum aestuvi L*, variety 232). For substrate preparation and growing oyster mushroom a standard technology was used (Heltay, 1979). A residual substrate and fruit bodies of oyster mushroom were directed to a worm cultivation bed.
- Operation - W_0 (10-11 weeks). A mixed culture of several hybrid lines close in characteristics to red Californian worms was grown in a mixture of residual substrate and fruit bodies of oyster mushroom. The worms were grown in a cultivation bed according to standard technology (*Bioconversion of* ..., 1990). No plants were grown on the bed. The biohumus containing worms was directed to a plant growing system. The mushroom growing system and the worm cultivation bed were the parts of bioregeneration module.

  The operations P_0, M_0, and W_0 were assigned to get an initial mass of biohumus for growing plants in the following operations.
- Operation - P_1. Wheat (*Triticum aestuvi L*, variety 232) was grown on biohumus placed in culture vessels. After 10-11 weeks a harvest (plant biomass) containing straw and a part of grains was

directed to a mushroom growing system. Another part of grains and a residual soil substrate (a residual biohumus in the culture vessels) were kept for using in operation P_2.

- Operations - M_1 and W_1 were carried out as M_0 and W_0. The biohumus formed was directed to the plant growing system.
- Operation - P_2 (10-11 weeks). The biohumus formed in W_1 was mixed with the residual soil substrate. The mixture was placed in the culture vessels and sown with wheat grains.

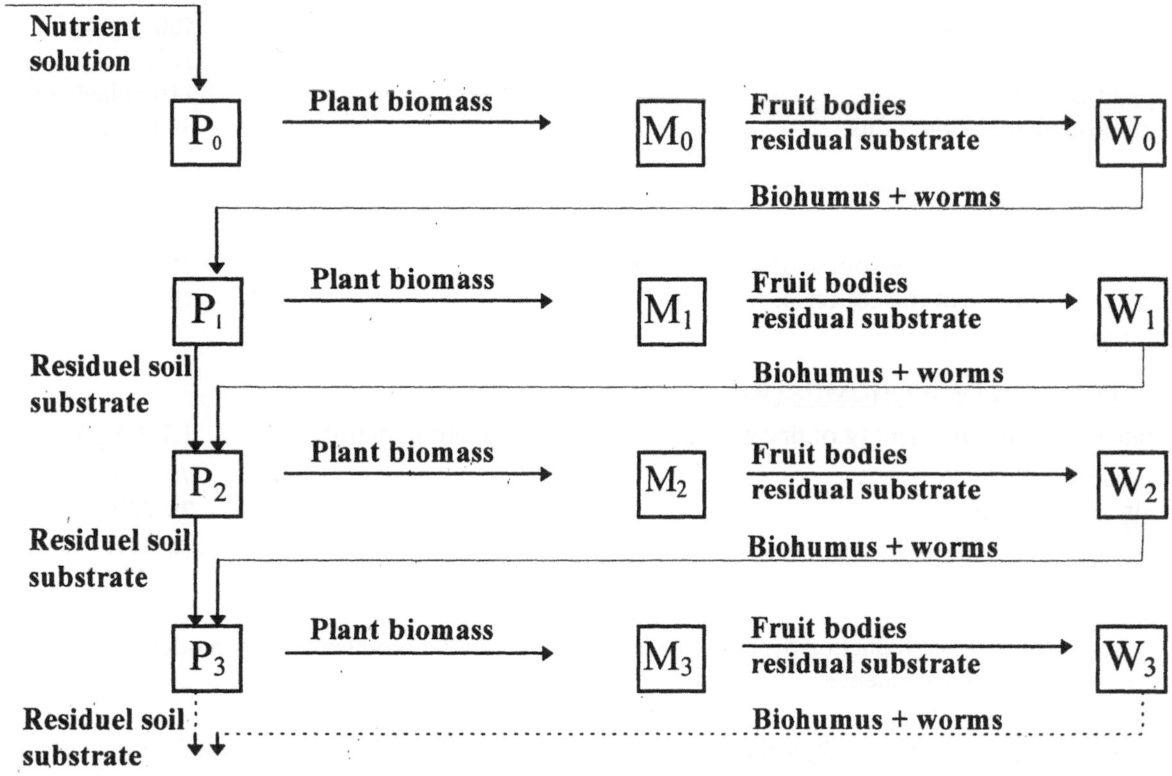

Fig. 1. Soil substrate regenerated from biomass of plants. P_i - growing plants, M_i - growing mushrooms, W_i - production of biohumus. The index (i) denotes the namber of the cycle.

The operation P_2 could be considered as the beginning of the following cycles: $P_2 \rightarrow M_2 \rightarrow W_2 \rightarrow P_3$ and $P_3 \rightarrow M_3 \rightarrow W_3$.. The third cycle was incomplete. Thus, the cycles were closed in terms of solid matter, and opened in terms of gaseous substances.

The parameters of growing plants and biomass conversion by means of oyster mushrooms and worms are given in Table 1. The yields of wheat on ESS and in the hydroponic system were compared. Under growing on ESS plants were watered once a day. The water used was not changed in the course of growing. Under growing in the hydroponic system a nutrient solution of Knopp was used.

The average total biomass production of wheat for one cycle on ESS including straw, grains and roots was 4156 g/m^2, the seed yield - 1531 g/m^2. Under wheat growing a hydroponic system the total biomass yield was 6636 g/m^2, the seed yield - 2444 g/m^2.

The additional investigations are needed to explain the decrease of yield on ESS. Grain productivity on ESS was 21.87 g m^{-2} day^{-1}. In the course of growing the mass of ESS lowered by 4-7 % because of forming carbon dioxide by soil reducers and absorption of biogenic elements by growing plants.

Table 1. Cycle component's parameters.

Parameters	Value
GROWING PLANTS (P_{1-3})	
vegetation period	70 days
filling rate of soil substrate (dry mass)	13.9 kg/ m^2
humidity of soil substrate	60-70 %
density of soil substrate	500-650 kg/ m^3
air temperature	22^{0}C
plant density	1100 plants/m^2
photosynthetic photon flux	500 μmol m^{-2} s^{-1}
GROWING MUSHROOM (M_{0-3})	
process time	60-70 days
air temperature in the course of spawn running	25-27^{0}C
air temperature in the course of fruiting	18-20^{0}C
humidity of substrate	70 %
PRODUCTION OF BIOHUMUS (W_{0-3})	
rate of worm settlement (g of dry weight per kg of dry weight substrate)	1.1-1.6 g/kg
process time	70 days
air temperature	20-22^{0}C
humidity of substrate	75-80%

We have assumed that reduction dynamics of plant biomass in the bioregeneration module is described by the exponential dependence

$$R = R_0 \cdot \exp\{-k_1 t\},$$

where: R is residual biomass; R_o is the initial plant biomass; t is time, k_1 is the decomposition constant. It has been found that the decomposition constant k_1 both for conversion of plant biomass by oyster mushroom and for convertion of spent substrate by worms is 0.01 day^{-1}

In the process of mushroom growing 1 kg of dry plant biomass was converted to 0.4-0.5 kg of residual substrate and 50-60 g of mushroom fruit bodies. Thereupon these products were converted to 0.2-0.3 kg of biohumus by means of worms.

DESIGN OF CES

The CES contains the following components: man, plants, soil substrate, bioregeneration module and aquaculture. A method of mass balance was used for calculating the matter turnover inside the CES. We calculated the following parameters:

- Average daily amount of substances passing into and out of the every component of the CES;
- Stationary masses of circulating matter in the components.

Under calculating these parameters we proceeded from the average indeces of human metabolism (*Closed system* ..., 1979). Mushroom fruit bodies (8.5 g of protein) and fish (38 g of protein) were included in the daily diet.

The daily amount of fish protein was a starting point for calculating the parameters of aquaculture. Under calculating the aquaculture we took into account the data of fish physiology (*Activity* ..., 1967;

Bioenergetics ..., 1979;). Specific growth rate of fish population was assumed to be 0.014 day^{-1}. Feed for fishes was planned to be provided with wheat grains, algal biomass and mushroom fruit bodies.

Taking into account the requirements of consumers (man and fishes) and a data on plant biology (*Closed system* ..., 1979) the parameters of plant culture were calculated. The specific growth rate of plants was assumed to be 0.03 day^{-1}, the rate of the edible part of plant biomass - 34 %.

The parameters of the bioregeneration module were calculated on the basis of the input of inedible plant biomass (1542 g/day), urine (80 g/day), feces (22 g/day) and wastes of aquaculture (234 g/day). Having calculated the stationary mass of matter in the bioregeneration module on the basis of our data mentioned above we assumed that a constant of matter decay was 0.01 day^{-1} and duration of decay was 120 days.

The stationary mass of ESS was calculated on the basis of the daily input of biohumus and the value of the decomposition constant of soil organic matter. The value of the decomposition constant was assumed to be 0.001 day^{-1} (Manukovsky *et al.*, 1996).

The data and assumptions mentioned above were used for developing a model of a 4 component CES (Fig. 2.).

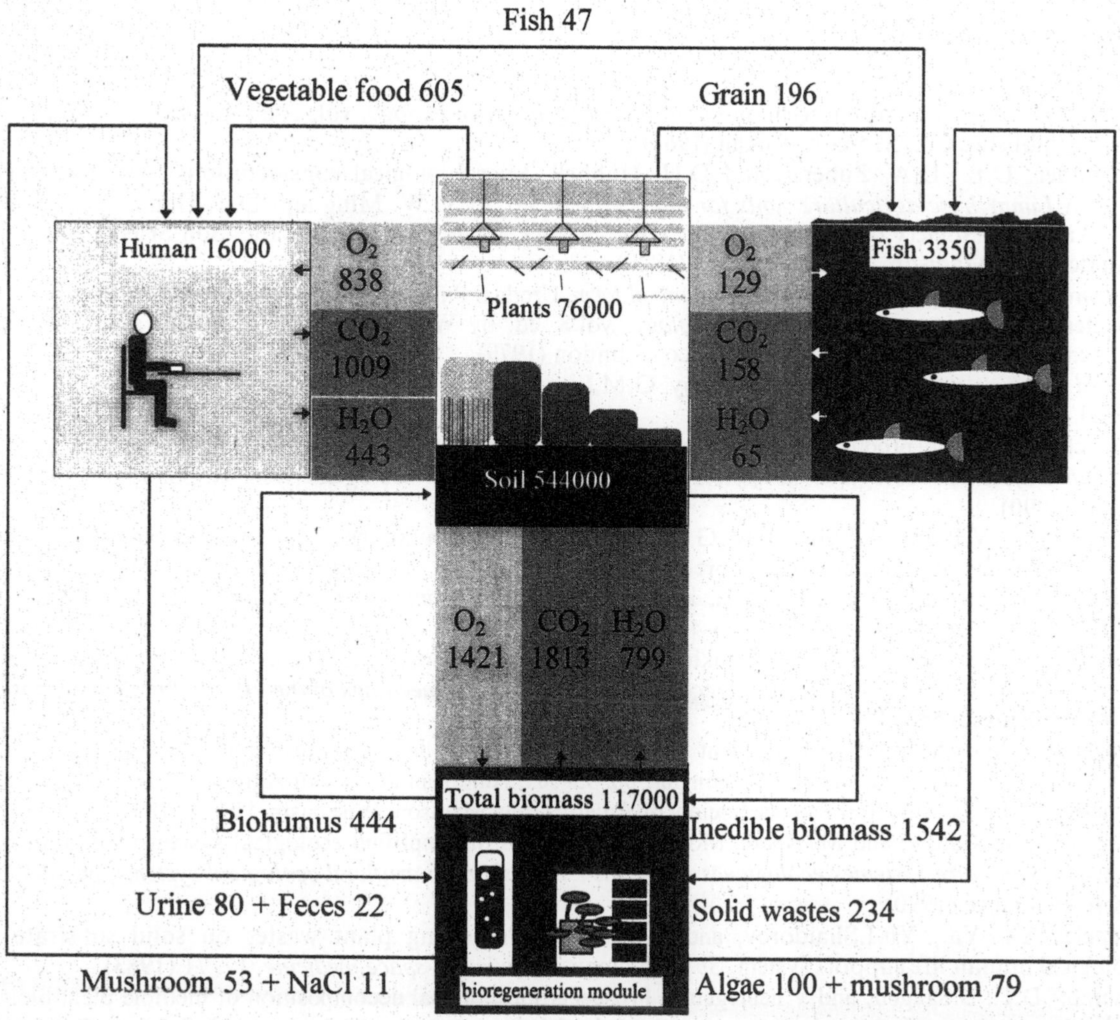

Fig. 2. Matter turnover in a four-component CES. Massflow rates of solid substances are expressed in grammes of dry weight/day. Massflow rates of gases are expressed in g/day. Only metabolic water is taken into account. Stationary biomasses of components are expressed in grammes of dry weight.

It has been calculated that with the water volume of 1-1.5 m^3 sufficient to sustain 13-14 kg of fish (living weight) an aquaculture may supply the human diet with 38 g of animal protein in daily consumption of 235 g of fish (living weight) in addition to 8.5 g of mushroom protein in daily consumption of 400 g of fresh mushrooms grown in the bioregeneration module on plant mass. To sustain the fish population it is necessary to spend daily 375 g of dry fodder involving 196 g of seeds, 100 g of algal biomass and 79 g of mushroom fruit bodies. As compared to a vegetarian version of nutrition this would require increasing the plant area by 24% and the stationary biomass in the bioregeneration module to 117 kg.

The bioregeneration module requires 60 % of oxygen produced by phytotron, the man - 35 % and the aquaculture - 5 %. The prevalence of the bioregeneration module in the consumption of oxygen is explained by the prevalence of inedible plant biomass in the crop.

It follows from Fig. 2, that the dry circulating mass of the CES is 756 kg, The soil substrate contains 72 % of this mass and the bioregeneration module - 16%. In case of using natural soil in a space station this data seem to be a disadvantage because of large transport expenditures. That is not quite true as regargds ESS.

The design of a CES is intended for the development of long-term space stations.

REFERENCES

Activity, habitability and biotechnology, in *Problems of space biology*, vol. 7, ed. by V.N. Chernigovsky, Nauka Publising Co., Moscow, Russia (1967).

Alexander, D.B., D.A. Zuberer, and D.H. Hubbell, Microbiological considerations for lunar-derived soils, in *Lunar base agriculture: soils for plant growth,* ed. by D.W. Ming and D.L. Henninger, pp. 245-255, American Society of Agronomy, Inc. Crop Science Society of America, Inc. Soil Science Society of America, Inc. Madison, Wisconsin, USA (1989).

Bioconversion of organic wastes in biodynamical farm, Urodgai Publishing Co, Kiev, USSR (1990).

Bioenergetics and growth, in *Fish physiology,* vol.8, ed. by W.S. Hoar, D.J. Randall, and J.R. Brett, Academic press, New York - San Francisco - London (1979).

Closed system: man-higher plants, ed. by G.M.Lisovsky, Nauka Publising Co., Novosibirsk, Russia (1979).

Daunicht H.J., Crop production systems in space: characteristics and further biological research needs, in *Seminaire systemes ecologiques artificiels / Workshop on artificial ecological systems*, pp. 75-90, Marseille, France (1990).

Gitelson J.I., V.Blum, A.I.Grigoriev, G.M.Lisovsky, N.S. Manukovsky, and et *al*., Biological-physical-chemical aspects of a human life support system for a lunar base, *Acta Astronautica,* 37, 385 (1995).

Heltay I., Industrieller Pleurotus Anbau, in *Proceedings of 10th international congress on the science and cultivation of edible fungi,* part 2, pp. 463-481, France (1979).

Manukovsky N.S., Kovalev V.S., Zolotukhin I.G., and V.Ye.Rygalov, Peculiarities of soil substrate transition to steady state at starting a CES, in *Correction of homeostasis, Proceedings of the 7th All-Russian symposium,* pp. 45-48, Krasnoyarsk, Russia (1996).

Ming D.W., Manufactured soils for plant growth at a lunar base, in *Lunar base agriculture: soils for plant growth,* ed. by D.W. Ming and D.L. Henninger, pp. 93-105, American Society of Agronomy, Inc. Crop Science Society of America, Inc. Soil Science Society of America, Inc. Madison, Wisconsin, USA (1989).

Saugier B., M. Caloin, and M. Andre, Modelling Dynamics of simplified ecological system based on higher plants, in *Seminaire systemes ecologiques artificiels / Workshop on artificial ecological systems*, pp. 191-202, Marseille, France (1990).

Shepelev Ye. Ya., Yu.I.Shaidorov, and V.V.Popov, Processing plant wastes on solid substrate for biological human life support system, *Space biology and aviation-space medicine*, 17, 71 (1983).

Smernoff D.T., B.Saugeir, and V.Fabregettes, Prospects for biolgical decomposition of inedible plant biomass on long-term space missions, in *Procceedings of the 5th European Symposium on Life Sciences Research in Space,* pp. 239-243, Arcachon, France (1993).

Pergamon

Adv. Space Res. Vol. 20, No. 10, pp. 1833–1843, 1997

Printed in Great Britain
0273-1177/97 $17.00 + 0.00

PII: S0273-1177(97)00849-1

INTEGRATION OF CROP PRODUCTION WITH CELSS WASTE MANAGEMENT

K. Wignarajah* and D. L. Bubenheim**

**Lockheed Martin Engineering and Science Corporation, NASA Ames Research Center, Space Technology Division, Regenerative Life Support Branch, Mail Stop 239A-3, Moffett Field CA 94035, U.S.A.*

***NASA Ames Research Center, Space Technology Division, Regenerative Life Support Branch, Mail Stop 239-15, Moffett Field CA 94035, U.S.A.*

ABSTRACT

Lettuce plants were grown utilizing water, inorganic elements, and CO_2 inputs recovered from waste streams. The impact of these waste-derived inputs on the growth of lettuce was quantified and compared with results obtained when reagent grade inputs were used. Phytotoxicity was evident in both the untreated wastewater stream and the recovered CO_2 stream. The toxicity of surfactants in wastewater was removed using several treatment systems. Harmful effects of gaseous products resulting from incineration of inedible biomass on crop growth were observed. No phytotoxicity was observed when inorganic elements recovered from incinerated biomass ash were used to prepare the hydroponic solution, but the balance of nutrients had to be modified to achieve near optimal growth. The results were used to evaluate closure potential of water and inorganic elemental loops for integrated plant growth and human requirements.

INTRODUCTION

Short duration space missions involving small crews can be undertaken with a supply of all expendable life support materials and the return of wastes to Earth. As mission duration increases and larger crew sizes become involved, the logistics of resupply of all consumables and storage of wastes becomes difficult. Resupply is costly /13/ and has a negative impact on the design of the spacecrafts /14/. Some level of self-sufficiency which provides for mass recycling and regeneration of life support materials will reduce logistics and increase crew safety.

Plants have demonstrated the ability to create a safe and habitable environment on Earth. The Controlled Ecological Life Support System (CELSS) attempts to exploit the food production, water purification, and air revitalization function of crops in a controlled manner to provide life support for future space missions /4/.

The consumable life support needs of individual crewmembers in space are estimated to be 0.62 kg of dry food per day (approximately 12 MJ $person^{-1}$ day^{-1}), 1.62 kg potable water per day, generation of 0.84 kg O_2 and removal of 1 kg of CO_2 /20/. The challenge is to recycle mass and regenerate resources for life support from human and habitation wastes.

The wastes generated from a human habitat include all three states of matter - solid, liquid, and vapor. In a fully functional lunar/Mars colony, producing its own food, liquid phase waste represents the biggest component on a mass basis - approximately 62% /20/. The inedible biomass and other solid wastes comprise 36% of the total waste stream, assuming a harvest index of 0.46 for crop production. The gaseous fraction consists chiefly of respired CO_2 while the liquid fraction is mainly wastewater from human consumption. From a chemical standpoint, the liquid waste stream is the smallest resource for elemental matter but the largest resource for water. Solid wastes are the major resource for inorganic elements and carbon for crop production. The solid wastes require processing for recovery and reuse of the resources.

In the investigations summarized here, we have attempted to demonstrate the ability to recover resources from wastes and regenerate life support materials using plants. A number of technologies are worth considering /5, 24/. We considered a limited number of waste treatment technologies to recover water, CO_2 and inorganic resources from liquid and solid waste streams. Liquid waste processing was limited to vapor compression distillation (VCD) and biological processing while incineration was selected for solid waste processing. VCD technology has produced high purity water and has attained high water recovery rates at a relatively high specific consumption rate /10/. Biological processing was chosen for wastewater processing since it is one of the foremost technologies currently in use. It provides considerable potential for use with naturally selected and genetically engineered microbes (GEM) for specifically targeting the biodegradation of defined chemicals in the wastewater. Dry incineration was used for solid waste processing because previous modeling studies showed that it is an excellent candidate technology for the processing of wastes /18/, and there is extensive industrial experience in processor design and operation. Our investigations have attempted to define the suitability of untreated wastes and recovered products from the selected treatment technologies as crop production inputs.

METHODS AND MATERIALS

Lettuce Production Protocol

Lettuce (*cv.* Waldmann's Green) seeds were germinated at 23°C under a 24 hour photoperiod at 85 µmol m^{-2} s^{-1} provided by fluorescent lamps. Initially nutrient levels were at one-fifth strength with the levels being increased to one-third and full strength by day 3 and 5, respectively (Table 1). Plants were transferred to the greenhouse at the 3-leaf stage which ranged from 6-10 days after seeding. The pH of the nutrient solution was maintained between 5.7-6.3; increases in pH were compensated for by the addition of 100 mM HNO_3. Electrical conductivity was maintained at 980 µS cm^{-1} by addition of complete nutrient solution. Plants were grown at a planting density of 36 plants m^{-2}. The temperature in the greenhouse was maintained between 21-24°C; natural photoperiod (maximum of 14 hours in the summer and a minimum of 11 hours in the fall and spring) and natural spectral quality of sunlight was provided.

Wastewater Stream Production

The major contaminant of hygiene wastewater stream is the surfactant employed for washing purposes. A number of formulations for space station soap were developed under NASA contract /9/. Human rating tests done at NASA Johnson Space Center using these formulations showed that shampoo formulation, No. 6503.45.4 was favored as the candidate soap /19/. This formulation has as its main ingredient an anionic surfactant, Igepon TC-42[1]. A simulated wastewater stream was generated by using a range of Igepon concentrations.

Water Recovery Techniques and Product Water Suitability

In these studies, gray water was pre-processed by either the physical process of vapor compression distillation /10/ or biological treatment /21/. The treated product water from these two processes were collected separately and used in phytotoxicity and crop productivity studies.

Vapor Compression Distillation – The VCD process is an energy efficient water distillation process; the design and function of the apparatus is described in another paper /10/. The wastewater stream was fed into the VCD processor which produces a distillate product stream of purified water and a bleed stream containing the waste products. The distillate product was collected and used for the preparation of the hydroponic solutions for crop production.

Table 1. Full strength nutrient solution for lettuce production - source and concentration of salts used

Nutrient Source	Solution Concentration
Macronutrients (mM)	
KNO_3	5.0
$Ca(NO_3)_2$	2.5
$MgSO_4$	1.5
K_2HPO_4	0.5
KH_2PO_4	0.5
Micronutrients (µM)	
H_3BO_3	46.6
$MnSO_4$	4.5
$ZnSO_4$	0.38
$(NH_4)_6 Mo_7O_{24}$	0.05
$CuSO_4$	0.15
$CoCl_2$	0.08
Fe^{3+} (Sequestrene 330)	96.0

The Bioreactor – The bioreactor was an aerobic, packed-bed column with extended surface packing for microorganism attachment /21/. The simulated shower water was recirculated through the reactor by pumping liquid from a reservoir to the top of the column and allowing the solution to flow down as a film over the packing surface. The bioreactor had been seeded previously by collecting roots from lettuce plants grown hydroponically in nutrient solution containing Igepon and the microorganisms cultured on agar plates /21/. Hygiene water processing was performed in a batch mode and the product water collected and used for the preparation of the hydroponic solutions for crop production.

Wastewater Phytotoxicity Determinations – Lettuce, a candidate CELSS crop was chosen for phytotoxicity studies. Studies were conducted using a short-term bioassay as well as long-term production studies using Igepon concentrations in a concentration range from 0-1000 mg l^{-1} to establish the dose response. The short-term bioassay was conducted according to the method of Berry /3/. Ten replicates were used per treatment. Whole plant production studies were carried out in the greenhouse under environmental conditions described in an earlier section. Production parameters were evaluated through measurements of fresh weights, leaf area and dry weights at regular time intervals during experimentation. Six replicates were used at each harvest per treatment.

Surfactant Analysis – Concentrations of the surfactant in the nutrient solution were measured by either monitoring GC-MS analysis of the fatty acid profile /1/ or the use of a surfactant electrode (Orion - Model No. 93-42) probe on the Orion 960 Autochemistry system. Sodium Lauryl Sulfate (SLS) was used as the standard.

---

[1]Igepon TC-42 is the major constituent of candidate surfactant (ECOLAB 6503.45.4) selected for use in the Space program. Chemically Igepon is N (coconut oil acyl) N-methyltaurine, sodium salt.

Incineration of Biomass and Product Suitability Studies

For purposes of recovering the inorganic elements in biomass, 25 g of ground wheat straw were placed in ceramic crucibles and incinerated at a temperature of 800°C for 1 hour. The ash was dissolved in a known amount of 4 M nitric acid. The amount of nitric acid was appropriate so that the final concentraton was 10 mM.

Lettuce plants were grown in one of four nutrient solutions: (1) incinerated ash, (2) simulated ash - prepared using reagent grade chemicals to provide an inorganic elemental mix identical in chemical composition to the ash, (3) ash supplemented to provide a full complement of nutrients equivalent to the control, and (4) the control - providing a full complement of nutrient solution (Table 1). The plants were grown for 42 days. Whole plant samples were harvested weekly; 2 harvests were done each week at 3- and 4-day intervals during the exponential phase of growth.

Incineration Product Gas Studies - Incineration was carried out at 800°C in a Lindberg Model 51142 box furnace, using air (21% O_2, balance N_2) as the oxygen source for combustion. The gas mixture evolved from the incineration of wheat straw was collected in a pressurized null voided cylinder. Final cylinder CO_2 concentration measured using GC was 10%. Measurements of CO, SO_2 and NO_x were made colorimetrically using Dräger™ tubes. Organic volatiles from the incinerator gas were measured using a GC-MS instrument. The incinerator product gas was used as the CO_2 source for lettuce production in a closed chamber. The gas was metered from the cylinder into the Plant Volatile Chamber (PVC) based on the need to maintain the CO_2 set point. The impact of incinerator-derived gases on the physiological processes of dark and light respiration, root respiration, photosynthesis and transpiration were quantified.

Inorganic Elemental Analysis - Inorganic elements in nutrient solutions were analyzed using inductively coupled plasma (ICP) and ion chromatography (IC). Solid samples were analyzed by thermal arc spectroscopy.

RESULTS AND DISCUSSION

Wastewater Suitability and Igepon Phytotoxicity

The phytotoxicity effects of Igepon are presented in Table 2. At a concentration of 250 mg l^{-1}, there was a 40% suppression in fresh weight production over a 36-day growing period. Higher concentrations resulted in greater suppression of growth with an 85% reduction of growth at 1000 mg l^{-1} Igepon. Direct utilization of simulated shower water in crop production is clearly a problem due to the presence of surfactants.

Table 2. Effect of Igepon concentration on lettuce production in greenhouse conditions over a 36-day growth period

Igepon Concentration mg l^{-1}	Fresh Head Yield g $plant^{-1}$	Yield Suppression %
0	79.95	
125	69.74	13%
250	47.83	40%
500	14.12	82%
1000	12.41	85%

Monitoring Igepon concentration on a daily basis after addition to the nutrient solution showed that within 3 days of addition, the Igepon in the hydroponic solution was destroyed. The destruction of Igepon by the microbial population is addressed in another paper /12/. The plants grown in Igepon, however, continued to show reduced growth rates for an additional 3-5 days. At the end of this period, the plants recovered and re-established growth rates similar to the control plants /23/. The severity of surfactant phytotoxicity and recovery following Igepon exposure is also evident through the measurement of the gas exchange in lettuce (Figure 1). Photosynthesis is suppressed after the addition of Igepon to the nutrient solution. An almost identical response pattern is true for transpiration response to Igepon (data not shown). In all cases, there is a recovery from acute phytotoxicity with degradation of the surfactant.

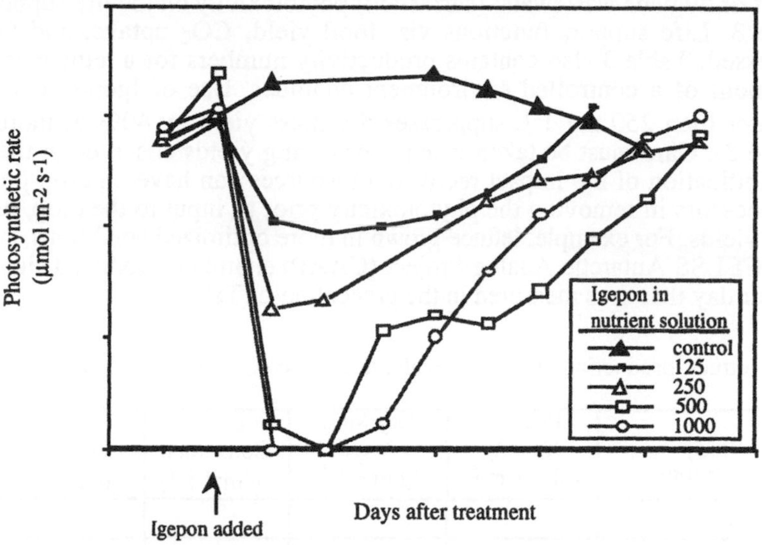

Fig. 1. Effect of surfactant (Igepon) concentration on photosynthesis of lettuce.

Wastewater Treatment and Product Suitability

Treatment of wastewater to eliminate phytotoxic effects prior to utilization may be appropriate to maintain a predictable, high level of crop performance. Two techniques were used for this purpose: (1) biological treatment and (2) vapor compression distillation (VCD). The product water of both liquid waste treatment systems provided water suitable for crop production with no phytotoxic effects (Figure 2).

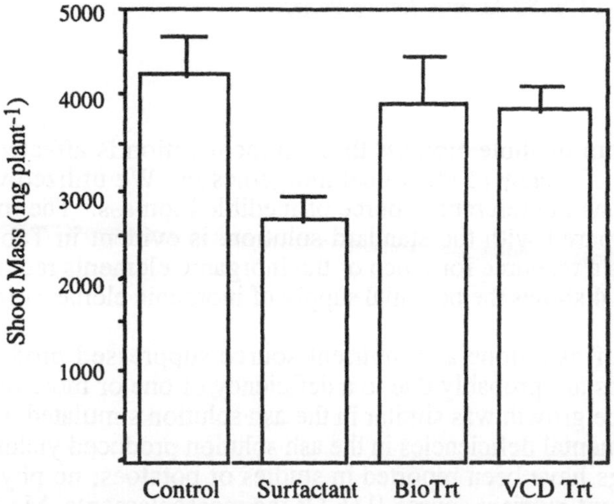

Fig. 2. Mass of 17-day-old lettuce plants grown in nutrient solution formulated with recovered water following biological or vapor compression distillation treatment.of wastewater containing 250 mg l^{-1} Igepon.

Growth of lettuce was almost identical in hydroponic solutions formulated with de-ionized water, VCD product water, or product water from the bioreactor. Neither the product water from bio-processing nor from VCD is of potable quality and thus it is available to the crops: with the phytotoxicity removed, it is an appropriate method for further purification to potable quality.

The potential impact of hygiene wastewater (surfactant) phytotoxicity on the life support function of plants is evident in Table 3. Life support functions viz. food yield, CO_2 uptake, and O_2 production were significantly suppressed. Table 3 also contains productivity numbers for a lettuce crop grown in a more optimized environment of a controlled environment chamber. Use of Igepon contaminated water, at concentrations greater than 250 mg l^{-1}, suppresses dry mass yield by 40% or more, depending on the concentration (Table 3). Care must be taken when considering yields observed in the greenhouse. These studies provide an indication of the impact recovered resources can have on crop production and on the effectiveness of processors in removing the phytotoxicity prior to input to the crop, however, they do not represent optimized yields. For example, lettuce grown in more optimized conditions, such as the controlled environment of the CELSS Antarctic Analog Project (CAAP) chamber, produced significantly more plant mass per unit area per day than was measured in the greeenhouse (Table 3).

Table 3. A lettuce production system to evaluate effectiveness in the CO_2/O_2 loop of CELSS

Study Conditions	Planting Density (plants m^{-2})	Dry Mass Yield (g m^{-2} d^{-1})	CO_2[1] Consumed (g m^{-2} d^{-1})	O_2[1] Produced (g m^{-2} d^{-1})
Greenhouse				
No Igepon	36	4.40	6.45	4.69
125 mg l^{-1} Igepon	36	3.84	5.63	4.09
250 mg l^{-1} Igepon	36	2.63	3.86	2.81
500 mg l^{-1} Igepon	36	0.77	1.13	0.82
CAAP[2]				
No Igepon	96	18.47	27.09	19.70

[1]CO_2 consumption and O_2 production values were computed from the biomass production values.
[2]CAAP stands for CELSS Antartic Analog Project.

Inorganic Recovery

Recovery of inorganics from inedible biomass through incineration is affected by both the quality of the biomass and the recovery efficiency of individual inorganics /6/. We utilized wheat straw, grown under a standard set of conditions, as the reference source of inedible biomass. The imbalance of nutrients in the ash-derived solution, compared with the standard solution, is evident in Table 4. Ash from incinerated wheat straw can be used as a resource for much of the inorganic elements required for the preparation of a hydroponic solution. Table 4 shows the potential supply of inorganic elements from solubilized ash.

Results showed that use of ash alone as a nutrient source suppressed productivity by more than 40% (Figure 3). The lower yields are probably due to a deficiency of one or more of the nutrients rather than to any phytotoxic effects, since growth was similar in the ash solution simulated with reagent grade chemicals /7/. Supplementing the elemental deficiencies in the ash solution produced yields comparable to the controls (Figure 3). Similar findings have been reported in studies of potatoes; no phytotoxic effects were shown when ash was used as a plant nutrient source /11/. Of the macroelements, Mg was present at only 24% of the concentration in the standard solution, while amongst the microelements, Cu, Fe and Zn were present in very low concentrations. Some of the deficiencies may be overcome by using more ash than the present 25 g per 100 *l* ratio, but the potential consequences of the imbalance, such as toxicity and/or inability to supply the required micronutrients, are still present.

Table 4. Supply of inorganic elements when 25 g of wheat straw ash was dissolved in 100 *l* of water*

Element	K	Ca	Mg	B	Cu	Fe	Mn	Mo	Zn	P
Standard Solution (mM)	5	2.5	1.5	0.047	0.00015	0.096	0.0045	0.0005	0.0038	1
Ash Solution (mM)	5	1.68	0.36	0.021	0.0000045	0.00672	0.00486	0.208	0.00003	0.6
Percentage supply of complete nutrient	101	67	24	44	3	7	108	416	8	60

*Ash was dissolved in a minimal amount of nitric acid prior to water addition.

Interestingly, attempts to utilize bioprocessor effluent to recycle inorganic nutrients from solid wastes showed phytotoxic effects on hydroponic potato production. Filtering the effluent through a 0.2 μM filter removed the phytotoxic component /15/. This suggests that it may be advantageous to use incineration technology for recovering resources from solid wastes.

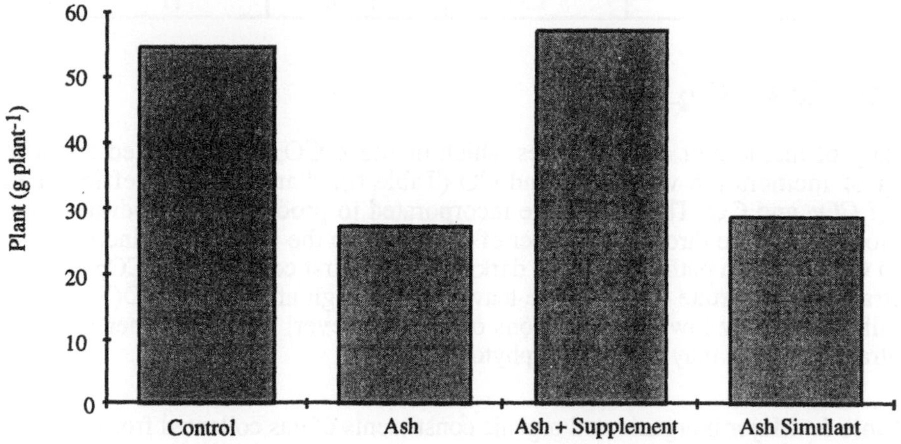

Fig. 3. Dry mass of 22-day-old lettuce plants as influenced by source of nutrients.

The quality of the waste stream to be processed, as well as the processing techniques, will influence the balance of nutrients in the final product. By looking at the quality of the waste streams anticipated for a human habitat, we can identify areas of potential deficiency or toxicity. The potential for using waste streams other than biomass is evident from an examination of the inorganic content of these waste streams (Table 5)

Table 5 shows that the use of all wastes, other than inedible biomass, potentially generates a nutrient solution which is slightly deficient in the major cations, K, Ca, Mg, but sufficient in most of the micronutrients, except for Fe, B, and S. The elemental composition of the solid wastes listed in Table 5 are from unprocessed wastes, and the effects of the processing on these wastes are undefined and require further investigation.

Table 5. Supply of inorganic elements from a mixture of liquid and solid waste streams other than biomass (feces, urine, hand and shower wastewater, urinal flush, sweat and metabolic water condensate) expected to be produced by a single crewmember per day /20/.

Element	Mass Available from Waste (mg)	Mass Required (mg)	Percentage Supplied (%)
K	3697	4887	75
Ca	1845	2505	74
Mg	572	911	63
Na	5907	--	--
P	1342	775	171
N	12143	3500	347
S	580	1200	49
Cl	2483	0.0071	--
Zn	17.26	0.63	2773.8
Cu	3.11	0.24	1296
Mn	10.6	6.19	171
Fe	27	134.4	20
B	1.38	12.58	11

Incinerator Product Gas as a CO_2 Source

The phytotoxicity of incinerator derived gases which included CO_2 are presented elsewhere /8, 16/. The major products of incineration were CO_2 and CO (Table 6). Plants do have efficient mechanisms for incorporation of CO_2 and CO. The former are incorporated to produce carbohydrates by photosynthesis, while CO fixation takes place through a number of pathways. In the light, CO is incorporated into serine or in C4 plants via CO_2 fixation pathways /2/. In darkness CO is first converted to CO_2 and then incorporated into malate, citrate and aspartate /17/. Lettuce leaves have a high affinity for CO(K_m= 7 nM), suggesting that the leaf will utilize very low concentrations of CO. However, there are other gases present in much smaller concentrations which may prove to be phytotoxic.

Table 6. Major inorganic and organic constituents of gas collected from incinerating dried wheat straw

Compound	Chemical Formula	Molar Concentration in Incinerator Gas (μmol mol^{-1})
inorganics		
Carbon dioxide	CO_2	106,000
Carbon monoxide	CO	8066
Sulfur dioxide	SO_2	<1.0*
Nitrogen oxides	NO_x	3-4
organics		
Benzene	C_6H_6	2.4
Toluene	$C_6H_5CH_3$	0.24
Xylene	$C_6H_4(CH_3)_2$	0.04

Both photosynthesis and transpiration were adversely affected by exposure to the incinerator-derived gases (Figure 4). Within 3 days of injection of the incinerator gas based on CO_2 supply needs, photosynthesis was suppressed by 30%; after 6 days following injection of suppression was approximately 80%. The source of this phytotoxicity has not been determined and is the subject of current study.

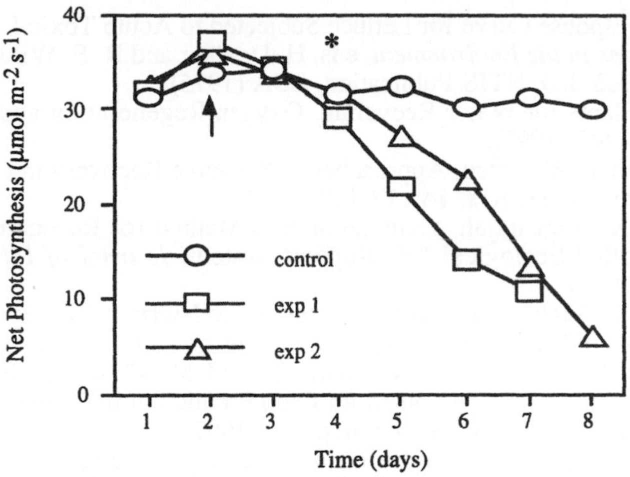

Fig. 4. The effect of incinerator gas on photosynthesis when used as a source of CO_2. The up arrow indicates the start of incinerator gas as a CO_2 source. The asterisk indicates when incinerator gas was removed and replaced with pure CO_2 input during experiment 2.

Significant control over the quality of gaseous product is possible via manipulation of incinerator combustion conditions /22/. Identification of the phytotoxic component(s) of incinerator product gas, characterization of the ability to control gaseous product quality through incineration conditions, and identification of practical post processing techniques should provide an avenue for effective CO_2 recovery.

SUMMARY AND CONCLUSIONS

Waste streams provide an accessible source of resources for crop production and maintenance of plant-based life support functions. Technologies for water and inorganic recovery were exercised and shown to be effective in providing resources for crop production from waste streams. Not unexpectedly some limitations were identified. Though use of incinerator ash did not prove to be phytotoxic, the elemental quality of waste and differential recovery of individual elements can lead to an imbalance in nutrients and the requirement for nutrient supplements. Removal of the phytotoxic components of wastewater resulting from the surfactant was demonstrated. Techniques for carbon recovery and utilization as a crop input requires further definition. Studies are currently being conducted to characterize the products of incineration gas and define the products from solid waste processing technologies. Control of combustion conditions and/or post processing of product gas to remove phytotoxic component(s) is promising. Our studies have dealt with representative waste streams and processing technologies. These data show that utilization of recovered water, carbon dioxide, and inorganics can fuel the production of life support materials by crops and provide a promising outlook for the development of fully regenerative life support systems.

ACKNOWLEDGEMENT

Thanks to Karen Bunn for the preparation of the manuscript.

REFERENCES

1. Belisle, W., Fatty Acid Compositions in Soap. Document ACL # 10020, Central Analytical Laboratories, NASA Ames Research Center, Moffett Field, CA, USA (1993).
2. Bidwell, R. G. S. and D. E. Fraser, Carbon Monoxide Uptake and Metabolism by Plants. *Can. J. Bot.* **50**:1435-1439 (1972).
3. Berry. W. L., Dose-response Curve for Lettuce Subjected to Acute Toxic Levels of Copper and Zinc, *Biological Implications in the Environment,* eds. H. Drucker and R. E. Wilding, ERDA Symp. Series, CONF-750929, pp. 365-369, NTIS Publication, USA (1975).
4. Bubenheim, D. L., Plants for Water Recycling, Oxygen Regeneration and Food Production, *Waste Manage. Res.* **9**:435- 443 (1991).
5. Bubenheim, D. L. and T. Wydeven, Approaches to Resource Recovery in Controlled Ecological Life Support Systems, *Adv. Space. Res.* **14**:113-123 (1994).
6. Bubenheim, D. and K. Wignarajah, Incineration as a Method for Resource Recovery from Inedible Biomass in a Controlled Ecological Life Support System, *Journal of Life Support and Biosphere Science* **1**:129-140 (1995).
7. Bubenheim, D. and K. Wignarajah, Recycle of Inorganic Nutrients for Hydroponic Crop Production Following Incineration of Inedible Biomass, *Advances in Space Research* (in press) (1997).
8. Bubenheim, D., M. T. Patterson, K. Wignarajah, and M. Flynn, Incineration of Biomass and Utilization of Product Gas as a CO_2 Source for Crop Production in Closed Systems: Gas Quality and Phytotoxicity, *Advances in Space Research* (in press) (1997).
9. Economics Laboratory Inc., A Study to Define a Set of Requirements for Cleansing Agents for Use in the Space Station Whole Body Shower. CR. No. 171910. NASA, USA (1985).
10. Flynn, M., Vapor Phase Catalytic Ammonia Reduction, SAE Tech. Paper No. 941398, 24th International Conference on Environmental Systems Conference (1994).
11. Garland, J. L., C. Mackowiak, and J. C. Sager, Hydroponic Crop Production Using Recycled Nutrients from Inedible Crop Residues. SAE Tech. Paper No. 932173, 23rd International Conference on Environmental Systems Conference, Colorado, USA. (1993).
12. Greene, C., D. Bubenheim, and K. Wignarajah, Significance of Plant Root Microorganisms in Reclaiming Water in CELSS, *Advances in Space Research* (in press) (1996).
13. Gustan, E. and T. Vinopal, Controlled Ecological Life Support System: A Transportation Analysis. NASA CR.-166420. Boeing Aerospace, Seattle. pp. 126 (1982).
14. MacElroy, R. D., H. P. Klein, and M. M. Averner, The Evolution of CELSS for Lunar Bases, *Lunar Bases and Space Activities of the 21st Century,* ed. W. W. Mendell. pp. 623-633, Lunar and Planetary Institute, Houston (1984).
15. Mackowiak, C. L., J. L. Garland, R. F. Strayer, B. W. Finger, and R. F. Wheeler, Comparison of Aerobically Treated and Untreated Crop Residue as a Source of Recycled Nutrients in a Recirculating Hydroponic System, *Adv. in Space Res.* **18**:281 (1996).
16. Patterson, M. T., K. Wignarajah, and D. Bubenheim, Biomass Incineration as a Source of CO_2 for Plant Gas Exchange: Phytotoxicity of Incinerator-derived Gas and Analyses of Recovered Evapotranspired Water. *Life Support and Biosphere Science* **3**: (In press) (1997).
17. Peiser, G. D., C. C. Lizada, and S. F. Yang, Dark Metabolism of Carbon Monoxide in Lettuce Leaf Disks. *Plant Physiol.* (1982)
18. Upadhye, R. S., K. Wignarajah, and T. Wydeven, Incineration for Resource Recovery in a Closed Ecological System, *Environ. Int.* **19**:381-392 (1993).
19. Verostko, C. E., R. Garcia, R. Sauer, R. P. Reysa, A. T. Linton, and T. Elms, Test Results on Reuse of Reclaimed Shower Water - a Summary, SAE Tech. Paper No. 891443, 19th International Conference on Environmental Systems Conference (1989).
20. Wieland, P. O., Designing for Human Presence in Space: An Introduction to Environmental Control and Life Support Systems, NASA RP No. 1324, Marshall Space Flight Center, AL, USA (1994).
21. Wisniewski, R., and D. Bubenheim, Aerobic Biological Degradation of Surfactants in Wastewater. Tech. Paper No. 93-4152. *Proceedings of the American Institute of Aeronautics & Astronautics,* Huntsville, AL, USA (1993).
22. Wignarajah, K., S. Pisharody, M. Flynn, D. L. Bubenheim, B. Potter, and C. Klein, Nitrogen Sources of NO_X Generated During Incineration of Inedible Plant Biomass. *Journal of Life Support and Biosphere Science* **4**: (In press) (1997).

23. Wignarajah, K., D. L. Bubenheim, and T. Wydeven, Performance of Lettuce in Gray Water Streams. NASA Tech. Memo. No. 103996, pp. 165-167, NASA Ames Research Center, Moffett Field, CA, USA (1992).
24. Wydeven, T., A Survey of Some Regenerative Physico-chemical Life Support Technologies. NASA Tech. Memo. No. 101004, NASA Ames Research Center, Moffett Field, CA, USA (1988).

Pergamon

Adv. Space Res. Vol. 20, No. 10, pp. 1845–1850, 1997
Published by Elsevier Science Ltd on behalf of COSPAR
Printed in Great Britain
0273-1177/97 $17.00 + 0.00

PII: S0273–1177(97)00850–8

INCINERATION OF BIOMASS AND UTILIZATION OF PRODUCT GAS AS A CO_2 SOURCE FOR CROP PRODUCTION IN CLOSED SYSTEMS: GAS QUALITY AND PHYTOTOXICITY

D. L. Bubenheim*, M. Patterson**, K. Wignarajah** and M. Flynn*

**NASA Ames Research Center, Space Technology Division, Regenerative Life Support Branch, Moffett Field CA 94035*
***Lockheed Martin Engineering and Science Corporation, Ames Research Center, Space Technology Division, Regenerative Life Support Branch, Moffett Field CA 94035*

ABSTRACT

This study addressed the recycle of carbon from inedible biomass to CO_2 for utilization in crop production. Earlier work identified incineration as an attractive approach to resource recovery from solid wastes because the products are well segregated. Given the effective separation of carbon into the gaseous product stream from the incinerator in the form of CO_2 we captured the gaseous stream produced during incineration of wheat inedible biomass and utilized it as the CO_2 source for crop production. Injection rate was based on maintenance of CO_2 concentration in the growing environment. The crop grown in the closed system was lettuce. Carbon was primarily in the form of CO_2 in the incinerator product gas with less than 8% of carbon compounds appearing as CO. Nitrogen oxides and organic compounds such as toluene, xylene, and benzene were present in the product gas at lower concentrations (<4 μmol mol^{-1}); sulfur containing compounds were below the detection limits. Direct utilization of the gaseous product of the incinerator as the CO_2 source was toxic to lettuce grown in a closed chamber. Net photosynthetic rates of the crop was suppressed more than 50% and visual injury symptoms were visible within 3 days of the introduction of the incinerator gas. Even the removal of the incinerator gas after two days of crop exposure and replacement with pure CO_2 did not eliminate the toxic effects. Both organic and inorganic components of the incinerator gas are candidates for the toxin.

INTRODUCTION

Recovery of resources from waste streams is essential for future implementation and reliance on a regenerative life support system. The purpose of the resource recovery component of such a system is to retrieve from the waste streams the raw materials that must be converted to forms appropriate for recycle to support the production of human consumables. Incineration is a highly-ranked candidate technology for use in recovering resources from waste streams in a Controlled Ecological Life Support System (CELSS) (Bubenheim and Flynn, 1996; Bubenheim and Wignarajah, 1995; Bubenheim and Wydeven, 1994; Dreshel et al, 1991; Wydeven, 1988). Inedible biomass will be the primary solid waste stream in a CELSS and represents a resource for carbon, water, and inorganics (Bubenheim, 1991; Wydeven, 1988). Incineration has been shown to be effective in oxidation of inedible biomass and recovery of inorganics (Bubenheim and Wydeven, 1994). The resulting gaseous and solid product streams from incineration are well segregated. The ash provides a suitable, although not complete, source for inorganic nutrient inputs to crop production (Bubenheim and Wignarajah, 1996). The suitability of incinerator gas as a source of CO_2 input to crop production has not been well defined (Bubenheim and Wignarajah, 1995; Upadhye et al, 1993).

Incineration produces a number of oxides of carbon, nitrogen, and sulfur and some hydrocarbons. The oxidation of inedible biomass during incineration should yield a CO_2 rich gas. When CO_2 generated by humans is combined with the CO_2 recovered from solid waste processing and delivered to the crop, assimilation by green plants via photosynthesis and the production of oxygen will complete the air revitalization step (Bubenheim, 1991). An incinerator was included as part of the Russian BIOS-3 study and implicated in visual symptoms of crop injury and suppression in yield (Gitelson et al, 1975). The addition of a post-incineration, catalytic oxidation unit reduced visible injuries. While the BIOS-3 work clearly shows that there can be problems associated with the inclusion of an incinerator in a closed system, the description of the incinerator, the degree of plant performance monitoring, and the specificity and consistency of the atmospheric analysis are such that they provide little or no insight into the identification of any phytotoxic component or it's source.

In closed systems, even trace gases may accumulate to acute phytotoxic levels. To date, very little work has addressed the effects of incinerator-derived gases on plant performance in closed systems. The objectives of the work presented here are to quantify any effects on whole plant gas exchange (photosynthesis and transpiration) resulting from exposure to incinerator-derived gas, identify the constituents of the incinerator product gas and determine their fate in crop production, and suggest the viability of using incinerator-derived gas as a CO_2 source for support of plant-based life support functions.

MATERIALS AND METHODS

Incinerator Gas Production and Measurement. Dried wheat straw was incinerated at 800 °C in a modified muffle furnace. A cylinder of synthetic air (21% O_2 in a balance of N_2; Scott Specialty Gases) was used as the oxygen source for combustion. Resultant gas from the wheat straw incineration was pulled from the incinerator with a vacuum/pressure pump and pressurized into a null voided cylinder. CO_2 concentration of the incinerator-derived gas was measured with a gas chromatography. Drager™ tubes were used to determine concentrations of SO_2 (1 - 25 μmol mol^{-1} range) and NO_x (0.5 - 10 μmol mol^{-1} range). Organic volatiles and carbon monoxide from both the incinerator gas and the plant chamber atmosphere were monitored with a GC-MS.

Plant Culture and Environmental Control. Six week old lettuce plants grown hydroponically in a greenhouse were transferred to a closed stainless steel plant chamber identified as the Plant Volatiles Chamber (PVC). The chamber has a volume of 0.52 m^3 and complete environmental control. A 40 L recirculating hydroponic system was isolated from the shoot zone by a sheet of ABS plastic through which the lettuce plants were planted in 1 cm holes, the stems wrapped in open-cell foam plugs to support the plants and maintain an effective barrier. Sixteen lettuce plants were used in each experiment with a total projected canopy area of 0.157 m^2. Photosynthetic photon flux (PPF) was maintained at 800 μmol m^{-2} s^{-1} with a 12-h photoperiod. Temperature was held constant at 23 °C. The chamber atmosphere was maintained at 350 μmol mol^{-1} CO_2 and 21% O_2 and a dew point of 10 °C (50% relative humidity).

Measurement of Photosynthesis and Transpiration. Net CO_2 assimilation was determined from the flow of CO_2 into the chamber required to maintain a constant CO_2 level. Photosynthesis was expressed as the rate of CO_2 assimilation per unit of lettuce canopy area (μmol CO_2 m^{-2} s^{-1}). Transpiration was measured from water collected on a dew point condenser at 10 °C. Transpiration was expressed as the rate of condensed water collected per unit canopy area (mmol H_2O m^{-2} s^{-1}).

Experimental Protocol. Lettuce cultures were placed in the plant chamber and allowed to equilibrate for a minimum of 24 hours prior to initiating any study. Instantaneous photosynthetic rates were monitored and baseline rate of both net photosynthesis and transpiration were determined. The CO_2 source for photosynthesis was initially supplied from pressurized cylinders of pure CO_2 (Scott Specialty Gas). After stable baseline measurements were determined, a cylinder of incinerator generated CO_2 was substituted for the pure CO_2 cylinder for 6 days. The experiment was repeated, this time exposing plants to the incinerator derived gas for only 2 days of the 6 day study, after which pure CO_2 was used as the CO_2 source for the remaining days. Chamber air was collected daily and analyzed. Controls were performed using only the pure CO_2 source but otherwise identical culture and environmental conditions.

RESULTS AND DISCUSSION

The product gas resulting from incineration of wheat straw contained organic as well as inorganic constituents (Table 1). The organic compounds identified in the product gas were primarily benzene, toluene and xylene. The inorganic components were CO_2, CO, SO_2 and NOx. The SO_2, NOx, benzene, toluene and xylene are all potentially phytotoxic materials (Darrall, 1989; Malhotra and Khan, 1984). Carbon dioxide concentrations of the incinerator product gas was approximately 10% on a molar basis. Carbon monoxide occurred in higher concentrations than reported for product gases leaving a microwave incinerator / catalytic oxidation combination (Srinivasan and Sun, 1995), but does not generally disrupt plant physiological processes even in fairly high concentrations (Naik et al, 1992).

Table 1. Major inorganic and organic constituents of the product gas collected from incineration of dried wheat straw.

Measured Compound	Chemical Formula	Incinerator Gas Concentration (μmol mol $^{-1}$)
inorganics		
Carbon dioxide	CO_2	106,000
Carbon monoxide	CO	8066
Sulfur dioxide	SO_2	<1.0
Nitrogen oxides	NO_x	3-4
organics		
Benzene	C_6H_6	2.4
Toluene	$C_6H_5CH_3$	0.24
Xylene	$C_6H_4(CH_3)_2$	0.04

Gas exchange processes were not affected during the initial 24 hours of exposure to incinerator-derived gases (Figures 1 & 2). This demonstrates that CO_2 derived from biomass incineration can support photosynthesis. However, there was a clear and dramatic decrease in photosynthesis on the third day of exposure to the incinerator-derived gas (Figure 1). A suppression in transpiration was evident on the second day of exposure (Figure 2). Visible injury symptoms became evident on the third day of exposure, the lettuce leaves began to yellow and the leaf margins had a general water-soaked appearance. The pattern of declining photosynthesis and transpiration was similar for both experiments. Even when the incinerator gas was replaced with pure CO_2 48 hours after injection of the incinerator gas was initiated (experiment 2), the decline in gas exchange parameters still remained. By the end of the experiment, photosynthetic rates had dropped over 70% and a widespread leaf necrosis was evident. This suggests that this acute injury occurred during the exposure period.

Volatile organics increased with time in the chamber when the plants were actively photosynthesizing. However, as the plants were slowing their photosynthetic rates and the injection of incinerator gas was decreased in response to CO_2 demand, concentrations of these organics began to drop (Figure 3). This indicates a continuing sink for these volatiles although it is not clear whether this results from a natural breakdown or the likely interaction of these volatiles with plant biomass. Toluene has been shown to be removed from the atmosphere by plants (Porter, 1994). The organic volatiles identified in the incinerator product gas could likely disrupt cellular membranes and thus disrupt normal function (Lacaze and Ducreux, 1987).

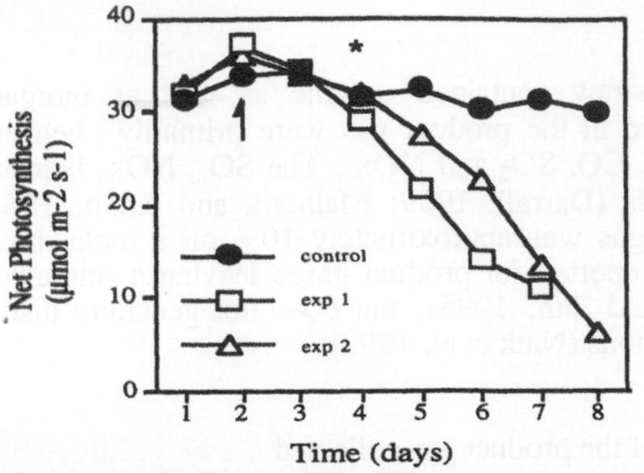

Fig. 1. The effect of incinerator gas as a source of CO_2 for photosynthesis. The up arrow indicates the start of incinerator gas as a CO_2 source. The asterisk indicates when incinerator gas was removed from experiment 2 and replaced with a pure CO_2 source.

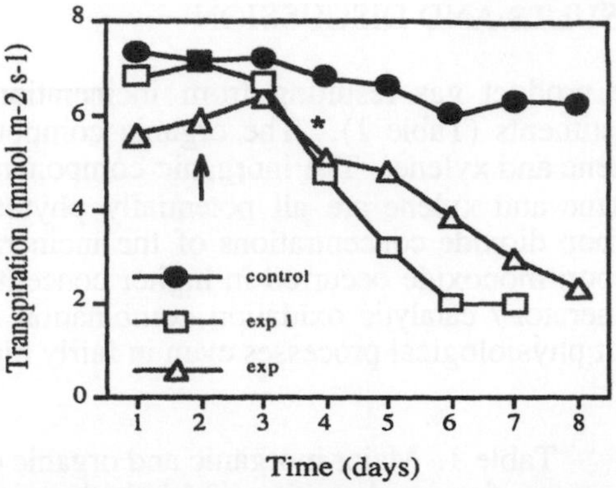

Fig. 2. The effect of incinerator gas as a CO_2 source on transpiration. The up arrow indicates the start of incinerator product gas as the CO_2 source. The asterisk indicates when the incinerator gas was removed from experiment 2 and replaced with a pure CO_2 source.

At no time during the studies could NOx, SO_2 or CO be measured in the closed chamber although these gases were clearly present in the incinerator product gas stream, further supporting the notion of the plant material as a sink. Movement of these gases into the plant via the open stomata, the same mechanism important to CO_2 movement into the plant, is the likely sink and avenue of exposure of internal plant tissues. Carbon monoxide can be fixed by plants (Bidwell and Bebee, 1974) but must exceed 5,000 µmol mol^{-1} before becoming phytotoxic effects to sweet pea (Crocker, 1988). While reports regarding the effects of oxides of nitrogen on plant growth are conflicting, NO_2 has been shown to be taken up through open stomata, dissolved in water and then incorporated into amino acids (Yoneyama and Sasakawa, 1979). The presence of NO_2 has also been implicated in stimulation of RuBP carboxylase activity (Malhotra and Khan, 1984), stimulation of nitrate reductase activity (Zeevart, 1974), increases in chlorophyll content and increases in shoot growth (Malhotra and Khan, 1984). There are also reports of detrimental effects of NO_2, especially in the dark and in combination with other pollutants (Hill and Bennett, 1970). The reported toxicity of nitric oxide (NO) has been partly attributed to low solubility in water (Hill and Bennett, 1970).

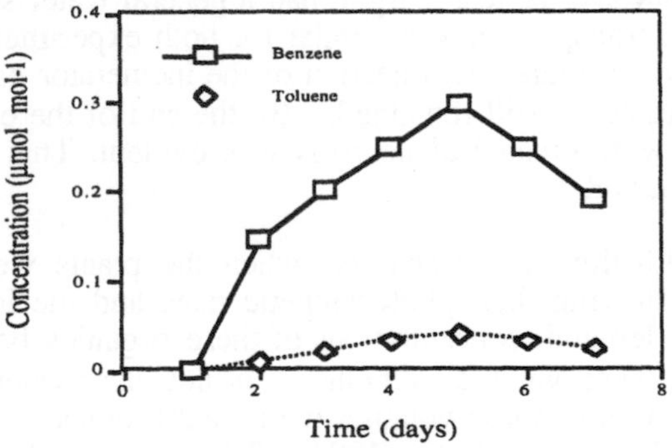

Fig. 3. Concentration of benzene and toluene in the plant chamber atmosphere over the course of the experiment. On day 2 pure CO_2 was replaced by incinerator-derived gas as the CO_2 source for photosynthesis. Other organic volatiles that showed similar patterns of concentration but at much lower concentrations include chloroform and xylene.

An important factor to be considered regarding potential phytotoxicity of oxides of nitrogen is their reactivity in the presence of oxygen and ultra-violet radiation. This situation can result in the production of ozone which is known to be phytotoxic at low concentrations (0.03 µmol mol^{-1} exposure for 4 hours) (Winner, 1994). While there is literature outlining the phytotoxic response to exposure to individual gaseous compounds such as NO_2, there is little understanding of mixed atmosphere effects. The presence of several organic and inorganic gaseous compounds in the incinerator-derived product gas makes identification of a specific toxic material in the growing atmosphere difficult. Inorganic oxidizing gases, such as ozone, have been shown to increase the threshold for toxicity responses due to hydrocarbon exposure (Winner, 1994). The simple presence of individual gases, such as oxides of nitrogen, previously identified as potentially phytotoxic is not adequate to assume their responsibility for observed effects of the incinerator derived gases on gas exchange processes in these studies.

Table 2. Maximum concentrations of inorganic and organic constituents of incinerator gas measured in a closed chamber over a 7 day experimental period of gas injection as the CO_2 Source.

Measured Compound	Chemical Formula	Maximum Conc. in Chamber (µmol mol^{-1})
inorganics		
Carbon dioxide	CO_2	350
Carbon monoxide	CO	-
Sulfur dioxide	SO_2	-
Nitrogen oxides	NO_x	-
organics		
Benzene	C_6H_6	0.300
Toluene	$C_6H_5CH_3$	0.041
Xylene	$C_6H_4(CH_3)_2$	0.003

These results should not be interpreted as justifying the elimination of incineration from the list of candidate technologies for resource recovery. Research has shown that effective control of incineration conditions can enable control of product gas quality (Wignarajah et al, 1996). The phytotoxic component, whether a direct incineration product or resulting from interaction of the gas constituents, must be identified before this ability can be utilized and any firm conclusions regarding the long-term usefulness of incineration determined.

CONCLUSIONS

Carbon dioxide rich gas resulting from the incineration of plant biomass can be used to support photosynthesis in closed systems. However, some phytotoxic component(s) in the incinerator-derived gas resulted in acute injury to plant tissues after only a few days. These phytotoxic components, still unidentified clearly, will need to be avoided through control of incinerator conditions or removed or reduced prior to utilization of incineration as a resource recovery technique for CELSS.

REFERENCES

Bidwell, R. G. S, and G. P. Bebee, Carbon monoxide fixation by plants, Can. J. Bot. 52, 1841 (1974)

Bubenheim, D.L, Plant for water recycling, oxygen regeneration, and food production, Waste Management and Research, 9, 435 (1991).

Bubenheim, D.L. and M.T. Flynn, The CELSS Antarctic Analog Project and validation of assumptions and solutions regarding regenerative life support technologies, 26th ICES, SAE Technical Paper No. 961589 (1996).

Bubenheim, D.L. and K. Wignarajah, Recycle of inorganic nutrients for hydroponic crop production following incineration of inedible biomass, (In this COSPAR - Advances in Space Research issue) (1996).

Bubenheim, D. and K. Wignarajah, Incineration as a method for resource recovery from inedible biomass in a controlled ecological life support system, Journal of Life Support and Biosphere Science 1, 129 (1995).

Bubenheim , D.L. and T. Wydeven, Approaches to resource recovery in controlled ecological life support systems, Advances in Space Research. 14, 113 (1994).

Crocker, W., Physiological effects of ethylene and other unsaturated carbon-containing gases, in *Growth of Plants*, Reinhold Publishing Corp., Ch. 4, New York, NY (1988)

Darrall, N.M, The effects of air pollutants on physiological processes in plants, Plant Cell and Environment 12, 1 (1989)

Dreshel, T.W., R.M. Wheeler, C.R. Hinkle, J.C. Sager, and W.M. Knott, Investigating combustion as a method of processing inedible biomass produced in NASA's biomass production chamber, NASA Tech. Memo. # 103821, Kennedy Space, FL (1991).

Gitelson, I.I., B.G. Kovrov, G. M. Lisovsky, Y.N. Okladnikov, M.S. Rerberg, F.Y. Sidko and I.A. Terskov, *Problems of space biology*, Volume 28, Experimental ecological systems including man, (NASA -TT-F-16993) (1975)

Hill, A. C, and J. H. Bennett, Inhibition of apparent photosynthesis by nitrogen oxides, Atmos. Environ. 4, 341 (1970)

Lacaze, J.K. and J.Ducreux, Toxicity of water-soluble aromatic hydrocarbons issuing from two types of oil surfaces and one type of hydrocarbon surface: effects on the photosynthetic activity of the marine diatom Phaedoctylum tricornatum and on the ingestion of this alga by the copepod Trigriopus brevicormis, Science De L'eau 6, 414 (1987).

Malhotra, S. S, and A. A. Khan, Biochemical and physiological impacts of major pollutants, in Air Pollution and Plant Life (Ed. M. Treshow), John Wiley and Sons Ltd., Ch. 7, USA (1984)

Naik, R.M., A.R. Dhage, S. V. Munjal, P. Singh, B. B. Desai, S. L. Metha, and M. S. Naik, Differential carbon monoxide sensitivity of cytochrome C oxidase in leaves of C-3 and C-4 plants, Plant Physiology 98, 984 (1992).

Porter, J. R., Toluene removal from air by Dieffenbachia in a closed environment. Adv. Space Res., 14, 99. (1994)

Srinivasan, V, and S. Sun, The application of microwave incinerator to regenerative life support. Proceedings of the International Microwave Power Conference, (1995).

Upadhye, R.S., K. Wignarajah and T. Wydeven, Incineration for resource recovery in a closed ecological life support system, Environ. Int. 19, 381 (1993).

Wignarajah, K., S. Pisharody, M. Flynn, D.L. Bubenheim, B. Potter and C. Klein, Nitrogen sources of NO_x generated during incineration of inedible plant biomass, Journal of Life Support and Biosphere Science , (1996).

Winner, W.E. Mechanistic analysis of plant responses to air pollution. Ecological Application 4:651-661, (1994).

Wydeven, T, A survey of some regenerative physico-chemical life support technologies, NASA Tech. Memo. No. 101004, NASA-Ames Research Center, Moffett Field, CA (1988).

Yoneyama, T, and H. Sasakawa, Transformation of atmospheric NO_2 absorbed in Spinach leaves, Plant and Cell Physiol. 20, 263 (1979)

Zeevart, A.J, Induction of nitrate reductase by NO_2, Acta Bot. Neerl., 23, 345 (1974).

Pergamon

Adv. Space Res. Vol. 20, No. 10, pp. 1851–1854, 1997

Printed in Great Britain
0273-1177/97 $17.00 + 0.00
PII: S0273-1177(97)00851-X

THEORETICAL AND PRACTICAL CONSIDERATIONS FOR STAGGERED PRODUCTION OF CROPS IN A BLSS

G. W. Stutte*, C. L. Mackowiak*, N. C. Yorio* and Wheeler**

**Dynamac Corporation, Kennedy Space Center, FL 32899, U.S.A.*
***Biomedical Operations and Research Office, Kennedy Space Center, FL 32899 U.S.A.*

ABSTRACT

A functional Bioregenerative Life Support System (BLSS) will generate oxygen, remove excess carbon dioxide, purify water, and produce food on a continuous basis for long periods of operation. In order to minimize fluctuations in gas exchange, water purification, and yield that are inherent in batch systems, staggered planting and harvesting of the crop is desirable. A 418-d test of staggered production of potato cv. Norland (26-d harvest cycles) using nutrients recovered from inedible biomass was recently completed at Kennedy Space Center. The results indicate that staggered production can be sustained without detrimental effects on life support functions in a CELSS. System yields of H_2O, O_2 and food were higher in staggered than batch plantings. Plants growing in staggered production or batch production on "aged" solution initiated tubers earlier, and were shorter than plants grown on "fresh" solution. This morphological response required an increase in planting density to maintain full canopy coverage. Plants grown in staggered production used available light more efficiently than the batch planting due to increased sidelighting.

INTRODUCTION

Crop production tests in the Biomass Production Chamber (BPC) at Kennedy Space Center (KSC) for the past eight years have characterized the growth and yield responses of candidate crops though a single life cycle (Wheeler et al., 1996). These studies have supplied information on crop performance under optimal conditions and permitted the characterization of plant responses to environmental parameters at different stages of development (Wheeler et al., 1996). In an operational Bioregenerative Life Support System (BLSS), production of water, oxygen, and food from a closed recirculating resource stream will be required (Drysdale et al., 1994). Two potential methods of production are a series of "batch" plantings, where an entire compartment is planted and harvested at the same time, and a "staggered" planting, where a portion of a compartment is harvested and replanted at a given time.

The potential advantage of continuous production is that the production of food, water, and oxygen should be at a more uniform rate than from a series of batch plantings, thus reducing the size of storage buffers to maintain a system. However, the impact of different aged plants sharing the same atmosphere and nutrient solution for long periods of time is not known. Recent tests in the BPC have concentrated on determining whether staggered production using recirculating, recycled nutrient solutions will have detrimental effects on crop performance in a closed atmosphere (Mackowiak et al., 1994).

A long-term BPC experiment was initiated in June 1994, with the objective being to determine the effect of staggered cropping of potato using potato bioreactor effluent as a primary source of nutrients. We report on the results of the 418-d staggered production experiment and their relationship to estimated production potential.

MATERIALS AND METHODS

Plant material Potato (*Solanum tuberosum* L. cv. Norland) plantlets were transplanted into the growing trays of the Biomass Production Chamber at Kennedy Space Center, FL. All plants were grown using recirculating nutrient film culture as previously described (Wheeler et al., 1990). The pH of the nutrient solution was controlled to near 5.8+/- 0.2 by additions of dilute (2.5%) HNO_3 and nutrient solution temperature was maintained at 18 C in all levels for the duration of the growout. Solution electrical conductivity was controlled near 1.2 dS m^{-1} by additions of a complete nutrient replenishment solution.

A 12-h light cycle was maintained with high pressure sodium (HPS) lamps for the duration of the experiment. The atmosphere was enriched and controlled to 1200 μmol mol^{-1} CO_2 during the light cycle. No effort was made to suppress CO_2 levels during dark cycles. Relative humidity (RH) was maintained at 85% during the first 10-d of the study, then controlled to 70%. Air temperature was maintained at 20 C during the light cycle and 16 C during the dark cycle.

Gas exchange The population gas exchange characteristics related to various environmental conditions were further characterized. The CO_2 concentration (μmol mol^{-1}), CO_2 added to maintain a setpoint of 1200 μmol mol^{-1} (L), and water collected from condensate system (L) were monitored at five-minute intervals. Population CO_2 exchange rates were calculated from the rate of CO_2 increase during the dark cycle and the rate of CO_2 decrease when the lights were turned on. Net assimilation rates were calculated based on the daily rate of CO_2 utilization, determined by the mass flow of CO_2. Transpiration rates were calculated from the rate of condensate collected from the air handling system.

Harvesting The batch production treatment was in the upper chamber of the BPC. All 32 trays of the compartment were planted and harvested at 104-d cycles. Thus, the plants were all the same age during development. The gas exchange and yield of the initial 104-d harvest was used as a control treatment to estimate CO_2 assimilation under staggered production.

The staggered production treatment was in the lower chamber of the BPC. This treatment involved the harvest of eight trays (25%) at 26-d intervals. The trays was replanted at a density of four plants tray^{-1}. The final density of plants tray^{-1} was determined at time of thinning. The nutrient solution was not changed following the harvests. Plants were harvested in two-tray units from each shelf in the lower chamber, from random, predetermined positions. Following harvest, the fresh mass of leaves, stems, roots and tubers was determined for each plant. The tissue was either freeze-dried or air-dried in a forced air oven at 70 C. When tubers were present, a 100-g subsample was used to determine percent dry mass. Plant height and canopy coverage were manually determined at 7-d intervals.

Table 1: Summary of the life support capabilities of potatoes cv. Norland grown in staggered production for 418 days

	BWP941 Total	Total Normalized	[z]Human Needs	Area for Human
Water[y]	13,552 L	3.2 L m^{-2} d^{-1}	19.0 L d^{-1}	5.9 m^2
CO_2[x]	173 kg	41.4 g m^{-2} d^{-1}	1.0 kg d^{-1}	24.2 m^2
O_2[w]	126 kg	32.5 g m^{-2} d^{-1}	0.83 kg d^{-1}	24.2 m^2
Food[v]	61 kg	14.5 g m^{-2} d^{-1}	0.62 kg d^{-1}	42.8 m^2

[z]Source: NASA SPP 30262 Space Station ECLSS Architectural Control.
[y]Water need excludes laundry/dish washing requirement.
[x]CO_2 value is the amount assimilated by photosynthetic tissues.
[w]O_2 value is derived from CO_2 and assumes a 1.00 conversion efficiency.
[v]Food values assume that potato tubers are 17% dry matter.

RESULTS AND DISCUSSION

A summary of the water purification, CO_2 removal, O_2 generation, and food production obtained from a 418-d trial of staggered production of potato using recirculating NFT hydroponics is given in Table 1. Yields of each component were higher in the staggered culture than the batch production (Stutte and Sager, 1995). These results indicate that human requirements for H_2O can be met with 5.9 m^2 of growing area, the atmospheric regeneration component required 24.2 m^2, and food (based on calories) can be achieved with 42.8 m^2 growing area. This is 10% less area than required to achieve food requirement with batch production (42.8 m^2 vs. 47.1 m^2).

Staggered production systems should provide a relatively constant output of O_2, H_2O, and food once "steady-state" has been achieved. This is in contrast to batch production systems where an entire crop is harvested at a single time. However, most of the data used to characterize the large scale performance characteristics of CELSS candidate crops have utilized batch production systems.

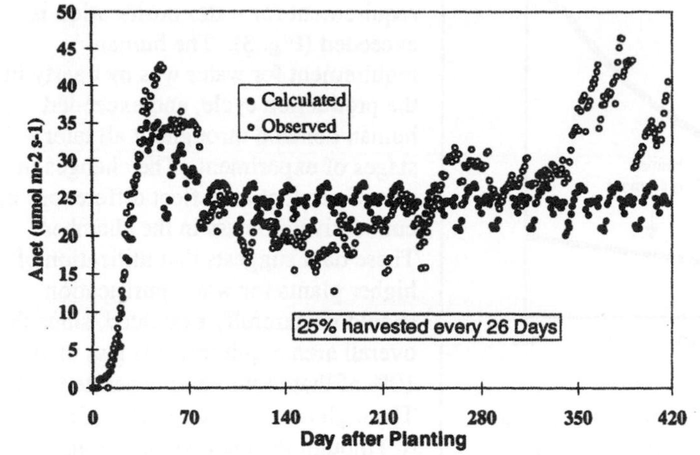

Figure 1: Comparison of calculated and observed CO_2 assimilation rates of potato cv. Norland grown under staggered NFT production.

We attempted to determine whether the information obtained from these batch production tests could be used to the simulate the productivity of crops during staggered production systems. In order to determine the "theoretical" CO_2 removal rates from the atmosphere, the developmental changes in CO_2 fixation obtained from the batch production treatment of BWP941 was assumed to be characteristic of all crops grown in the chamber. The curve was offset at regular 26-d intervals, and appropriate adjustments to planting area were made. The calculated cumulative CO_2 fixation curves were then used as the estimated output of a staggered production system. This "calculated" production curve was compared to the actual results obtained from a 418-d staggered production treatment (Fig. 1). The results indicate that the actual production closely approximated the estimated production rates for most of the growout. During the middle portion of the experiment, the production is reduced because of lower PPF at the canopy level. The reduced PPF is a result of shorter plants that resulted from potatoes sharing a common nutrient solution. Environmental changes which offset this effect were experienced during the final cycle of the experiment, and this is reflected than higher than expected gas exchange rates.

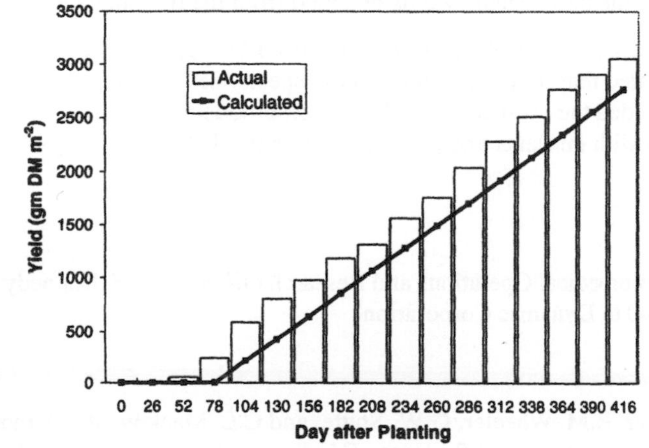

Figure 2: Comparison of calculated and actual cumulative tuber yields, expressed as dry mass, of potato cv. Norland under staggered production.

A similar analysis was conducted with yield, using the initial harvest of the batch production system as the actual data to generate a "theoretical" yield response. As with the gas exchange data, the actual yields exceed the estimates of yield by 8% over the life of the experiment (Fig. 2). This is not to suggest that yields from the batch production were constant, but that the fluctuations tended to balance. Two factors contributed to this increase in yield: The first is that staggered production system opens up the potato canopy, exposing additional leaf area to light. The increase in photosynthetic surface was between 10 and 25% of the total growing area, depending upon height of the plants and height of plants in adjacent trays. This increased light, provides additional energy for fixation of carbon. The second factor is that tubers were initiated earlier on potatoes transplanted into "aged" nutrient solution resulting in a higher harvest index than for plants grown in batch culture on fresh solution for most of the study.

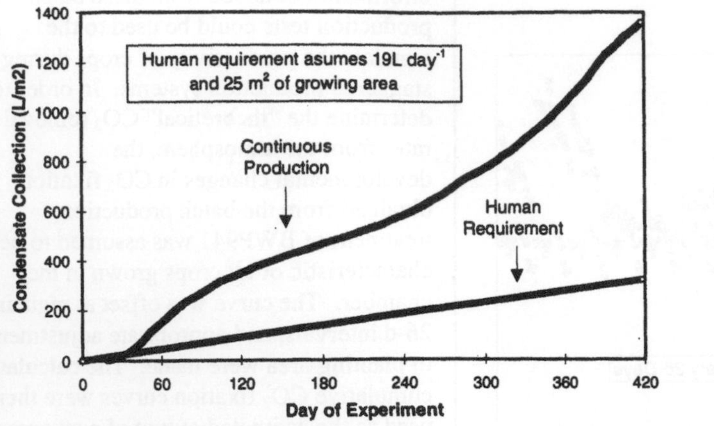

Figure 3: Cumulative water purification capability of potato cv. Norland grown in staggered production for 418-d on an area required to meet human CO_2 removal requirements (See Table 1).

This experiment also indicated that if the atmospheric regeneration requirements are achieved, the human requirement for water purification is exceeded (Fig. 3). The human requirement for water was met early in the production cycle, and exceeded human demand throughout all later stages of experiment. The changes in rate of production reflect differences in cumulative biomass in the chamber. These data suggests that utilization of higher plants for water purification should be carefully evaluated, since the overall area requirement is less than 10% of that required to generate food. The high rate of water cycling (3 L m^{-2} d^{-1}) though the plant system is an important consideration, since water is the greatest weight and mass requirements in a BLSS.

CONCLUSIONS

The BLSS Breadboard Project at KSC has demonstrated that potatoes can be produced continuously using recycled water and nutrients recovered from biological processing of inedible plant residues. The staggered production system has been able to sustain a relatively stable production of fundamental life support activities (CO_2 removal, O_2 production, water purification) that will meet or exceed the requirement for a single individual using 25 m^2 of growing area. These staggered production system conditions have not been optimized. However, the overall yield per unit area has been approximately 8% higher than the best "batch" yields obtained from the BPC to date.

This experiment has demonstrated the feasibility of using higher plants to function as a life support system for an extended period of time. This experiment has indicated that for potato, under these specialized growing conditions, that actual performance of the crop exceeded the estimates based on batch production systems. However, this experiment also revealed interactions with nutrients and plant age which needed to be managed throughout the experiment.

ACKNOWLEDGMENTS

This research were conducted under the auspices of Biomedical Operations and Research Office, John F. Kennedy Space Center, FL under NASA contract NAS10-12180 to Dynamac Corporation.

REFERENCES

Drysdale, A.E., H.A. Dooley, W.M. Knott, J.C. Sager, R.M. Wheeler, G.W. Stutte, and C.L. Mackowiak. A more completely defined CELSS. 24th Intl. Conf. on Environmental Systems. Friedrichshafen, Germany. *SAE Tech. Paper 941292* (1994)

Mackowiak, C.L., J.L. Garland, and G.W. Stutte. Growth regulator effects of water soluble materials from crop residues for use in plant hydroponic culture. *Proc. Plant Growth Regul. Soc. Amer.* **22:** 233-239 (1994)

Stutte, G.W. and J. C. Sager. Biological considerations in the design of continuous potato production systems. *ASAE paper no. 9555654* (1995)

Wheeler, R.M, C.L. Mackowiak, J.C. Sager, W.M. Knott, and C.R. Hinkle. Potato growth and yield using nutrient film technique. *Amer. Potato J.* **67**: 177-178 (1990)

Wheeler, R.M., C.L. Mackowiak, G.W. Stutte, J.C. Sager, N.C. Yorio, L.M. Ruffe, R.E. Fortson, T.W. Dreschel, W.M. Knott, and K.A. Corey. NASA's Biomass Production Chamber: A testbed for bioregenerative life support studies. Adv. Space Res. 18: 215-224 (1996)

Pergamon

Adv. Space Res. Vol. 20, No. 10, pp. 1855–1860, 1997

Printed in Great Britain
0273-1177/97 $17.00 + 0.00

PII: S0273-1177(97)00852-1

DYNAMIC OPTIMIZATION OF CELSS CROP PHOTOSYNTHETIC RATE BY COMPUTER-ASSISTED FEEDBACK CONTROL

C. Chun and C. A. Mitchell

NASA Specialized Center of Research and Training in Bioregenerative Life Support, Purdue University, West Lafayette, IN 47907-1165, U.S.A.

ABSTRACT

A procedure for dynamic optimization of net photosynthetic rate (Pn) for crop production in Controlled Ecological Life-Support Systems (CELSS) was developed using leaf lettuce as a model crop. Canopy Pn was measured in real time and fed back for environmental control. Setpoints of photosynthetic photon flux (PPF) and CO_2 concentration for each hour of the crop-growth cycle were decided by computer to reach a targeted Pn each day. Decision making was based on empirical mathematical models combined with rule sets developed from recent experimental data. Comparisons showed that dynamic control resulted in better yield per unit energy input to the growth system than did static control. With comparable productivity parameters and potential for significant energy savings, dynamic control strategies will contribute greatly to the sustainability of space-deployed CELSS.

INTRODUCTION

Crop-production systems for Controlled Ecological Life-Support Systems (CELSS) must reliably supply edible biomass and oxygen to crews throughout long-term space missions, with sustainability of the overall life-support system as the most important goal (Mitchell, 1993; Salisbury and Clark, 1996). High sustainability is favored by precise prediction and satisfaction of oxygen and edible biomass needs for the system, as well as by economical usage of limited resources. To achieve these goals, an energetically friendly strategy for crop production in CELSS is needed.

Chun and Mitchell (1996) proposed dynamic control of photosynthetic photon flux (PPF) for crop production in CELSS. According to that strategy, setpoints for PPF would change at different stages of crop development and/or at different times of day to produce a desired amount of oxygen, to transpire a desired amount of water, or to produce a desired amount of edible biomass. In that approach, fixed setpoints were used for all other environmental parameters. Although increasing PPF equates to an energy penalty for the entire system, CO_2 concentration management may be treated differently in CELSS. For example, CO_2 enrichment for crop production might benefit the overall system when quick removal of CO_2 from the human habitat is needed, whereas, in most cases, CO_2 still will be a limited and valuable resource for crop production in CELSS. In the present study, dynamic optimization of both PPF and CO_2 were tested simultaneously, and biomass productivity and energy-use efficiency were compared with the traditional strategy of using fixed setpoints for both environmental inputs. Leaf lettuce was used as a model salad crop because of its rapid production cycle, its adaptability to controlled-environment cultivation, and its relatively uncomplicated growth pattern (Hoff *et at.*, 1982; Wheeler *et al.*, 1994).

MATERIALS AND METHODS

Plant Culture

'Waldmann's Green' leaf lettuce (*Lactuca sativa* L.) canopies were grown from seed in two Minitron II plant-growth/canopy gas-exchange systems (Knight et al., 1988). During the first 7 days of growth, seedlings were nourished with half-strength Hoagland's no. 1 nutrient solution (Hoagland and Arnon, 1950). After day 8, single-strength solution was provided, including 2.5 mol$\cdot$m^{-3} NH_4NO_3. Solution pH and electrical conductivity were maintained at 5.5 ± 0.3 and 2.4 ± 0.2 mS$\cdot$cm^{-1}, respectively, throughout the cropping period.

Seedlings were germinated and grown for the first two days under a PPF of 3 µmol$\cdot$m$^{-2}\cdot$s^{-1}. PPF was raised to 200 µmol$\cdot$m$^{-2}\cdot$s^{-1} from days 2 to 10, and the day/night CO_2 concentration was 400/400 µmol$\cdot$mol^{-1} through day 10. From day 11, PPF and CO_2 concentration were manipulated to achieve a target Pn determined by an algorithm. The photo/nycto periods were 20/4 h, respectively, and shoot-zone temperature was maintained at 25 °C. Electrical power consumption from the lamp bank was measured with an appliance meter (Model ML-985, Duncan Electric Co., Lafayette, IN). At day 23, the canopy was harvested, divided into various plant parts, and dried at 70 °C for 3 days. Edible yield rate (EYR) was calculated as edible biomass per growth area per cropping time. An energy conversion efficiency (ECE=EYR/kWh) was calculated to evaluate the energy efficiency for crop production using different control strategies as described in the Control Algorithm section.

Minitron II Plant-Growth/Canopy Gas-Exchange System

Two Minitron systems (Chun and Mitchell, 1996) were installed in a controlled environment room, and their environmental sensors were interfaced with Optomux digital and analog input/output controllers (Opto 22, Temecula, CA). A control program was developed using Paragon 500 industrial control software (Intec Controls, Walpole, MA). The Minitron system utilizes a slow turnover gas-flow system with CO_2 injection to maintain a given chamber CO_2 level during gas exchange. CO_2 homeostasis is accomplished by a series of mass-flow control valves (MKS, models 1259A and 1259C, Andover, MA) for air and CO_2 inlet streams to each chamber. The computer receives signals from infrared gas analyzers (Infrared, model IR-705, Santa Barbara, CA and Horiba, model PIR-2000, Irvine, CA) measuring CO_2 in the outlet atmosphere from each Minitron chamber. Based upon voltage output from the analyzers, the computer provides feedback signals to mass-flow valves to increase or decrease in aperture, thereby changing the flow rate of CO_2 into a Minitron. Thus, a stable CO_2 environment is maintained during photosynthetic CO_2 fixation, while the system measures Pn by means of voltage signals from CO_2 injection valves to the computer (Mitchell, 1992).

Control Algorithm

The dynamic optimization of crop growth was strategically tested to combine aspects of phasic control gleaned from previous data together with computer-controlled modulation of environmental parameters during specific phases of crop development. Electrical energy consumption for crop production was measured for each combination of environmental inputs. Pn was the target of the dynamic control system, and environmental variables were manipulated to achieve target Pn, because it is the parameter that most quickly and directly indicates canopy ability to produce oxygen and biomass. Pn not only reflects immediate plant response to temperature, CO_2, and PPF, but can be controlled directly by manipulating these variables.

The purpose of the "Dynamic control strategy" is not to maximize canopy Pn but to optimize Pn to meet the overall system needs. Target Pn should be decided from the amounts of oxygen or biomass the entire CELSS system needs in a certain period. In the present study, however, to test this dynamic control strategy both upward and downward, the target Pn on a given day was set at 80 % of maximum Pn for that day, and the maximum canopy Pn was estimated during a preliminary experiment using high CO_2 (1100/400 $\mu mol \cdot mol^{-1}$ during day/night periods) plus high PPF (1000 $\mu mol \cdot m^{-2} \cdot s^{-1}$) at 25 °C. A logistic function was generated from lettuce canopy Pn data using curve fitting to determine maximum Pn during each day of the cropping cycle.

During the lag phase of growth (through day 10), PPF and CO_2 concentration were kept at 200 $\mu mol \cdot m^{-2} \cdot s^{-1}$ and 400 $\mu mol \cdot mol^{-1}$, respectively, but after day 11, the canopy was challenged hourly with changes in PPF and CO_2 concentration. From days 11 to 18 of exponential growth, initial values for the first hour of each photo-period were set at 80 % of maximum PPF and the same CO_2 concentration that gave the highest Pn on the previous day. Comparing target and measured Pn during each hour of a given day, setpoints for PPF and CO_2 concentration were adjusted for the next hour. Upward test increments for CO_2 were 30 $\mu mol \cdot mol^{-1}$ at a time, whereas CO_2 decrements of 15 $\mu mol \cdot mol^{-1}$ were used to test downward. For PPF, increments and decrements in this period were 50 and 25 $\mu mol \cdot m^{-2} \cdot s^{-1}$, respectively. During days 19-23 of the plateau phase, canopies were challenged with 20 $\mu mol \cdot mol^{-1}$ upward or 10 $\mu mol \cdot mol^{-1}$ downward for CO_2 concentration, and with 30 $\mu mol \cdot m^{-2} \cdot s^{-1}$ upward/15 $\mu mol \cdot m^{-2} \cdot s^{-1}$ downward for PPF.

RESULTS AND DISCUSSION

Eq. 1 was formulated by curve fitting Pn response data against saturating levels of CO_2 and PPF inputs to decide maximum Pn for a given day, when x equals the number of days after sowing.

$$Pn(x) = \frac{32.941}{1+e^{-0.568(x-15.908)}} - 0.708 \qquad (1)$$

This equation of a sigmoid curve reflects the fact that canopy Pn is very low through day 10, after which there is an exponential increase until day 18, after which, Pn curves over toward a plateau phase. This gas-exchange pattern suggests that environmental resources can be conserved when the canopy is unresponsive to optimizing conditions (eg., lag and plateau phases), and permits phasic control to be included in dynamic optimization strategies. Figure 1 illustrates dynamic control of target Pn (19.63 $\mu mol \cdot m^{-2} \cdot s^{-1}$ on day 18) achievement by manipulating PPF and CO_2 concentration. With the present version of the Minitron gas-exchange system, it is feasible to challenge lettuce canopies with higher or lower increments of PPF every 8 min, but it takes about 30 min for Pn to re-equilibrate in response to a change in CO_2 concentration. Thus, the control strategy is to first change PPF at a certain CO_2 concentration, and then change the CO_2 at constant PPF. In the present study, a combination of empirical mathematical models plus the use of rule sets developed from previous experimental data was adopted as an approach to decision making. The rules for establishing initial setpoints on a given day, the time interval for the next decision point, the order of changing multiple variables, and the increments of PPF/CO_2 concentration changes were decided empirically and programmed.

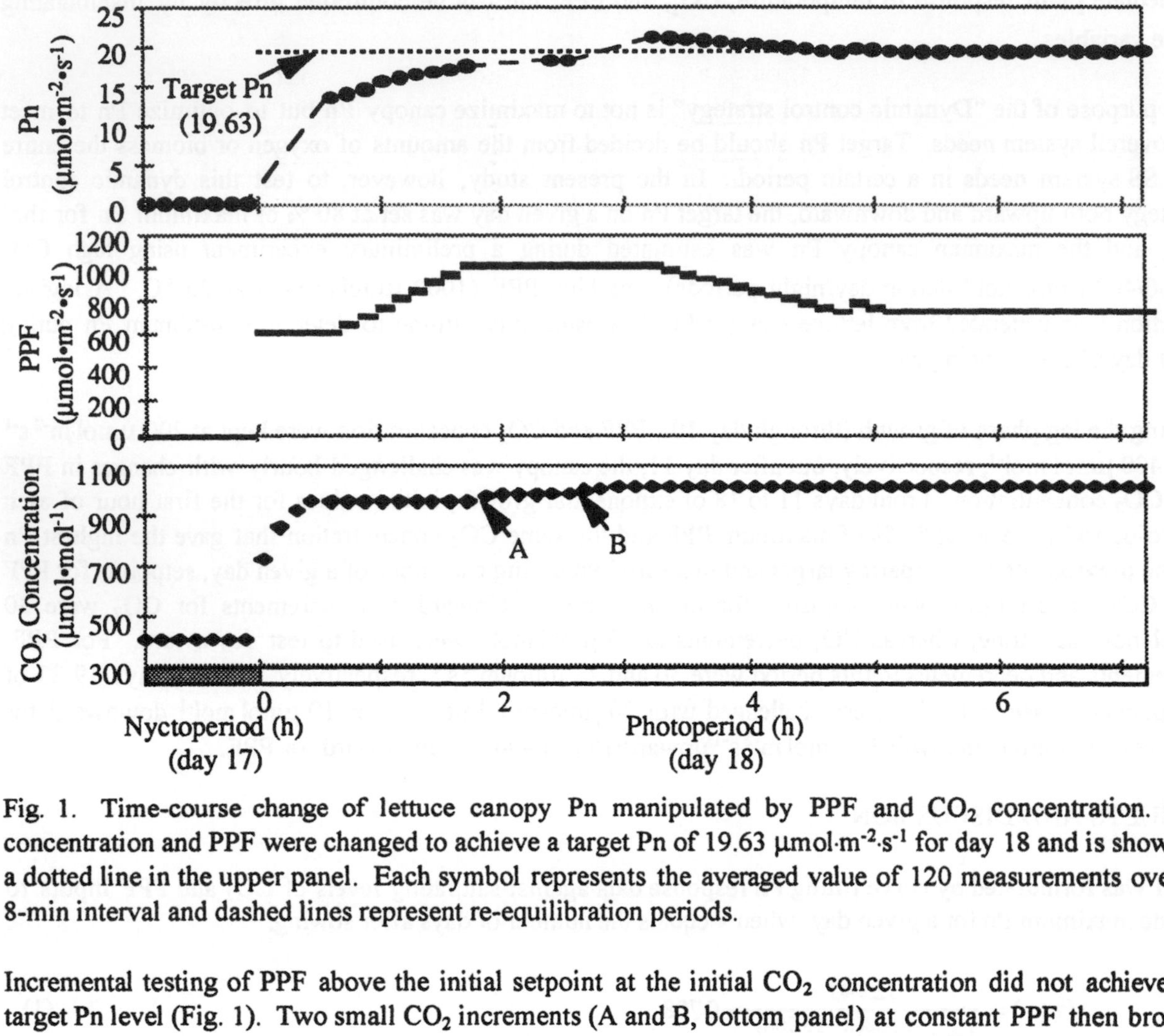

Fig. 1. Time-course change of lettuce canopy Pn manipulated by PPF and CO_2 concentration. CO_2 concentration and PPF were changed to achieve a target Pn of 19.63 $\mu mol{\cdot}m^{-2}{\cdot}s^{-1}$ for day 18 and is shown as a dotted line in the upper panel. Each symbol represents the averaged value of 120 measurements over an 8-min interval and dashed lines represent re-equilibration periods.

Incremental testing of PPF above the initial setpoint at the initial CO_2 concentration did not achieve the target Pn level (Fig. 1). Two small CO_2 increments (A and B, bottom panel) at constant PPF then brought Pn above the target level, and subsequent decrements of PPF at constant high CO_2 then brought Pn to the target by hour 5, at which point it could be held constant for the remainder of the photoperiod. At the highest CO_2 concentration tested, PPF could be brought down from 1027 to 752 $\mu mol{\cdot}m^{-2}{\cdot}s^{-1}$ without great degradation of Pn, and further demonstrating that a dynamic control strategy can save energy without disruption of plant growth. These results suggests that plant canopy Pn or plant growth rate can be optimized by manipulating both PPF and CO_2 and that plant-production units in CELSS can produce specific amounts of oxygen and/or biomass that the entire system requires by changing the target Pn value. Switching to higher capacity mass-flow valves (10 $L{\cdot}min^{-1}$ in the present study) would shorten Minitron re-equilibrium time when CO2 setpoints are changed, permitting incremental testing to proceed even faster.

Table 1 presents several CELSS-relevant crop yield parameters such as edible yield rate (EYR) and daily energy conversion efficiency (ECE). These parameters were compared with those from previous studies using dynamic control of PPF at a constant CO_2 concentration of 1000 μmol·mol^{-1} (Chun and Mitchell, 1996), as well as those from static control using 1000 μmol·mol^{-1} CO_2 and a constant PPF of 895 μmol·m^{-2}·s^{-1} (Knight and Mitchell, 1988). EYRs for dynamic and static control were comparable for all treatments except for static low PPF. Daily ECE indicates that energy-use efficiency was greater for dynamic control of both PPF and CO_2 than for other treatments. These findings suggest that dynamic control has potential for energy savings in resource-limited environments such as CELSS. Further improvement of decision making will lead to still better energy efficiency and sustainability of the overall system.

Table 1. Comparison of Crop Yield Parameters between Dynamic and Static Control Strategies. Units for CO_2 and PPF are μmol·mol^{-1} and μmol·m^{-2}·s^{-1}, respectively.

Efficiency Parameters	[CO_2] / PPF			
	Dynamic	Dynamic PPF	Static	
	(var / var)	(1000 / var)	1000 / 430	1000 / 895
Edible yield rate (EYR) (g ed·m^{-2}·d^{-1})	8.05	8.77	4.09	8.26
Daily energy consumption (kWh·m^{-2}·d^{-1})	32.54	38.85	27.53	44.14
Daily energy conversion efficiency (Daily ECE) (g ed·kWh^{-1})	0.25	0.23	0.15	0.19

In conclusion, test of a dynamic control strategy for both PPF and CO_2 based upon rapid crop response shows that it has potential to conserve energy during the production of a vegetative crop and perhaps during at least ßthe vegetative phase of any CELSS crop. Future experiments will develop even more economical crop-production protocols, more precise control algorithms, and eventually achieve full computer-directed automation of environmental control.

ACKNOWLEDGMENT

This research was sponsored in part by NASA grant NAGW 2329.

REFERENCES

Chun, C. and C.A. Mitchell, Dynamic Control of Photosynthetic Photon Flux for Lettuce Production in CELSS, *Acta Hort.*, **440**, 7-12 (1996).

Hoagland. D.R. and D.I. Arnon, The Water-Culture Method for Growing Plants without Soil, *California Agricultural Experiment Station Circular*, **347** (1950).

Hoff, J., J. Howe, and C. Mitchell, Nutritional and Cultural Aspects of Plant Species Selection for a Regenerative Life-Support System, *NASA Contractor Report* **166324**, NASA Ames Research Center, Moffett Field, California (1982).

Knight, S., C. Akers, S. Akers, and C. Mitchell, The Minitron II System for Precise Control of the Plant Growth Environment, *Photosynthetica* **22**, 90-98 (1988).

Knight, S. and C. Mitchell, Effects of CO_2 and Photosynthetic Photon Flux on Yield, Gas Exchange and Growth Rate of *Lactuca sativa* L. 'Waldmann's Green', *J. Exp. Bot.* **30**, 317-328 (1988).

Mitchell, C.A., Measurement of Photosynthetic Gas Exchange in Controlled Environment, HortScience, **27(7)**, 764-767 (1992).

Mitchell, C.A., The Role of Bioregenerative Life-Support Systems in a Manned Future in Space, *Trans. Kansas Acad. Sci.* **96(1-2)**, 87-92 (1993).

Salisbury, F. and M. Clark, Choosing Plants to be Grown in a Controlled Environment Life-Support System (CELSS) based upon Attractive Vegetarian Diets, *Life Support & Biosphere Sci.* **2**,169-179 (1996).

Wheeler, R.M., C.L. Mackowiak, J.C. Sager, N.C. Yorio, and W.M. Knott, Growth and Gas Exchange by Lettuce Stands in a Closed, Controlled Environment, J. Amer. Soc. Hort. Sci. **119(3)**, 610-615 (1994).

Pergamon

Adv. Space Res. Vol. 20, No. 10, pp. 1861–1867, 1997

Printed in Great Britain
0273-1177/97 $17.00 + 0.00

PII: S0273–1177(97)008853–3

ATMOSPHERIC LEAKAGE AND METHOD FOR MEASUREMENT OF GAS EXCHANGE RATES OF A CROP STAND AT REDUCED PRESSURE

K. A. Corey*, D. J. Barta**, M. A. Edeen** and D. L. Henninger**

**Department of Plant & Soil Sciences, University of Massachusetts, MA 01003, U.S.A.*
***NASA, Johnson Space Center, Houston, TX 77058, U.S.A.*

ABSTRACT

The variable pressure growth chamber (VPGC) was used in a 34-day functional test to grow a wheat crop using reduced pressure (70 kPa) episodes totalling 131 hours. Primary goals of the test were to verify facility and subsystem performance at 70 kPa and to determine responses of a wheat stand to reduced pressure and modified partial pressures of carbon dioxide and oxygen. Operation and maintenance of the chamber at 70 kPa involved continuous evacuation of the chamber atmosphere, leading to CO_2 influx and efflux. A model for calculating CO_2-exchange rates (net photosynthesis and dark respiration) was developed and tested and involved measurements of chamber leakage to determine appropriate corrections. Measurement of chamber leakage was based on the rate of pressure change over a small pressure increment (70.3 to 72.3 kPa) with the pump disabled. Leakage values were used to correct decreases and increases in chamber CO_2 concentration resulting from net photosynthesis (Ps) and dark respiration (DR), respectively. Composite leakage corrections (influx and efflux) at day 7 of the test were 9 % and 19 % of the changes measured for Ps and DR, respectively. On day 33, composite corrections were only 3 % for Ps and 4 % for DR. During the test, the chamber became progressively tighter; the leak rate at 70.3 kPa decreasing from 2.36 chamber volumes/day pretest, to 1.71 volumes/day at the beginning of the test, and 1.16 volumes/day at the end of the test. Verification of the short-term leakage tests (rate of pressure rise) were made by testing CO_2 leakage with the vacuum pump enabled and disabled. Results demonstrate the suitability of the VPGC for conducting gas exhange measurements of a crop stand at reduced pressure.

INTRODUCTION

Reduced pressure atmospheres will be employed in extraterrestrial habitats in order to minimize the pressure gradient between the atmosphere and the external environment. Since structural requirements and atmospheric leakage are both directly proportional to the pressure gradient, considerable mass and energy savings to launch or produce mass (e.g. extract oxygen from lunar regolith) will be realized using reduced pressures (McCarthy and Green, 1991). In habitats employing plants as a means of air revitalization, food production, and water purification, there may be additional benefits of reduced pressure related to water flux (Andre and Massimino, 1992; Daunicht and Brinkjans, 1996; Ohta *et al.*, 1993) and enhanced photosynthesis (Corey *et al.*, 1996a; Corey *et al.*, 1996b). A long term objective of advanced life support testing is to determine responses of crop stands to sustained reduced pressure. A component of such testing involves the measurement of gas exchange rates of entire crop stands to assess overall crop health and performance and to track oxygen production. Measurements of photosynthetic carbon dioxide uptake in growth chambers has typically involved leakage corrections to account for diffusive leakage. Methods for making diffusive leak rate corrections to gas exchange measurements have been developed by several workers (Acock and Acock, 1989; Hand, 1973; Kimball, 1990).

In small chamber testing, it is usually possible to minimize chamber leakage by minimizing the number of penetrations to the chamber. Several research groups are involved with reduced pressure plant growth work using small growth areas and chamber volumes (Andre and Massimino, 1992; Daunicht and Brinkjans, 1996; Ohta *et al.*, 1993; Corey *et al.*, 1996a; Schwartzkopf *et al.*, 1995). One set of chambers has the capability of sustaining pressures down to 50 mb (5 kPa) with leak rates of only 24 % of the chamber volume per day (Schwartzkopf *et al.*, 1995). A chamber designed for measurement of the nature and quantity of plant volatiles has been reported to have leak rates at 50 kPa of only 0.003 % of the chamber volume per day (Bates *et al., 1994)*. On a larger scale, Kennedy Space Center's biomass production chamber has typical diffusive leak rates of 9.8 % of the chamber volume per day (Wheeler *et al., 1991)*. During temperature changes such as occurs during a day/night temperature transition, pressure gradients of about 0.5 kPa develop and result in mass leakage (2 % of chamber volume during a 6 C event) in or out of the chamber depending on the direction of the pressure gradient (Wheeler, 1992).

Johnson Space Center's Variable Pressure Growth Chamber (VPGC) is a large, closed, controlled environment chamber designed for testing of human life support in concert with a combination of bioregenerative and physicochemical systems (Barta and Henninger, 1996). The VPGC has the capability of operation at 10.2 psi (70 kPa), but had not been tested for plant growth operations at reduced pressure until the current study. In large chambers with numerous penetrations and seals such as the VPGC, leakage at reduced pressure may be high enough to preclude accurate measurements of gas exchange rates. Also, the leakage characteristics of the chamber may change over time, necessitating periodic checks to apply appropriate corrections to gas exchange measurements. The objectives of this test were to verify facility and subsystem performance of the VPGC at 70 kPa and to determine leakage characteristics of the chamber for application to testing responses of a wheat stand to reduced pressure and modified partial pressures of oxygen and carbon dioxide.

MATERIALS AND METHODS

Test Description

Wheat seed was planted in the VPGC on October 18, 1995 according to protocol established previously (Edeen and Barta, 1996). Baseline conditions for atmospheric control and other environmental conditions are summarized in Table 1. Seeding rate was estimated to be 2800 seed/m^2 based on 3 measurements of

Table 1. Nominal environmental conditions used for the reduced pressure functional test.

Variable	Designation	Values
Atmospheric Pressure	P_t	70 & 101 kPa, alternating
Carbon Dioxide Partial Pressure	$ppCO_2$	121.6 Pa
Oxygen Partial Pressure	ppO_2	
70 kPa		21.2 kPa
101 kPa		14.7 kPa
Relative Humidity	RH	70 %
Light Intensity	PPF	1200 $\mu mol \cdot m^{-2} \cdot s^{-1}$
Canopy Temperature	T_c	23 C
Volumetric Air Flow	V_a	
70 kPa		51.0 $m^3 \cdot min^{-1}$
101 kPa		53.0 $m^3 \cdot min^{-1}$

counts from seeding 3, half-trays (0.175 m^2/half-tray). Lights were set to one-third the maximum voltage at day 3, were increased to two-thirds of maximum on day 7, and finally to 100 % on day 11. The plant growth trays were lowered several times during growth of the wheat stand to give a distance of about 20 cm between the light bank barrier and the estimated top of the plant canopy. This distance decreased after day 26 when the limit of tray lowering was reached. Reduced pressure episodes were conducted during regular business hours and generally lasted between six and seven hours. Twenty-one reduced pressure episodes totalling 130.7 hours were used in alternation with ambient periods (Figure 1). During reduced pressure episodes, the CO_2 setpoint was elevated from 1200 ppm at ambient pressure to 1729 ppm in order to maintain a constant $ppCO_2$ of 121.6 Pa for both pressures. Initial depressurization operations involved continuous pumpdown to 70.3 kPa, but were adjusted to a slower stepdown procedure when the lower growth trays were observed to fill and overflow onto the floor of the VPGC. The problem was caused by a

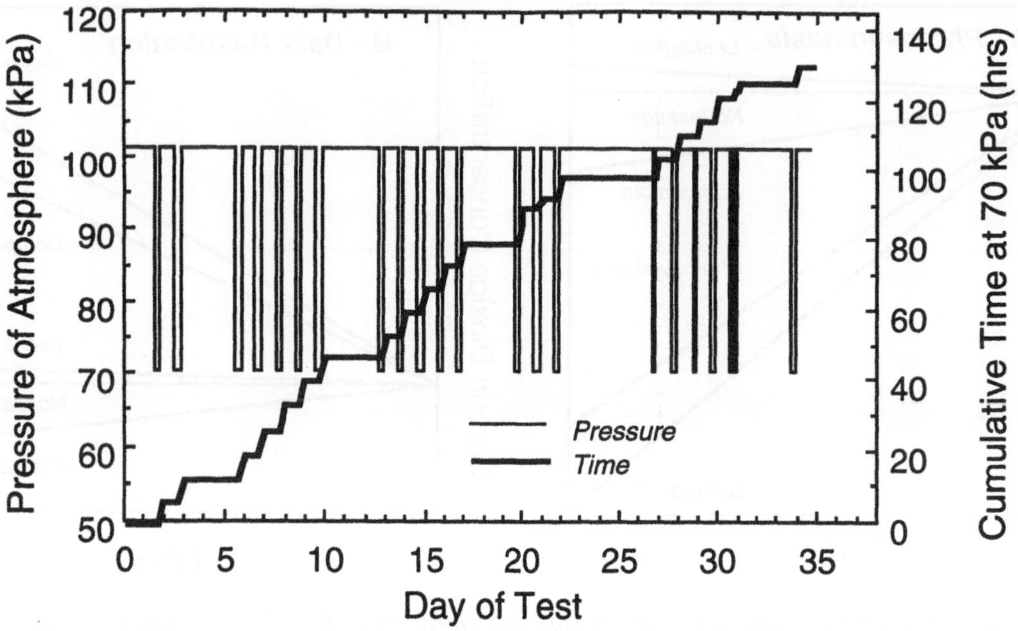

Fig. 1. Atmospheric pressure and cumulative time at reduced pressure in the VPGC during the 34-day wheat test.

pressure differential between the VPGC air volume and the air volume in the nutrient feed tank during depressurization. Ramping the pressure in increments of about 3.45 kPa (0.5 psi) over a total period of about 30-min eliminated the tray overflow problem.

Gas Exchange Model

At ambient pressure and isothermal conditions, CO_2 leakage from the chamber is predominantly by diffusion, is negligible over the time frame used for gas exchange measurements (about 1.9 % of chamber volume/day), and therefore may be ignored for gas exchange calculations made in less than an hour. However, measurements of the rates of change in CO_2 concentration with time during maintenance of 70.3 kPa necessitate corrections that account for CO_2 pumped out (L_o) and CO_2 leaked in/diluted (L_i) from the ambient environment. In the absence of metabolic activity, and given a chamber of fixed volume, constant temperature, and sustained reduced pressure, CO_2 concentration in the chamber with a high initial value (e.g. 1730 µmol/mol) will tend to decrease over time as it is pumped from the atmosphere (Figure 2A). A small offset of the decrease is attributable to the CO_2 leakage into the chamber as external air moves via mass flow into the chamber. With a stand of plants in the chamber, the lights on, and CO_2 injection disabled, the decline in CO_2 concentration is a composite of three processes; that which is pumped out, that which leaks in from the atmosphere outside the chamber and becomes diluted, and that net amount which is fixed by the plants photosynthetically (Phs). Conversely, when the lights are off and CO_2 injection disabled, evolution of CO_2 by the plant stand during dark respiration (DR) would cause the CO_2 concentration in the chamber to increase (Figure 2B). To calculate only the quantity of CO_2 fixed or evolved by the plant stand, it is necessary to make corrections to the composite measurements by estimating the leakage components. The leakage rates shown in Figure 2 are relative rates used to illustrate the carbon dioxide exchange dynamics of the chamber applied to measurement of metabolic rates of plant stands, but may also be applied to the calculation of evolution rates of other gases such as ethylene. The proportion of the composite gas exchange (measured changes in CO_2) attributable to the plants (i.e. CO_2 uptake or evolution) depends on the actual chamber leak rate as it may vary with time, partial pressures of CO_2 inside and outside the chamber, and stage of growth of the crop. It is important for absolute leakage corrections to be low (less than a few percent) in order to make comparisons of plant gas exchange responses at reduced pressure with those made at ambient pressure.

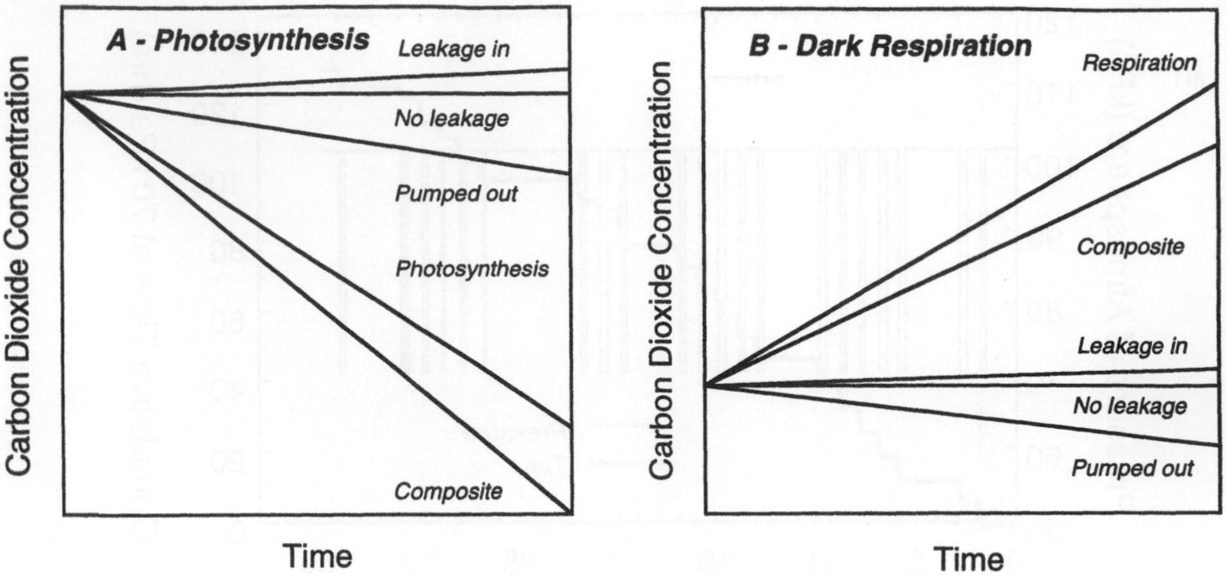

Fig. 2. Leakage model for changes in carbon dioxide concentration in the VPGC atmosphere used to calculate rates of net photosynthesis (A) and dark respiration (B) during maintenance of reduced pressure. See text for assumptions and explanation.

Leakage Determinations

Six days prior to the 34-day test, a complete leak test was conducted following a depressurization, by disabling the pump and allowing the pressure to increase to ambient pressure. The same leak test was also conducted following the wheat harvest (Figure 3). Since reduced pressure episodes involved sustained operation of a constant reduced pressure, it was of interest to determine leakage rates at 10.2 psi (70.3 kPa). Pressure changes over a small change in pressure (e.g. 10.2 to 10.5 psi) were linear, so periodic short leak checks during the test were achieved by disabling the pump and determining the change in pressure after 30 minutes. One day after the test (day 35), two additional leak tests were conducted with the initial CO_2 concentration close to 1700 ppm. One test was conducted to determine CO_2 leakage when the pump was enabled to maintain 10.2 psi and the other was done with the pump disabled to determine CO_2 leakage in and its dilution, accompanied by a gradual increase in atmospheric pressure.

Crop Gas Exchange Measurements

Photosynthesis (Pn) and dark respiration (DR) measurements were made daily at both pressures. Carbon dioxide analysis was conducted on samples of chamber atmosphere pulled continuously through a series of gas analyzers (Beckman Industrial, model 880) prior to return to the chamber. Carbon dioxide injection was disabled at or slightly above the setpoint and allowed to decrease for 10 to 15 minutes. Dimmers were then set to 0 and the CO_2 allowed to increase for 15 minutes, followed by manual injection to bring the CO_2 concentration back to the setpoint. It was necessary to increase the setting on the heating coils during the lights off portion of the measurements to maintain a constant temperature. Slopes of linear decreases and increases in CO_2 concentration were calculated from 8 to 10 minute stretches, and the rates of Pn and DR determined. The molar volumes at 23 C for the VPGC used in the calculations at 101.3 kPa and 70.3 kPa were 1111.2 moles and 774.2 moles, respectively. Leak rates obtained from the most recent leakage determination were used for corrections to the slopes of the reduced pressure CO_2 drawdowns. For calculation of Phs rate, L_o was subtracted from and L_i was added to the composite measurement. The converse was done to correct measurements of rate of DR.

RESULTS AND DISCUSSION

Based on measurements of gas exchange rates from previous tests, the leak rate at the beginning of the test

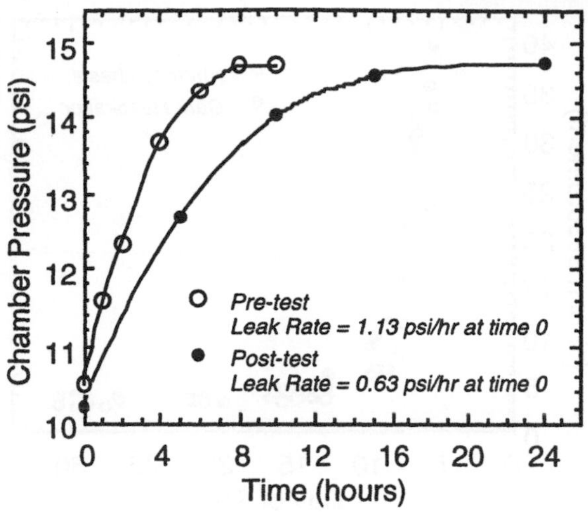

Fig. 3. Repressurization due to leakage of the VPGC atmosphere from 10.2 psi to ambient pressure prior to and following the 34-day reduced pressure test.

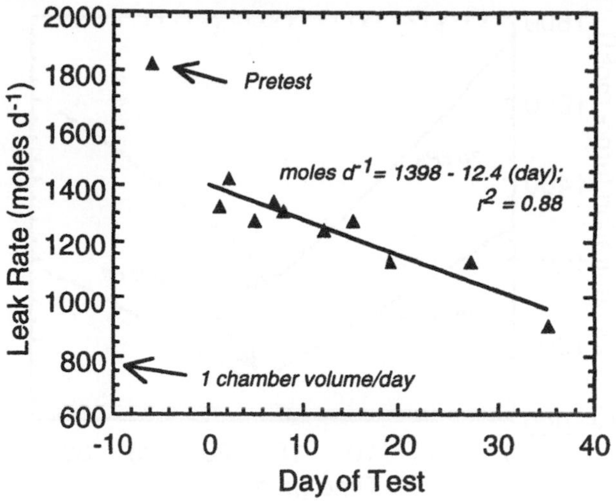

Fig. 4. Changes in atmospheric leakage of the VPGC at 70 kPa during the 34-day wheat test.

was estimated to be less than 10 % of expected changes in CO_2 concentration for an early vegetative canopy (day 10). An increase in the magnitude of CO_2 concentration changes (higher rates of Pn and DR) was expected to lower the % of the measurement attributable to leakage. Prior to the test, VPGC leakage was fairly high with a pressure loss of 1.13 psi/hr at 10.5 psi (Figure 3). A number of leaks were detected in the plumbing of the fluid delivery system which were reduced greatly by replacement of valve caps. The first short leak test (rate of pressure rise) on day 1 showed a substantial reduction in leakage from the pretest point (Figure 4). During the test, the leak rate declined linearly, indicating a considerable tightening of the chamber during closure. Perhaps repeated depressurizations and repressurizations led to clogging of pores in seals and joints by microparticulates, leading to reduced leakage rates. By the end of the test, the leak rate at 10.2 psi was only 1.16 reduced pressure chamber volumes/day.

Representative patterns of the uncorrected CO_2 drawdowns and increases for measurements of Pn and DR at 70 and 101 kPa are illustrated for day 19 (Figure 5). Calculations of rates were made as described in the procedures. Rates of Pn and DR increased rapidly with canopy growth approaching a maximum between 15 and 20 days from seeding, the time of full canopy development (Wheeler *et al.*, 1993). Leakage corrections made to measurements early in the test decreased from nearly 40 % of the measurement on day 5 to less than 10 % by day 9 (Figure 6). Leakage corrections were generally less than 5 % of the measurement after day 10. After the wheat was harvested, carbon dioxide leakage when the pump was enabled to sustain 70.3 kPa, resulted in a rate of decline of 0.65 µmol/mol·min (Figure 7A). This rate reflects a combined effect of removal of chamber CO_2 by evacuation during reduced pressure maintenance and the dilution of the chamber CO_2 concentration with ambient air having a much lower CO_2 concentration (i.e. 350 to 420 ppm). In order to isolate just the L_i component, the pump was disabled and the CO_2 monitored. The rate of decline was 0.66 µmol/mol·min, essentially the same as with the pump enabled (Figure 7B). To verify that this latter rate was attributable to dilution of chamber air with ambient air, predicted CO_2 concentrations were calculated assuming the same initial concentration, the slope of the pressure versus time function through time 0 as 0.0082809 psi/min, an average ambient CO_2 concentration during the test of 379 µmol/mol, and the actual pressure at a given time. The predicted decline in CO_2 concentration was not appreciably different from those actually measured (0.651 ppm/min compared with 0.657 µmol/mol·min). Since the rate of CO_2 decline was also the same whether the pump was enabled or disabled, the leakage value served as a confirmation of making leak corrections based on the rate of pressure increase over a small range of pressure (to ensure linearity) with the pump disabled. This method represents a straightforward means of determining leakage corrections to gas exchange measurements of crops made during the test, particularly if pre- and post-test leakage measurements do not accurately reflect the actual leakage near the time of the gas exchange measurements, as was the case in this study. A disadvantage of

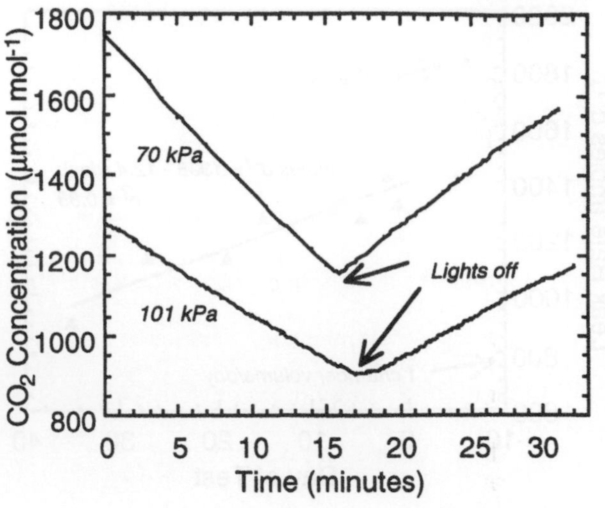

Fig. 5. *Patterns of changes in carbon dioxide concentration of the VPGC atmosphere used for calculations of net photosynthesis (Pn) and dark respiration (DR) of a 19-day old wheat stand during episodes of reduced (70 kPa) and ambient (101 kPa) pressures.*

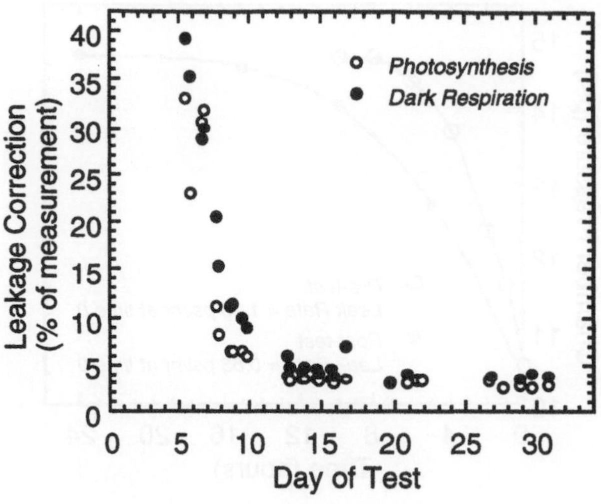

Fig. 6. *Leakage corrections made to gas exchange measurements during growth of a wheat stand.*

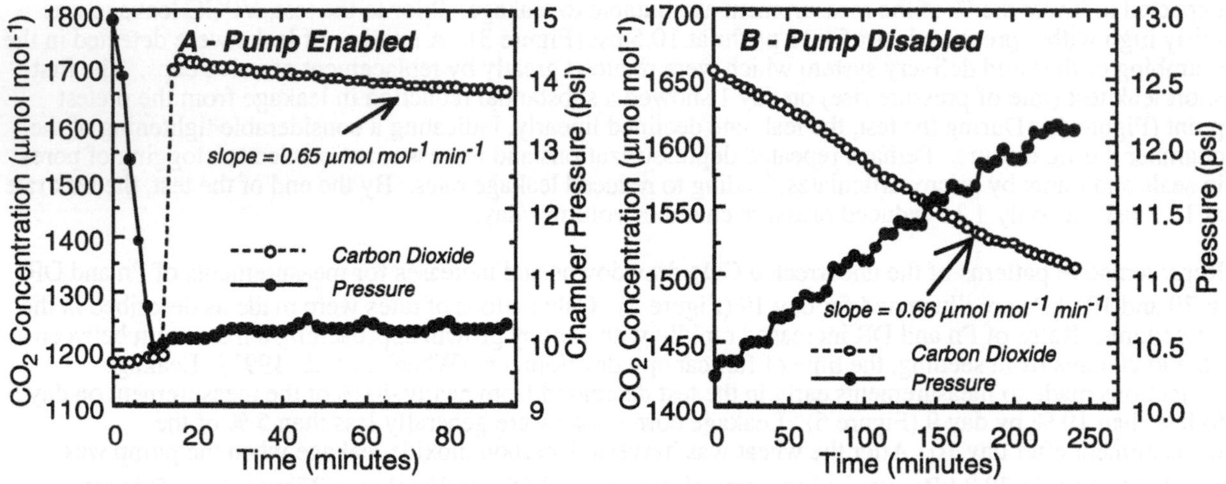

Fig. 7. *Decay of carbon dioxide concentration of the VPGC atmosphere due to evacuation and dilution with the pump enabled (A) and due to dilution with the pump disabled (B).*

the approach is that the pump must be disabled for about 30 minutes resulting in a pressure transient. Another drawback of this procedure is that the correction becomes a larger proportion of the measured change when low partial pressures of CO_2 (e.g. < 35 Pa) are used for the generation of photosynthetic response functions. Below ambient ppCO_2 (35-40 Pa), $L_i > L_o$, and therefore composite corrections become negative. Another approach for measuring leakage during the test and still allow the pump to be enabled would be to measure the decay of a nonmetabolic gas such as argon.

In reduced pressure research involving measurements of gas exchange rates of plants, it is necessary to minimize and update the status of chamber leakage in order to make appropriate corrections. With a densely planted crop cultured for high rates of stand photosynthesis, leakage rates at reduced pressures may preclude accurate measurements of crop gas exchange rates early in growth but become small enough (< 5 % of measurement) later in development for accurate measurements of gas exchange rates. The use of crops which establish and develop slowly, or ones that are planted at relatively low densities (e.g. lettuce), may not enable measurements of gas exchange rates until late in development. Despite its size and the number of penetrations for subsystem components, results of this work demonstrate the suitability of the VPGC for conducting measurements of gas exchange rates of a large crop stand at reduced pressure.

ACKNOWLEDGMENTS

The authors would like to acknowledge the work of Ed Mohr, Don Overton, Tim Monk, Don Kilpatrick, and John Gibbs with mechanical, instrumentation, and computer control subsystems. Research by the senior author was supported in part by a grant from the National Research Council.

REFERENCES

Acock, B. and M.C. Acock, Calculating air leakage rates in controlled-environment chambers containing plants, *Agron. J.*, 81, 619-623 (1989).

Andre, M. and D. Massimino, Growth of plants at reduced pressures: Experiments in wheat-technological advantages and constraints, *Adv. Space Res.* 12 (5), 97-106, (1992).

Barta, D.J. and D.L. Henninger, Johnson Space Center's regenerative life support systems test bed, *Adv. Space Res.* 18, 211-221, (1996).

Bates, M.E., K.D. Wilson, D. Bubenheim, S. Cronin, R. Averill, D. Sedlak, and K. Wagenbach, A closed plant growth chamber for volatiles research, *Amer. Soc. of Agric. Eng.* Paper No. 944078, (1994).

Corey, K.A., D.J. Barta, and D.L. Henninger, Photosynthesis and respiration of a wheat stand at reduced atmospheric pressure and reduced oxygen, Abstract for paper presented at 31st Committee for Space Research Scientific Assembly held 14-21 July 1996, Birmingham, UK, (1996).

Corey, K.A., M.E. Bates, and S.L. Adams, Carbon dioxide exchange of lettuce plants under hypobaric conditions, *Adv. Space Res.* Vol 18, (4/5), 265-272, (1996).

Daunicht, H.J. and H.J. Brinkjans, Plant responses to reduced air pressure: advanced techniques and results, *Adv. Space Res.* 18 (4/5), 273-281, (1996).

Edeen, M. and D. Barta, Early human testing initiative phase I. Final report. JSC-33636, (1996).

Hand, D.W, Techniques for measuring CO_2 assimilation in controlled environment enclosures, *Acta Hort.* 32, 133-147, (1973).

Kimball, B.A., Exact equations for calculating air leakage rates from plant growth chambers, *Agron. J.*, 82, 998-1003, (1990).

McCarthy, K. B. and J. A. Green, The effect of reduced cabin pressure on the crew and the life support system. *American Society of Agricultural Engineering* Paper No. 911331, (1991).

Ohta, H., E. Goto, T. Takakura, F. Takagi, Y. Hirosawa, and K. Takagi, Measurement of photosynthetic and transpiration rates under low total pressures, *American Society of Agricultural Engineering* Paper No. 934009, (1993).

Schwartzkopf, S.H., J.R. Grote, and T.L. Stroup, Design of a low atmospheric pressure plant growth chamber. *Society of Automotive Engineers* Technical Paper Series No. 951709, Warrendale, PA, (1995).

Wheeler, R.M., K.A. Corey, J.C. Sager, and W.M. Knott, Gas exchange characteristics of wheat stands grown in a closed, controlled environment, *Crop Science* 33, 161-168, (1993).

Wheeler, R.M., Gas exchange measurements using a large, closed plant growth chamber, *HortScience* 27, 777-780, (1992).

Wheeler, R.M., J.H. Drese, and J.C. Sager, Atmospheric leakage and condensate production in NASA's Biomass Production Chamber: Effect of diurnal temperature cycles, John F. Kennedy Space Center. *NASA Tech. Mem.*, 103819, (1991).

Pergamon

Adv. Space Res. Vol. 20, No. 10, pp. 1869–1877, 1997

Printed in Great Britain
0273-1177/97 $17.00 + 0.00

PII: S0273-1177(97)00854-5

PHOTOSYNTHESIS AND RESPIRATION OF A WHEAT STAND AT REDUCED ATMOSPHERIC PRESSURE AND REDUCED OXYGEN

K. A. Corey*, D. J. Barta** and D. L. Henninger**

**Department of Plant & Soil Science, University of Massachusetts, Amherst, MA 01003, U.S.A.*
***NASA, Johnson Space Center, Houston, TX 77058, U.S.A.*

ABSTRACT

A 34-day functional test was conducted in Johnson Space Center's Variable Pressure Growth Chamber (VPGC) to determine responses of a wheat stand to reduced pressure (70 kPa) and modified partial pressures of carbon dioxide and oxygen. Reduced pressure episodes were generally six to seven hours in duration, were conducted at reduced ppO_2 (14.7 kPa), and were interrupted with longer durations of ambient pressure (101 kPa). Daily measurements of stand net photosynthesis (Pn) and dark respiration (DR) were made at both pressures using a $ppCO_2$ of 121 Pa. Corrections derived from leakage tests were applied to reduced pressure measurements. Rates of Pn at reduced pressure averaged over the complete test were 14.6 % higher than at ambient pressure, but rates of DR were unaffected. Further reductions in ppO_2 were achieved with a molecular sieve and were used to determine if Pn was enhanced by lowered O_2 or by lowered pressure. Decreased ppO_2 resulted in enhanced rates of Pn, regardless of pressure, but the actual response was dependent on the ratio of $ppO_2/ppCO_2$. Over the range of $ppO_2/ppCO_2$ of 80 to 200, the rate of Pn declined linearly. Rate of DR was unaffected over the same range and by dissolved O_2 levels down to 3.1 ppm, suggesting that normal rhizosphere and canopy respiration occur at atmospheric ppO_2 levels as low as 11 kPa. Partial separation of effects attributable to oxygen and those related to reduced pressure (e.g. enhanced diffusion of CO_2) was achieved from analysis of a CO_2 drawdown experiement. Results will be used for design and implementation of studies involving complete crop growth tests at reduced pressure.

INTRODUCTION

Sustainable human presence in bases constructed in extreme environments will require bioregenerative components to human life support systems where plants are used for generation of oxygen, food, and water. Pressure and composition of atmospheres in future life support systems such as lunar and martian bases will likely differ from those of a sea-level, earth-based environment. Hypobaric pressures may be used to reduce the pre-breathe time for astronauts to conduct extra-vehicular activity in low-pressure space suits (4.3 psi) and to decrease the mass and engineering requirements for establishing and sustaining life support systems at extraterrestrial outposts (Corey *et al.*, 1995; McCarthy and Green, 1991). The National Aeronautics and Space Administration is considering 10.2 psi (70 kPa) as a candidate atmospheric pressure for space-based life support systems. However, few studies have assessed the metabolic and developmental responses of plants to reduced pressure atmospheres with or without modifications of oxygen partial pressure. There is a near absence of reduced pressure studies conducted on a large scale under tightly controlled conditions.

Recent studies suggest that plant growth at reduced pressures may be enhanced when combined with decreased partial pressures of oxygen or unaffected when the partial pressure of oxygen is held at 21 kPa (Andre and Massimino, 1992; Corey *et al.*, 1996b; Daunicht and Brinkjans, 1992, 1996; Ohta *et al.*, 1993; Rule and Staby, 1981). At present, there has been no clear separation of variables to explain effects of hypobaric pressure on plant processes. However, there are two possibilities based on well established physical principles and the current state of knowledge in plant metabolism. First, the diffusion coefficient of gases in air increases at atmospheric pressures below those of ambient sea level. For example, the diffusion coefficients for CO_2 and H_2O are 1.44 times greater at 0.7 atm than at 1.0 atm total atmospheric pressure (Corey, 1995). Increased rates of CO_2 diffusion will result in a more rapid rate of transport to the site of photosynthesis assuming that stomatal conductance and leaf boundary layer resistance also remain unaffected (Gale, 1972). Water diffusion at reduced pressures should also be enhanced and will lead to greater rates of transpiration (Daunicht and Brinkjans, 1992; Gale, 1973, Ohta *et al.*, 1993).

A second possibility for reduced pressure effects relates to the effects of O_2 on plant metabolic processes. A well established effect of decreased oxygen partial pressure is the decrease in activities of ribulose bisphosphate carboxylase-oxygenase (RUBISCO-oxygenase) and glycolic acid oxidase. The net effect of the activities of these photorespiratory enzymes in C-3 plant species is an inhibition of Pn with increasing ppO_2 accompanied by an increased carbon loss (Bjorkman, 1966; Forester *et al.*, 1966; Gerbaud and Andre, 1980, 1989; Parkinson *et al.*, 1974; Poskuta, 1968; Siegel, 1961; Siegel *et al.*, 1963). The literature on photorespiratory pathways has been reviewed (Ehleringer, 1979; Govindjee, 1983, Jackson and Volk, 1970).

The contribution of enhanced CO_2 diffusion at reduced pressure is expected to depend upon the partial pressure gradient of CO_2, which will in part depend on the molecular ratio of O_2 and CO_2 and their binding to RUBISCO. The overall subcellular gradient will be determined by the combined activities of CO_2- and O_2- binding enzymes. Recent results suggest that the primary effect of reduced atmospheric pressure on the rate of photosynthesis of whole lettuce plants is attributable to the accompanying reduction in oxygen partial pressure, since holding oxygen constant over a range of total pressures resulted in similar rates of carbon dioxide uptake (Corey *et al.*, 1996b). In addition, there is an effect of lowered oxygen on dark respiration (DR) (Chun and Takakura, 1993; Corey *et al.*, 1996b) which may further increase the rate of biomass production through decreased carbon loss. Thus, optimizing productivity of plants also may involve a combination of reduced pressures and lowered partial pressures of oxygen; yet few studies have assessed the metabolic and developmental responses of large crop stands to those conditions.

The development of Johnson Space Center's Variable Pressure Growth Chamber (VPGC), a large, closed plant growth chamber rated for 0.7 atm, has the capability for conducting large scale studies on crop responses to low pressure and varied atmospheric compositions (Barta and Henninger, 1996; Corey, 1995). A goal of the human testing programs at Johnson Space Center is to grow crop plants under reduced pressure (70 kPa) and quantify the mass balances of metabolic products for meeting human requirements. The objectives of this study were to determine responses of a wheat stand to reduced pressure and varied partial pressures of carbon dioxide and oxygen. In the testing of plant responses it was of interest to answer the following question: Does reduced atmospheric pressure alter the rates of gas exchange (i.e. photosynthesis, dark respiration, and evapotranspiration) of a wheat stand and if so, what roles do altered partial pressure of oxygen and enhanced diffusivity of carbon dioxide have in the responses?

MATERIALS AND METHODS

Gas Exchange Measurements

The test and the leakage model used for determination of crop gas exchange rates at reduced pressure have been described (Corey *et al.*, 1998b - current issue). The $ppCO_2$ was maintained at 122 Pa for both atmospheric pressures except during short term rate determinations. Measurements of rates of photosynthesis (Pn) and dark respiration (DR) were made daily. Periods of reduced pressure (70 kPa) were generally six to seven hours in duration. Changes in carbon dioxide concentration at reduced pressure were corrected for leakage components using results of the most recent leakage measurement based on rate of pressure increase (Corey *et al.*, 1998b - current issue).

Carbon Dioxide Partial Pressure Manipulations

A complete carbon dioxide drawdown was conducted on day 16. Carbon dioxide concentrations were increased to 100 to 200 ppm above the setpoints for the two pressures, injection disabled, and the CO_2 concentration allowed to decrease to near the compensation point. Numerous short drawdowns were conducted on days 30 to 33 to determine the photosynthetic responses of the wheat stand to $ppCO_2$ and ppO_2 at ambient and reduced pressures during an experiment where ppO_2 was reduced to various levels.

Oxygen Partial Pressure Manipulations

On day 30, oxygen was scrubbed down to 15.5 % (15.7 kPa) using a molecular sieve. Measurements of Pn and DR were conducted at the lowered ppO_2, the atmosphere was pumped down to 70.3 kPa (ppO_2 = 11.1 kPa), gas exchange measurements taken again, and the door opened to bring the oxygen level back to ambient for a final set of gas exchange measurements. The second oxygen scrub was initiated at the end of day 30 and continued through day 33. Gas exchange measurements were made by short duration drawdowns near the setpoints. The second scrub brought the % O_2 from 20.6 to 9.7 %. One reduced pressure episode occurred during this scrub occurring on day 33; the ppO_2 reaching a low of 7.1 kPa (approximately one-third of ambient ppO_2).

Water Use

The air stream in the VPGC return ducts passes through a variable speed blower and then across a condensing heat exchanger. Condensate is directed to tanks for each chamber side outside the chamber. When the tanks are full (side A = 1.25 liters & side B = 1.27 liters), they are drained and the time is computer logged. Rates of water use were calculated for time increments ranging from several hours for reduced pressure episodes to a complete day when no reduced pressure episodes were used.

Harvest Measurements

The experiment was terminated after 34 days for analysis of biomass production. The shoots were trimmed at the surface, roots and shoots were separated, bagged, fresh weights taken, and then all biomass dried at 70 C and reweighed. Intact plant samples (about 20 plants/pot) were removed from one tray and transplanted to 15-cm pots containing calcined clay and placed in a growth chamber at 23 C to mature. Plants were cultured in the growth chamber at a photosynthetic photon flux of about 400 $\mu mol/m^2 \cdot s$ with no carbon dioxide enrichment. After an additional 49 days (total of 83 days from seed), grain was harvested to assess the degree of seed set. Seed were germinated on moistened filter paper in petri dishes to evaluate viability.

RESULTS AND DISCUSSION

Photosynthesis and Dark Respiration

Rates of Pn and DR increased rapidly with canopy growth approaching a maximum between 15 and 20 days from seeding, the time of full canopy development (Figure 1). During the test, trays were lowered periodically to maintain a near constant distance of 20 cm between the light barrier and canopy top. These events are often reflected by small decreases in the measured rates of photosynthesis. Even though CO_2 concentration was targeted to be near saturation for Pn, there was photosynthetic enhancement at reduced pressure. At reduced pressure, Pn was consistently higher than at ambient pressure averaging 14.6 % higher for the entire test (Figure 1). This is likely attributable in large part to the reduction in ppO_2 (14.7 kPa from 21.2 kPa) that accompanies the reduced pressure episodes.

Rate of DR was apparently unaffected by reduced pressure with pairwise comparisons over the entire test averaging only 1.2 % lower (Figure 1). This suggests that normal background respiration for the crop stand (shoots and roots) is saturated at 14.7 kPa. It is possible that differences in DR attributable to reduced pressure are not detectable during the vegetative stages, but may occur at other developmental stages. For example, a comparison of the rate of DR on days 26 to 30 reveals that DR was reduced by an average of 6.2 % from ambient rates. It was speculated that given the density of the root mass that develops within the trays, oxygen availability within the root mass might limit cytochrome oxidase and overall respiration would decrease. This does not seem to be the case during the vegetative stage of development despite the large reduction in dissolved oxygen that accompanies reduced pressure. For example, upon depressurization, the dissolved oxygen in the nutrient solution declines from 6.6 ppm at 101 kPa to 4.5 ppm at 70 kPa, a predictable decline based on the reduced ppO_2 of the atmosphere at 70 kPa. Root mass is a relatively small proportion of the total standing biomass (15 - 20 %). Therefore, reductions in root respiration attributable to a decline in DO may be overwhelmed by a lack of difference in shoot respiration.

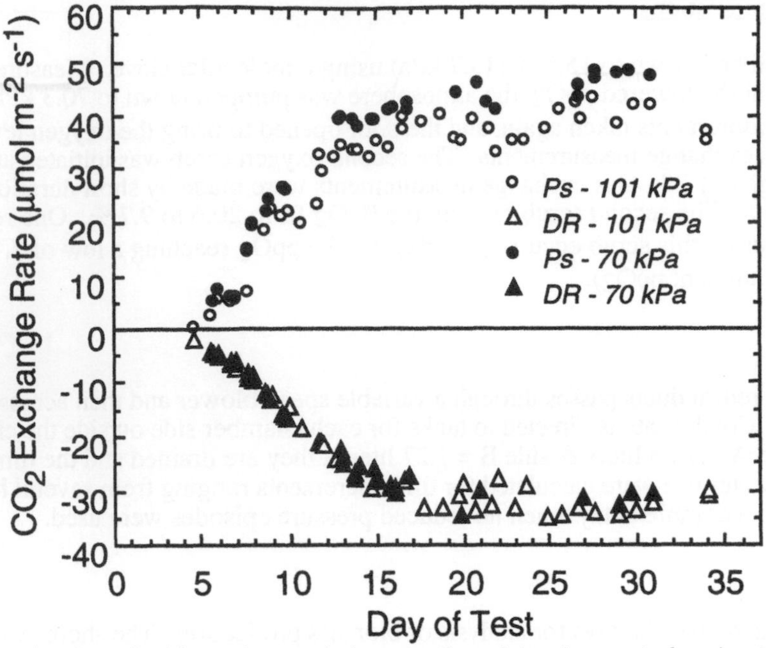

Fig. 1. Rates of carbon dioxide exchange during growth and development of a wheat stand during 70 kPa and 101 kPa pressure episodes.

Since we could not separate the metabolic activity of the roots from the shoots, measurements of stand DR would not necessarily reveal changes in respiratory activity of the root system.

Absolute values for Pn and DR at ambient pressure were quite similar to those obtained from the EHT1-2 test for the first 16 days of the test, indicating reproducibility of a large scale growth test (Edeen and Barta, 1996). Possible cumulative effects of sustained reduced pressure and ethylene scrubbing (by evacuation at reduced pressure) are not known. It will be important to conduct tests with complete growouts to determine if differences in gas exchange translate into differences in biomass production.

Rates of Pn and DR were on the order of 60% and 150 % higher in this test than in Kennedy Space Center's Biomass Production Chamber (Wheeler *et al.*, 1993). Higher net photosynthesis is probably mainly attributable to the higher light intensity used in this study (average PPF = 690 $\mu mol/m^2 \cdot s$ at KSC and ~ 1200 $\mu mol/m^2 \cdot s$ at JSC). The higher rates of DR indicate an overall lower carbon use efficiency; an important consideration if a dark period is used. Lower carbon use efficiency may be due to the higher stand density used in this study.

<u>Responses to Carbon Dioxide Partial Pressure</u>

Photosynthetic responses to $ppCO_2$ were derived from CO_2 drawdowns at both pressures (Figure 2). Carbon dioxide partial pressure declined more rapidly at 70 kPa than at 101 kPa (Figure 2A). Within the 2-hour time frame of the drawdown, the CO_2-compensation point was more closely approached at 70 kPa than at 101 kPa with values of approximately 10 and 15 Pa for 70 kPa and 101 kPa, respectively. Both values are somewhat higher than the 5 to 8 Pa $ppCO_2$ generally observed for CO_2-compensation points for single leaves (Araus and Tapia, 1987), whole plants (Gifford, 1977), and large stands (Wheeler *et al.*, 1993). Rates of photosynthesis were derived from successive 2-point slopes on the drawdown curves. Leakage corrections made to the rates of CO_2 uptake were a large proportion of the measurement at low CO_2 concentrations (< 200 µmol/mol) and became less substantial at higher concentrations; 2 to 4 % over the CO_2 concentration range of 200 to 2400 µmol/mol. Below ambient CO_2 concentration, corrections were opposite in sign because there was a net flux of CO_2 into the chamber. Neither response function reached a plateau even at $ppCO_2$ levels as high as 196 Pa. This suggests that photosynthesis of the stand is not saturated at the nominal $ppCO_2$. Perhaps the stand density was high relative to other tests, complete

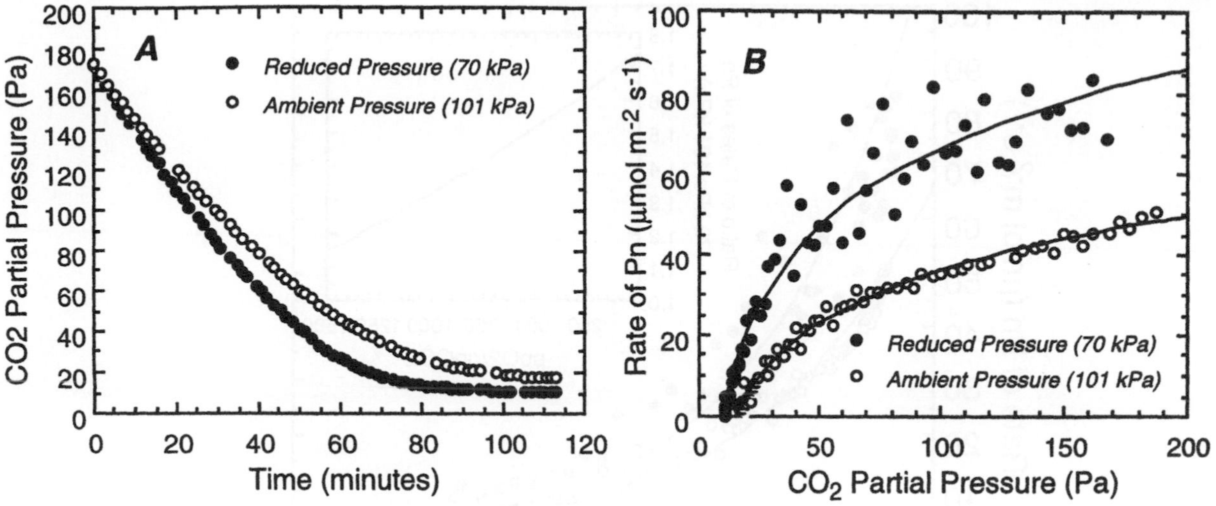

Fig. 2. Photosynthetic carbon dioxide drawdown experiment of a wheat stand on day 16 conducted at 70 and at 101 kPa (A) and the photosynthetic response functions (B) derived from A.

canopy mixing was not achieved, and localized pockets of lower $ppCO_2$ occurred within the crop canopy, leading to a small overestimate of the rate of CO_2 uptake. Inadequate canopy mixing may be attributable to insufficient blower speed or an air circulation pattern that did not result in complete canopy mixing.

Rates of Pn at 70 kPa were substantially higher than at 101 kPa over the entire range of $ppCO_2$ levels (Figure 2B). This is likely in part attributable to the reduction in ppO_2 that accompanies reduced pressure. If the actual ppO_2 values at the two pressures are used to compute a ppO_2 to $ppCO_2$ ratio and the rate of Pn expressed as a function of this ratio (Figure 3), effects attributable to reduced pressure should be removed and the remaining difference between 70 and 101 kPa should be attributable to reduced pressure effects alone (i.e. diffusion).

At low $ppO_2/ppCO_2$ differences in Pn are greatest and decline with increasing $ppO_2/ppCO_2$ with convergence of the two curves at very high ratios. Low ratios favor increased carboxylase activity of RUBISCO and high ratios favor the cumulative activities of oxygenases. The near linear decline in the ratio of the rate of Pn at 70 kPa with increasing $ppO_2/ppCO_2$ (Figure 3 - inset) is suggestive of a reduced pressure effect attributable to diffusion.

Responses to Oxygen Partial Pressure

When the VPGC atmosphere is depressurized to 70 kPa, the mole fraction of all the gases remains the same but the partial pressures decrease proportionately and the resulting ppO_2 is 14.7 kPa. Decreased ppO_2 should result in a proportionate reduction in the dissolved O_2 concentration of the nutrient solution in accordance with Henry's law. A typical sequence of depressurization and repressurization curves with the accompanying changes in dissolved oxygen concentration on day 26 are illustrated (Figure 4). This particular depressurization episode was conducted over a 30 minute period. The decline in dissolved oxygen paralleled the decline in pressure achieving about 4.7 ppm in 60 minutes. Thereafter, dissolved oxygen declined slowly until an equilibrium value of 4.5 ppm was reached. The time lag for approaching the return to equilibrium following repressurization to 101 kPa (dissolved oxygen concentration of 6.35 ppm) was shorter than for depressurization. Equilibrium values at a given pressure may change with growth and develoment of the wheat stand depending on the extent of root mass development and therefore the rate of depletion of dissolved oxygen.

On days 30 to 33, the molecular sieve was used to reduce the ppO_2 of the VPGC (Figure 5). The two oxygen scrubs and reduced pressure episodes decreased the ppO_2 down to 15 and 7 kPa. Slight changes in total chamber pressure accompanied the reductions in ppO_2. The rate of Pn declined linearly with increasing

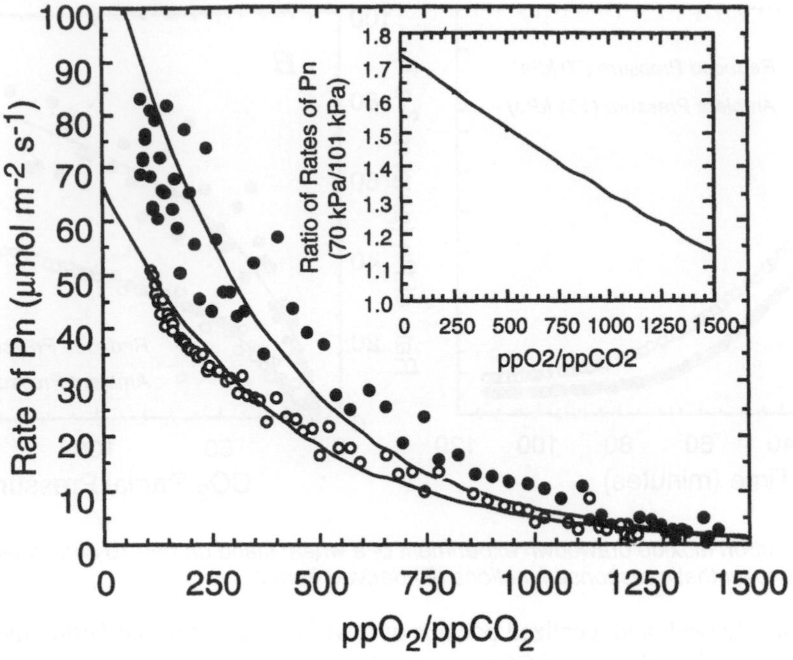

Fig. 3. Relationships of the rate of Pn with the ratio of ppO2 to ppCO2 at 70 (•) and 101 (o) kPa. Inset shows the functional relationship of the ratio of the rates of Pn at 70 kPa with those at 101 kPa.

$ppO_2/ppCO_2$ (Figure 6), but unlike the drawdown experiment (Figure 3), differences attributable to reduced pressure were not discerned. This may have been due to insufficient data points and to the extended time of the experiment. Rates of DR were not different over the entire range of $ppO_2/ppCO_2$, even though dissolved oxygen levels varied over the range of 3.1 to 6.2 ppm (data not shown).

Evapotranspiration

Water use by the entire wheat stand increased rapidly approaching a maximum in the range of 11 to 14 liters/m^2·day (Figure 7). These values compare closely with those measured in a previous test (Edeen and Barta, 1996), and are approximately twice those measured in the biomass production chamber at Kennedy Space Center (Wheeler *et al.*, 1993). The latter effect is likely due to the much higher stand density used in this study. The curves for the two pressures indicate little difference early in the test with divergence occurring after day 10. Averaged over the entire test, there was a 15 % increase in water throughput at 70 kPa compared to water use at 101 kPa. Despite fluctuations in temperature and relative humidity that occur as a result of the control systems, the vapor pressure deficits at the two pressures were comparable (i.e. for 23 C and relative humidity of 70 %, the VPD = 0.8427). Thus, differences in water throughput at 70 and 101 kPa are not attributable to differences in VPD. Increased water throughput at reduced pressure is likely due to enhanced diffusivity of water vapor since the diffusion coefficient of water at 70 kPa increases by a factor of 1.44 relative to ambient pressure. It is not known whether reduced pressure affects stomatal aperture and what proportion of enhanced water throughput is attributable to altered resistance of the stomates.

Plant Appearance and Biomass Production

After 34 days, the plants were in the early stages of anthesis. While the plant stand appeared vigorous and healthy, signs of stress were exhibited in the form of mild chlorosis and leaf curling. These symptoms have been observed in other large scale closed chamber growth tests and may in part be attributable to crop produced ethylene, particularly at critical times such as anthesis, seed set, and seed development. In previous tests with the VPGC, ethylene concentrations were quite high (Corey *et al.*, 1996a) and may have resulted in poor seed set and development. To determine if the plants in this test were predisposed to poor seed set, transplants were grown to maturity in a growth chamber. While light and CO_2 levels were not the

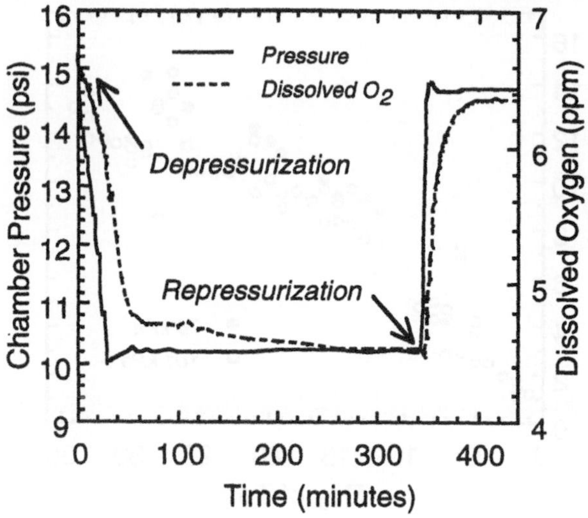

Fig. 4. Changes in dissolved oxygen of the nutrient solution following depressurization and repressurization events.

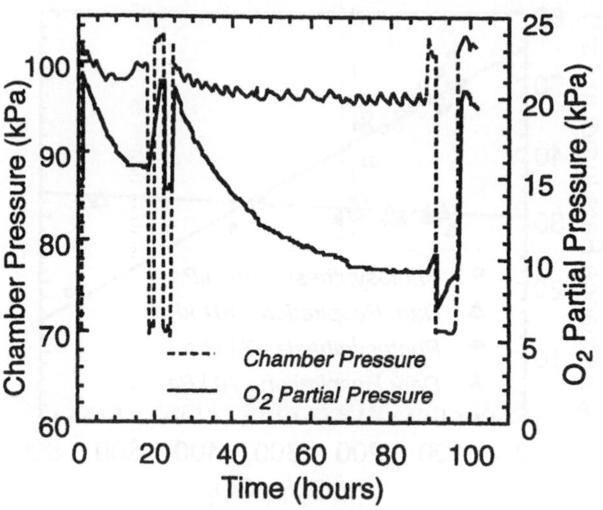

Fig. 5. Changes in ppO_2 and pressure of the VPGC atmosphere during depressurization and repressurization episodes and scrubbing of oxygen with the molecular sieve on days 30 to 33.

same as the conditions used in the test, it was possible to establish that the plants were still capable of seed set and development (Table 1). Perhaps the most critical indicator was the number of seeds set/head. While 8 seed/head is substantially higher than the seed set observed in previous VPGC growth tests of wheat, it was only about a third the seed set obtained by Bugbee *et al.* (1988). Future tests will either involve evacuation or scrubbing of ethylene from the chamber atmosphere and should reveal whether ethylene accumulation is an implicating factor involved in poor seed set.

Table 1. Yield components of wheat plants transplanted 34 days after harvest of the reduced pressure test and grown for an additional 49 days.

Measurement	Mean ± standard deviation[1]
Total Dry Biomass (g/pot)	65.0 ± 10.0
Primary Heads (#/pot)	19.7 ± 3.5
Grain Yield (# seeds/pot)	150 ± 7.4
Seed Weight (g/pot)	6.9 ± 0.9
Seeds/head	7.8 ± 1.1
Seed Weight (mg/seed)	45.6 ± 4.7

[1]Values represent means of 3 replications.

CONCLUSIONS

Results suggest that enhancing the growth environment for plants in a future life support facility operating at reduced pressure could also entail a reduction in ppO_2 that would still permit human entry without much

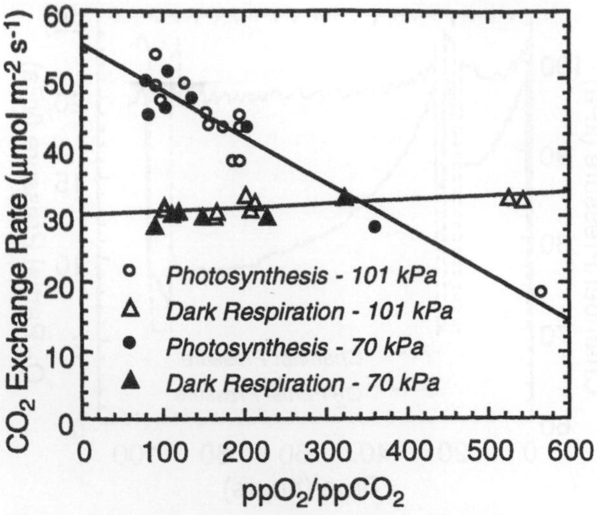

Fig. 6. Carbon dioxide exchange rate at 70 and 101 kPa measured during the oxygen scrub experiment on days 30 to 33.

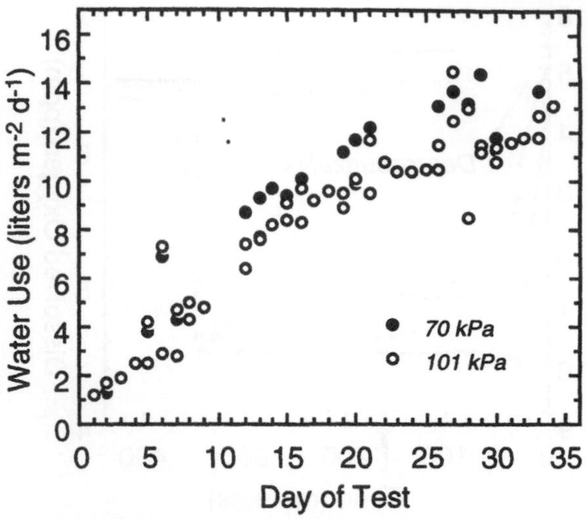

Fig. 7. Water use of a wheat stand at reduced and ambient pressures during growth and development.

discomfort. The ppO_2 accompanying 70 kPa would be comparable to that experienced at an altitude of 3000 meters. Alternatively, rates of photosynthesis can be maximized by minimizing the $ppO_2/ppCO_2$ ratio of the atmosphere, which could involve atmospheric isolation from humans at low ppO_2. The latter option should also yield increased rates of oxygen production, thereby improving the air revitalization capacity for human life support. Further testing will be required to determine if higher rates of photosynthesis translate into increased rates of biomass production at reduced pressure and to evaluate the roles of ppO_2 and increased CO_2 diffusion in gas exchange responses.

ACKNOWLEDGMENTS

The authors would like to acknowledge the work of John Gibbs, Keith Henderson, Don Kilpatrick, Ed Mohr, Tim Monk, and Don Overton with planting, mechanical, instrumentation, and computer control subsystems. Research by the senior author was supported in part by a grant from the National Research Council and by National Aeronautics & Space Administration Grant NAG-9-848.

REFERENCES

Andre, M., and D. Massimino, Growth of plants at reduced pressures: experiments in wheat-technological advantages and constraints, *Adv. Space Res.*, 12, 97-106, (1992).

Araus, J.L. and L. Tapia, Photosynthetic gas exchange characteristics of wheat flag leaf blades and sheaths during grain fill, *Plant Physiol.*, 85, 677-683, (1987).

Barta, D.J. and D.L. Henninger, Johnson Space Center's regenerative life support systems test bed, *Adv. Space Res.*, 18, 211-221, (1996).

Bjorkman, O., The effect of oxygen concentration on photosynthesis in higher plants, *Physiol. Plant.*, 19, 618-633, (1966).

Bugbee, B.G. and F.B. Salisbury, Exploring the limits of crop productivity, *Plant Physiol.*, 88, 869-878, (1988).

Chun, C. and T. Takakura, Control of root environment for hydroponic lettuce production - rate of root respiration under various dissolved oxygen concentrations, *American Society of Agricultural Engineering* Paper No. 934040, (1993).

Corey, K.A., D.J. Barta, M.A. Edeen, and D.L. Henninger, Atmospheric leakage and method for measurement of gas exchange rates of a crop stand at reduced pressure, *Adv. Space Res.*, current issue, (1998).

Corey, K.A., D.J. Barta, M.A. Edeen, D.L. Henninger, and R.G. Williamson, Ethylene evolution by wheat stands in a variable pressure growth chambner with human integration. Abstact for paper presented at 31st Committee for Space Research Scientific Assembly held 14-21 July 1996, Birmingham, UK, (1996).

Corey, K.A., M.E. Bates, and S.L. Adams, Carbon dioxide exchange of lettuce plants under hypobaric conditions, *Adv. Space Res.*, Vol 18, No. 4/5, 265-272, (1996).

Corey, K.A., Design of plant gas exchange experiments in a variable pressure growth chamber, NASA/ASEE Summer Faculty Fellowship Final Report, Contractor Report NGT-44-001-800, 18 pages, (1995).

Daunicht, H.J. and H.J. Brinkjans, Gas exchange and growth of plants under reduced air pressure, *Adv. Space Res.,* 12, 107-114, (1992).

Daunicht, H.J. and H.J. Brinkjans, Plant responses to reduced air pressure: advanced techniques and results, *Adv. Space Res.* 18 (4/5), 273-281, (1996).

Edeen, M. and D. Barta, Early human testing initiative phase I, Final report, JSC-33636, (1996).

Ehleringer, J.R., Photosynthesis and photorespiration: Biochemistry, physiology, and ecological implications, *HortScience* 14, 217-222, (1979).

Forrester, M.F., G. Krotkov, D.D. Nelson, Effect of oxygen on photosynthesis, photorespiration, and respiration in detached leaves: I. Soybean, *Plant Physiol.,* 41, 422-427, (1966).

Gale, J., Availability of carbon dioxide for photosynthesis at high altitudes: theoretical considerations. *Ecology* , 53, 494-497, (1972).

Gerbaud, A. and M. Andre, Effect of CO_2, O_2, and light on photosynthesis and photorespiration in wheat, *Plant Physiol.,* 66, 1032-1036, (1980).

Gerbaud, A. and M. Andre, Photosynthesis and photorespiration in whole plants of wheat, *Plant Physiol.,* 89, 61-68, (1989).

Gifford, R.M., Growth pattern, carbon dioxide exchange and dry weight distribution in wheat growing under differing photosynthetic environments, *Aust. J. Plant Physiol.*, 4, 99-110, (1977).

Govindjee, ed., *Photosynthesis, Vol II: Development, Carbon Metabolism and Plant Productivity*, Academic Press, Inc., New York, (1983).

Jackson, W.A. and R.J. Volk, Photorespiration, *Annu. Rev. Plant Physiol.,* 21, 385-432, (1970).

McCarthy, K. B. and J. A. Green, The effect of reduced cabin pressure on the crew and the life support system, *American Society of Agricultural Engineering* , Paper No. 911331, (1991).

Ohta, H., E. Goto, T. Takakura, F. Takagi, Y. Hirosawa, and K. Takagi, Measurement of photosynthetic and transpiration rates under low total pressures, *American Society of Agricultural Engineering* , Paper No. 934009, (1993).

Parkinson, K.J., H.L. Penman, and E.B. Tregunna, Growth of plants in different oxygen concentrations, *J. Expt. Bot.,* 25, 135-145, (1974).

Poskuta, J., Photosynthesis, photorespiration, and respiration of detached spruce twigs as influenced by oxygen concentration and light intensity, *Physiol. Plant.,* 21, 1129-1136, (1968).

Rule, D.E. and G.L. Staby, Growth of tomato seedlings at sub-atmospheric pressures, *HortScience* , 16, 331-332, (1981).

Siegel, S.M., Effects of reduced oxygen tension on vascular plants, *Physiol. Plant.,* 14, 554-557, (1961).

Siegel, S.M., L.A. Rosen and C. Giumarro, Plants at sub-atmospheric oxygen-levels, *Nature* , 198, 1288-1289, (1963).

Wheeler, R.M., K.A. Corey, J.C. Sager, and W.M. Knott, Gas exchange characteristics of wheat stands grown in a closed, controlled environment, *Crop Science* 33, 161-168 (1993).

Pergamon

Adv. Space Res. Vol. 20, No. 10, pp. 1879–1889, 1997

Printed in Great Britain
0273-1177/97 $17.00 + 0.00

PII: S0273–1177(97)00855–7

STORAGE STABILITY OF SCREWPRESS-EXTRACTED OILS AND RESIDUAL MEALS FROM CELSS CANDIDATE OILSEED CROPS

S. D. Stephens, B. A. Watkins and S. S. Nielsen

Department of Food Science, Purdue University, West Lafayette, Indiana 47907-1160. U.S.A.

ABSTRACT

The efficacy of using screwpress extraction for oil was studied with three Controlled Ecological Life-Support System (CELSS) candidate oilseed crops (soybean, peanut, and canola), since use of volatile organic solvents for oil extraction likely would be impractical in a closed system. Low oil yields from initial work indicated that a modification of the process is necessary to increase extraction efficiency. The extracted oil from each crop was tested for stability and sensory characteristics. When stored at 23°C, canola oil and meal were least stable to oxidative rancidity, whereas peanut oil and meal were least stable to hydrolytic rancidity. When stored at 65°C, soybean oil and canola meal were least stable to oxidative rancidity, whereas peanut oil and meal were least stable to hydrolytic rancidity. Sensory evaluation of the extracted oils used in bread and salad dressing indicated that flavor, odor intensity, acceptability, and overall preference may be of concern for screwpress-extracted canola oil when it is used in an unrefined form. Overall results with screwpress-extracted crude oils indicated that soybean oil may be more stable and acceptable than canola or peanut under typical storage conditions.

INTRODUCTION

Soybean (*Glycine max*), peanut *(Arachis lypogaea)*, and canola (dwarf, rapid cycling *Brassica napus*) are three major oilseed crops under consideration for use in a Controlled Ecological Life-Support System (CELSS) (Hoff *et a*l. 1982; Salisbury and Bugbee, 1988; Frick *et al.*, 1994). The most common commercial method used to obtain oil from oilseeds is solvent extraction (Watkins *et al.*, 1996). Only in less industrialized countries or for specialized health food products are oils extracted without organic solvents. However, a simple physical method such as screwpress extraction may have the greatest application in a CELSS. Oil recovery is not typically high with simple screwpress extraction, but the method has been preceded by extrusion to obtain a net oil recovery of about 70% for soybeans (Nelson *et al.* 1987). Relatively low oil yields may not be a serious problem in CELSS, because both the oil and the residual high-oil meal presumably would be utilized.

Prior to consumption, food-grade oils are refined by a process that includes degumming, alkali refining, bleaching, and deodorization (Jung *et al.*, 1989). The chemicals and equipment required for these procedures would make extensive oil refining impractical in CELSS, creating the need to utilize crude oils in the diet. While refining steps remove compounds that improve the color and flavor of oil, crude oils actually are more stable to oxidative rancidity than are refined oils (Jung *et al.*, 1989; Kwon *et al.*, 1984). However, high-oil meal may have reduced shelf-life due to fatty acid oxidation.

Data are needed on the stability and acceptability of crude oils and their residual meals to properly evaluate the dietary uses of these products. Therefore, soybean, peanut, and canola were screwpress extracted, the crude oils and residual meals were evaluated for stability over time at two temperatures, and the crude oils were used in bread and salad dressing and subjected to sensory evaluation.

Table 1. Extraction Parameters for Screwpress Extraction of Oilseeds

Type	Trial #	Pretreatment Temp (°C)	Moisture Content (%)	Extraction Pressure[1]	Screw Speed (RPM)	Extraction Yield (%)
Canola	1	57	6.0	23	49	28
	2	60	5.7	22	24	30
	3	68	4.4	20	49	33
Peanut	1	56	4.2	20	39	23
	2	49	4.2	20	39	22
	3	49	4.4	20	39	21
Soybean	1	88	6.3	20	35	13
Century	2	79	6.3	14	49	13
84	3	82	6.2	16	39	7
Soybean	1	82	6.0	10	41	12
Century	2	88	5.8	14	39	21
-L_2L_3	3	88	6.1	15	44	10

[1]Pressure was measured using a gauge (no units of measurement) on the screwpress.

MATERIALS AND METHODS

Sources of Oilseeds

Canola seeds and shelled peanuts, provided by Central Soya, Decatur, Illinois, were aspirated and cleaned prior to extraction. Soybean cultivars 'Century 84' and 'Century - L_2L_3' (genetically modified from 'Century 84' to remove the L_2 and L_3 lipoxygenase isozymes) (Davies *et al.*, 1987) were purchased from the Agricultural Alumni Seed Improvement Association in Romney, Indiana.

Pretreatment of Oilseeds

The four oilseeds were screwpress extracted, in three trials each of 22-45 kg seed, to obtain the extracted oil and protein meal. Pretreatment and extraction conditions are given in Table 1. Prior to expelling, the seeds were heated (Table 1) in a steam-jacketed kettle (Groen, Model DN/RA-15 SP, Groen, Elk Grove, Illinois). The seeds were stirred for even heating and to prevent burning. Samples of the seeds from each trial were collected both prior to heating and expelling, to be used for moisture determination. Moisture analysis was done to determine if any moisture was removed in the heating process.

Extraction Parameters of the Screwpress

Oilseeds were extracted using a screwpress provided by BAR North America, Inc., Urbana, Illinois. Each seed type was extracted in three replicate trials and the resulting oil and meal were collected. The press had a variable-speed drive with a range of 14 to 75 rpm. The press also allowed variable pressures, but the only method for measuring this parameter was by the depth of the screw, based on a built-in gauge on the machine. The gauge depth of the screw ranged from 0 to 30 units. The barrel length of the press was 15.2 cm with a diameter of 6.1 cm and a screw diameter of 5.7 cm, with a constant pitch of 2.2 cm. The 12 cast-iron plates that made up the barrel were separated by thin metal "shims." Shim thickness varied from 0.0013 to 0.0051 mm. The varying shim thickness between each plate in the barrel was arranged in different order for each seed type to increase oil extraction efficiency and reduce the number of "foots", or protein meal, in the extracted oil. The feed rate for the machine was not measured, but the device was kept full of seed during all extraction procedures.

Handling Extracted Oil and Meal

The protein meal from each extraction trial was mixed thoroughly and representative samples were taken. The oil was collected and allowed to settle at room temperature for 24 hrs to remove some of the "foots". Oil samples were centrifuged at 10,000 x g for 7 min., and the supernatant oil was decanted and filtered through Whatman #1 filter paper. The oil and meal samples then were divided into 10 representative samples and placed into 250 ml Nalgene amber storage bottles and stored at 23°C (room temperature) or 65°C (accelerated storage temperature). Samples from each bottle stored at 23°C were analyzed for peroxide value (PV), free fatty acid (FFA) content, and sensory quality of oil samples at 0, 2, 4, 6, or 8 weeks, with separate bottles used each time. Samples stored at 65°C were analyzed for PV and FFA content at 0, 1, 2, 3, 4 or weeks, with separate bottles used for each time period.

Determination of Oil Yield

Oil yield was calculated after measuring the fat content of the seeds prior to extraction and the fat content of the respective meal after extraction: % Oil Yield = 100% - [(% fat in meal ÷ % fat in seed) x 100].
Determination of yield was calculated on the basis of oil loss instead of mass loss, due to the amount of "foots" present, especially in peanut oils. The "foots" were filtered out as much as possible, but they were still noticeable upon visual inspection of the peanut oil.

Extraction of Residual Oil from Meal

The stability of screwpress-extracted meal was determined by measuring the PV and FFA content of oil removed from the meal by extraction with hexane. The solvent/oil mixture was filtered through Whatman #1 filter paper. The solvent was removed by heating at 55°C under vacuum with a rotary evaporator (Rotovapor-RKRVR 65145, Brinkmann Instruments, Inc., Westbury, New York).

Chemical Analyses

Raw oilseeds and meals were analyzed for moisture content by vacuum-oven drying (AOAC Method 925.09) (AOAC, 1990) so that all compositional data could be expressed on a dry weight basis. Samples were analyzed for ash content using a muffle furnace (AOAC Method 923.03) (AOAC, 1990); fat content by Soxhlet extraction with petroleum ether (AOAC Method 920.39B) (AOAC, 1990); and protein content by the micro-Kjeldahl procedure (AOAC Method 960.52) (AOAC, 1990). Total carbohydrate content was calculated by difference: % carbohydrate = 100% - (% ash + % fat + % protein). To assess oxidative rancidity of oil samples, PV was determined using American Oil Chemists' Society (AOCS) (AOCS, 1990) Method Cd 8b-90. To assess hydrolytic rancidity of oil samples, free fatty acid (FFA) content was measured as described by Pandurang Rao *et al.* (1972). Tocopherol content of crude oil samples and oil extracted from meal samples was determined by HPLC (Pascoe *et al.*, 1987). Color analysis of the oils was performed with a Hunter Lab Colorimeter (Model D25-DC2, Hunter Assoc. Lab., Inc., Reston, Virginia), to obtain L (lightness - darkness), a (redness - greenness), and b (blueness - yellowness) values.

Sensory Analysis of Oil

Sensory evaluation panels were used to 1) determine if changes in the sensory attributes of taste and odor occurred during storage of the oil samples, and 2) compare oils from various oilseeds with regard to acceptability and preference. Sensory evaluations were done with bread and salad dressing samples prepared with the various oils.

Panel description and development. Sensory evaluation procedures were based on information from Meilgaard *et al.* (1990). The sensory panel consisted of 15 to 25 individuals, ages 22 to 55, who were presented samples of white bread and a vinegar/oil salad dressing over lettuce. Both products contained the oils extracted from the four oilseeds and stored at 23°C for various periods of time. The two products were evaluated in separate trials. The order of presentation of the oil types and storage times was re-randomized for each panelist. Tests were performed to compare among oil types for each time period (0, 2, 4, 6 or 8 weeks), and to compare within each oil type over time. Attributes measured on a 9-point hedonic scale included overall odor intensity, overall odor acceptability, overall flavor intensity, and overall flavor acceptability. The samples in each trial were ranked also in order of overall preference.

Bread samples. Bread was baked in an automated breadmaker (Betty Crocker "Bake-It-Easy" Automatic Breadmaker, Model BC-1691, High Performance Appliances, Inc., Danbury, Connecticut), with the BASIC K setting on the breadmaker. The following ingredients were used in a recipe for white bread: 6.4-6.5 g salt, 3/4 cup water, 19.4-19.5 g sugar, 10 g oil sample being tested, 2 cups of Gold Medal "Better for Bread" flour, 4.4-4.5 g non-fat dry milk, and 3.5 g dry yeast. The baked bread (0.68 kg) was split into portions of even size (approximately 5 x 5 cm) and crust amounts, and served to each panelist.

Salad dressing. A vinegar-and-oil salad dressing was prepared by combining the following: 30 ml oil test sample, 15 ml white distilled vinegar, and 1 teaspoon water. The ingredients where mixed thoroughly and then poured over shredded iceberg lettuce. Subsamples of approximately 28 g from one large salad were then served to each panelist.

Statistical Design and Analysis

The extraction process was performed as a randomized complete block design with the four oilseed types extracted in a random order as one block. Triplicate trials of the blocks were made by re-randomizing each time. Oil samples and meal samples from each trail were collected and stored in separate containers for each storage time tested, with three subsamples from each container being analyzed. Thus, the experimental design was a split-split plot (Box *et al.*, 1978), with screwpress trials as blocked replicates, seed type as whole units, material (oil, meal) as subunits, and storage time as sub-subunits. Each storage temperature was analyzed separately.

Statistical analysis was performed using Statistical Analysis Software (SAS, 1989). Analysis of variance (ANOVA) was performed on PV and FFA levels, with the storage time effect and its interactions fully partitioned into orthogonal polynomial contrasts (Steel and Torrie, 1980). Due to lack of homogeneity of variances (according to Bartlett's test), in the storage study under both temperature conditions, data transformations were necessary (Steel and Torrie, 1980). Using the Box-Cox method for determining transformations (Box *et al.*, 1978), data from the study with accelerated temperatures (65°C) were transformed using a $\log_{10}$ transformation for both PV and FFA data. For the 23°C studies, PV data were transformed using a square root transformation, and the FFA data were transformed using a $\log_{10}$ transformation. Following ANOVA, regression analysis was performed to determine the treatment response of both temperatures as a function of storage time for the four seed types and for both the oil and meal. All statistical analyses were performed on the transformed data, and the data were then back transformed for presentation.

The sensory evaluation study was a randomized complete block design with the individual panelist considered a blocked replicate. ANOVAs were performed followed by mean separation tests using Student-Newman-Keuls at the 95% level. Tests to determine significant differences in preference rankings were performed according to Kahan *et al.* (1973) using non-parametric analysis.

RESULTS AND DISCUSSION

Proximate Composition and Oil Yield

Proximate compositions of raw oilseeds reported in Table 2 were similar to literature values (Bell, 1989). However, both soybean varieties were somewhat higher in protein content and lower in fat content than is typical for soybean. This inverse relationship between the content of protein and fat often is observed in oilseeds

Table 2. Proximate Composition (means ± standard deviation, n = 3) of Raw Oilseeds and Meals from Screwpress-Extracted Oilseeds

Sample	% Moisture	%, dry weight basis Protein	Fat	Ash	Carbohydrate
Oilseed					
Canola	8.2 ± 0.2	24.4 ± 0.5	45.9 ± 0.8	4.2 ± 0.1	25.5
Peanut	5.4 ± 0.0	31.6 ± 0.4	51.8 ± 1.4	2.5 ± 0.1	14.1
Soybean Century-84	6.6 ± 0.1	52.3 ± 0.3	19.2 ± 0.0	4.9 ± 0.1	23.6
Soybean Century-L_2L_3	6.9 ± 0.1	52.4 ± 0.6	19.2 ± 1.5	5.2 ± 0.1	23.2
Meal					
Canola	7.7 ± 1.4	28.3 ± 0.6	31.9 ± 1.2	4.9 ± 0.1	34.9
Peanut	6.1 ± 0.1	35.8 ± 1.0	40.4 ± 0.5	2.9 ± 0.1	20.9
Soybean Century 84	6.5 ± 0.5	53.5 ± 0.5	17.2 ± 0.6	4.9 ± 0.1	24.4
Soybean Century-L_2L_3	5.0 ± 1.7	54.2 ± 1.2	16.5 ± 1.1	5.4 ± 0.2	23.9

(Hanson *et al.*, 1961). As compared to the raw oilseeds, the increased protein and decreased fat contents of the meal were quite small, due to low extraction yields (Table 2). Oil yields for canola, peanut, and soybean were 30%, 25%, and 10 to 20%, respectively. This was due in part to the fact that screwpresses generally are more efficient after repeated uses, and this one had not been used previously (Boodram, personal communication).

Oil Color

Hunter L, a, and b values of the filtered oils indicated differences in the colors (Table 3). Visually, both soybean oil samples had a dark deep amber color typical of crude or unbleached oil. Peanut oil was lightest in color (highest L value), but was cloudy due to small particulates that passed through the filter. The crude canola oil had a deep green-black color (lowest L value; highest a value).

Tocopherol Content

Tocopherol content of screwpress-extracted oils and meals varied greatly among the three crops (Table 4). Total tocopherol concentrations were highest for soybean oil, but α-tocopherol, which has been shown to have the most antioxidant activity (Jung and Min, 1990), was highest in canola and peanut oils. Tocopherol levels in residual oil extracted from the meal were not as high as tocopherol levels in the oil expelled from the seeds by the screwpress. This may have been due to low extraction yield, but spiked samples showed over 95% recovery.

Storage Stability of Oils and Meals

Hydrolytic rancidity. Statistical analysis of the free fatty acid results required transformation of the data for both storage conditions to $\log_{10}$ values to achieve homogeneity of error variance. Actual FFA data are presented in Figure 1, but the following discussion is based on statistical analysis of $\log_{10}$ values, within a particular storage condition. FFA contents of soybean oil and meal at time 0 and throughout storage were significantly lower than those of canola and peanut oils and meals, whether stored at 23°C for up to 8 weeks or at 65°C for up to 4 weeks. There was no significant difference between the two soybean cultivars in FFA content of the oil or meal at either storage temperature. With the exception of meal from one extraction trial stored at 65°C, peanut oil and meal from all time periods at both storage conditions had significantly higher FFA content than soybean or canola oil and meal. The meal FFA values for all three oilseed types increased with storage time faster than did oil FFA values.

Oxidative rancidity. Statistical analysis of PV required transformation of the data for 23°C storage to square root values and data for 65°C storage to $\log_{10}$ transformation to achieve homogeneity of variance. Actual PV data are

Table 3. Hunter Color Values (mean ± standard deviation, n = 3) for Screwpress-Extracted Oils

	Hunter Value		
Sample	L	a	b
Canola	5.2 ± 0.8	2.9 ± 0.7	2.1 ± 0.8
Peanut	21.4 ± 2.8	-0.9 ± 1.7	9.9 ± 0.5
Soybean Century 84	10.5 ± 1.3	0.6 ± 1.4	6.8 ± 0.6
Soybean Century-L_2L_3	10.7 ± 1.2	0.5 ± 1.1	6.4 ± 0.2

Table 4. Tocopherol (δ, γ, α, and total) Concentrations (ppm) (means and standard deviation in parentheses below, n = 3) of Oils and Meals from Screwpress Extraction of Oilseeds

	Oil				Meal			
Sample	δ	γ	α	Total	δ	γ	α	Total
Canola	32 (23)	457 (22)	137 (12)	626 (22)	2.6 (0.6)	173 (18)	79 (5)	255 (23)
Peanut	25 (2)	238 (17)	138 (6)	401 (19)	3 (1.2)	74 (2)	89 (12)	166 (15)
Soybean Century 84	317 (36)	863 (80)	41 (7)	1221 (122)	39 (9)	173 (24)	11 (4)	223 (34)
Soybean Century-L_2L_3	326 (6)	1003 (48)	47.6 (3)	1376 (56)	46 (9)	347 (45)	31 (2)	424 (54)

presented in Figure 2, but the following discussion is based on statistical analysis of transformed data, within a particular storage condition. Data from the two temperature storage conditions were not compared statistically, because it is well documented that measurements of rancidity at low and high temperatures are difficult to compare statistically (Labuza, 1971). The PVs of canola oil and meal samples were much higher than those from peanut and soybean at time 0 and throughout storage at 23°C. The same was true for canola meal but not for oil throughout storage at 65°C. The higher PVs for canola suggest the presence of a pro-oxidant, perhaps chlorophyll (Niewiadomski, 1970). Oil samples from all three oilseeds developed high PVs upon storage at 65°C, with soybean PVs being the highest. A PV above 10 indicates sensory problems (Warner, private communication). Of the meal samples, only canola stored either at 23°C or 65°C developed a PV above 10. In contrast to FFA values for which meals were more affected by heat, temperature appeared to have a larger effect on PVs of oils than meals.

Sensory Analysis

Effect of oilseed type on sensory attributes. Results from the sensory evaluation of breads containing crude oils (Table 5) showed that, in general, odor of bread made with canola oil was less acceptable than that of bread made with peanut or soybean oils. There were minimal differences in odor acceptability of breads made with the two soybean oils compared to peanut oil. For week 2 oil used in the bread, the odor intensity value for 'Century 84' soybean oil was higher than those for the other oils, which indicates lower odor perceptibility. However, this trend did not continue for weeks 4, 6, and 8 oils used in the bread. For flavor acceptability, panelists again showed a strong dislike for canola oil in the bread compared to the other oils.

Regarding the effect of oilseed type on the sensory attributes of salad dressings made with crude oils (Table 5), at week 8 of storage, the odor of canola oil salad dressing was more intense than that for the other oilseeds. There was a trend for less acceptable odor in all stored canola oil salad dressing compared to salad dressings made with the other oils, but only the 6-week-canola-oil salad dressing was significantly less acceptable. There was a trend

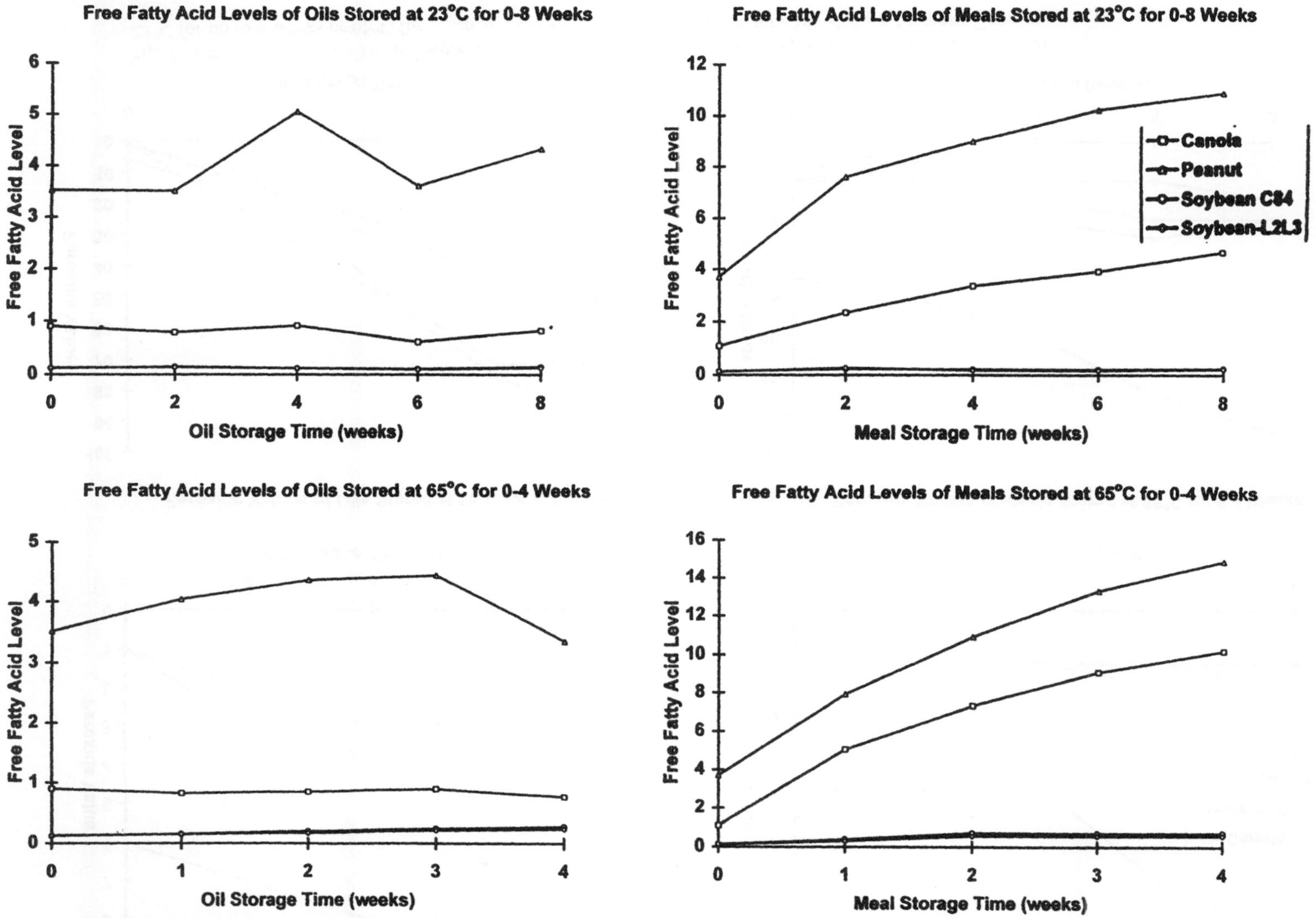

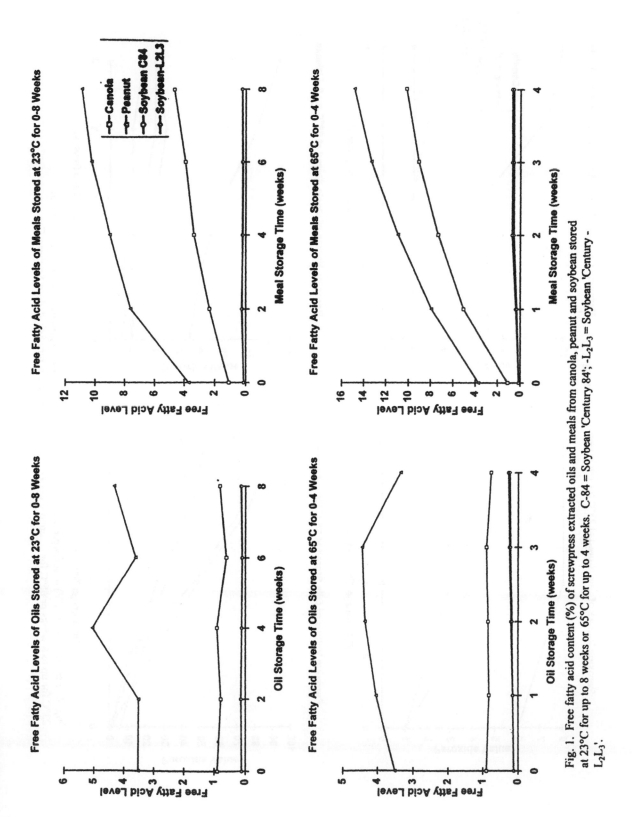

Fig. 1. Free fatty acid content (%) of screwpress extracted oils and meals from canola, peanut and soybean stored at 23°C for up to 8 weeks or 65°C for up to 4 weeks. C-84 = Soybean 'Century 84'; -L_2L_3 = Soybean 'Century - L_2L_3'.

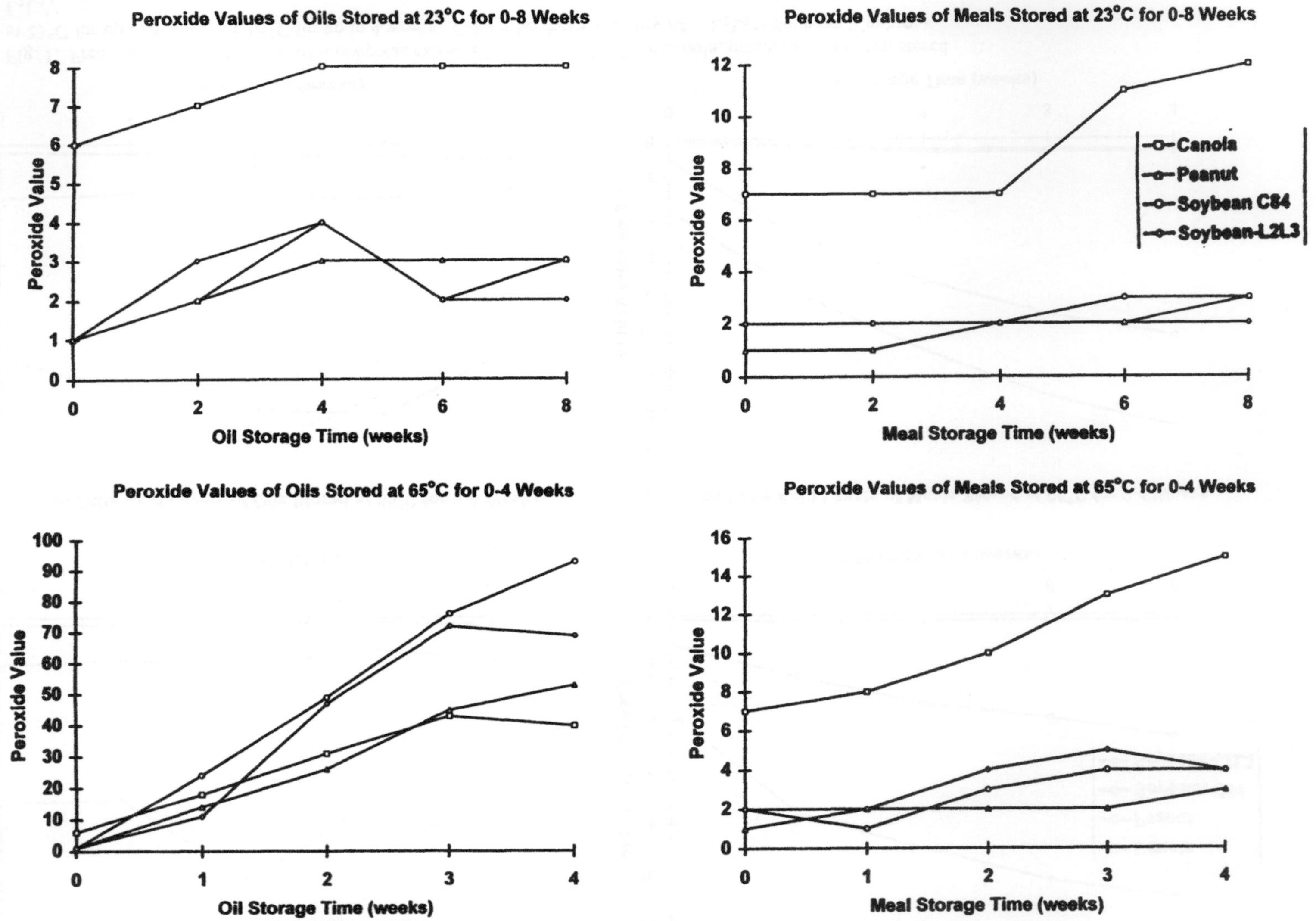

Fig. 2. Peroxide value (PV) of screwpress extracted oils and meals from canola, peanut, and soybean stored at 23°C for up to 8 weeks or 65°C for up to 4 weeks. C-84 = Soybean 'Century 84'; -L_2L_3 = Soybean Century - L_2L_3'.

Table 5. Effect of Oilseed Type on the Sensory Characteristics of Bread and Salad Dressing Made Using Screwpress-Extracted Crude Oils from Canola (C), Peanut (P), and Two Soybean Cultivars [Century 84, (S1) and Century-L_2L_3, (S2)] Stored at 23°C for up to 8 Weeks[1]

Time	Odor Intensity[2]				Odor Acceptability[3]				Flavor Intensity[2]				Flavor Acceptability[3]			
(wk)	C	P	S1	S2	C	P	S1	S2	C	P	S1	S2	C	P	S1	S2
Bread																
0	5.2	4.6	4.6	4.3	4.3^{ab}	4.8^{b}	3.6^{a}	3.6^{a}	4.2	4.8	4.8	4.6	4.6^{b}	4.3^{b}	3.1^{a}	3.1^{a}
2	4.1^{a}	4.0^{a}	5.4^{b}	4.5^{a}	5.4^{b}	3.9^{a}	3.8^{a}	3.8^{a}	4.0^{a}	4.3^{ab}	5.7^{c}	5.0^{bc}	5.4^{b}	3.8^{a}	3.4^{a}	3.2^{2}
4	5.0	4.8	4.5	4.4	5.0^{c}	4.3^{bc}	3.2^{a}	3.6^{ab}	5.7	5.4	5.3	5.0	5.0^{b}	4.8^{a}	3.4^{a}	3.6^{a}
6	4.4	4.5	4.7	4.7	4.7^{a}	4.6^{a}	3.3^{b}	4.3^{a}	4.8	5.1	4.9	5.3	4.3	4.4	4.0	4.2
8	4.4	4.9	5.3	5.0	5.3^{c}	4.3^{b}	3.2^{a}	3.7^{ab}	5.1	5.0	5.3	5.4	5.4^{b}	4.2^{a}	3.0^{a}	3.0^{a}
Salad Dressing																
0	4.6	5.0	5.4	5.2	5.7^{b}	4.0^{a}	4.8^{ab}	4.8^{ab}	3.2	3.6	3.7	3.2	7.6^{b}	4.8^{a}	5.1^{a}	5.9^{a}
2	4.4	5.1	4.6	4.6	5.0	4.4	4.7	4.6	3.3^{a}	4.4^{c}	3.9^{ab}	4.3^{bc}	6.3^{b}	4.8^{a}	5.2^{ab}	5.3^{ab}
4	5.4	5.5	5.2	4.8	5.4^{b}	4.1^{a}	4.7^{ab}	4.7^{ab}	3.3	3.5	4.0	3.6	7.0^{b}	5.6^{a}	5.0^{a}	5.3^{a}
6	4.0	4.3	4.7	4.4	6.1^{b}	4.6^{a}	4.6^{a}	5.0^{a}	2.9	3.9	3.7	3.5	7.1^{b}	5.6^{a}	5.4^{4}	5.4^{a}
8	4.0^{a}	5.2^{b}	4.6^{b}	4.6^{b}	6.3	4.9	4.5	4.9	3.1	3.9	4.2	4.4	6.8^{b}	5.4^{a}	5.0^{a}	5.0^{a}

[1]Means within each product, sensory attribute (i.e. odor intensity, odor acceptability, etc.), and time having a common superscript letter are not significantly different by Student-Newman-Keuls mean separation test ($p>0.05$).

[2]Intensity scale: 1 = very perceptible, 5 = moderately perceptible, 9 = not perceptible.

[3]Acceptability scale: 1 = like extremely, 2 = like very much, 3 = like moderately, 4 = like slightly, 5 = neither like nor dislike, 6 = dislike slightly, 7 = dislike moderately, 8 = dislike very much, 9 = dislike extremely.

for the same pattern with week 0, 2, 4, and 8 oils. There were significant differences in flavor acceptability between salad dressing samples for all oil storage time periods. Canola oil salad dressing had a significantly less acceptable flavor than all other oils at weeks 0, 4, 6, and 8, and was less acceptable than peanut oil salad dressing at week 2. This indicates a strong dislike for the flavor of salad dressing made using crude canola oil.

Preference ranking for breads made with the crude oils (Table 6) indicated that at weeks 0, 4, and 8, soybean 'Century 84' was the most preferred oil for use in a bread type product, with canola and peanut oils being significantly less preferred. At weeks 2 and 8, canola oil preference was lowest. Overall preference rankings for salad dressings made with the crude oils (Table 6) were similar to the rankings of breads, i.e., canola oil generally was less preferred. Though week 2 oils showed no differences in the salad dressings, weeks 0, 4, 6, and 8 data indicated that canola oil was significantly less preferred than were other oils in the salad dressing.

Effect of storage time on sensory attributes. The effect of oil storage time on overall preference of breads and salad dressings made with the crude oils was analyzed in a quantitative manner. Regression equations (not shown) indicated the following: 1) overall odor acceptability of bread made with canola oil decreased with oil storage time, 2) overall odor intensity of salad dressing made with either canola or peanut oil decreased with oil storage time, and 3) overall odor acceptability and flavor acceptability of salad dressing made with peanut oil decreased over oil storage time.

Using preference ranking to determine if any effect occurred within an oilseed type over time, no differences were noted with the breads (data not shown). For the salad dressings, only for the 'Century 84' soybean oils were the preference rankings significantly different over storage time of the oil (data not shown), with week 2 being the highest and week 4 the lowest.

Table 6. Effect of Oilseed Type on the Preference of Bread and Salad Dressing Made Using Screwpress-Extracted Crude Oils from Canola, Peanut, and Two Soybean Cultivars (Century 84 and Century - L_2L_3) Stored at 23°C for up to 8 Weeks[1,2]

	Storage Time (weeks)				
Sample	0	2	4	6	8
Bread					
Canola	1.9^a	1.8^a	2.0^a	2.4^a	1.6^a
Peanut	2.0^a	2.8^b	1.9^a	2.4^a	2.4^b
Soybean Century 84	3.1^c	2.3^b	3.2^c	2.9^a	3.3^c
Soybean Century-L_2L_3	2.9^b	2.9^b	2.9^b	2.4^a	2.9^b
Salad Dressing					
Canola	1.7^a	2.1^a	1.8^a	1.6^a	1.7^a
Peanut	3.0^b	2.9^a	2.5^b	2.7^b	2.7^b
Soybean Century 84	2.9^c	2.8^a	2.9^b	2.9^b	2.8^b
Soybean Century-L_2L_3	2.3^c	2.1^a	3.0^b	2.7^b	2.8^b

[1]Scale = 1-4, with 4 being highest preference.

[2]Values reported are means from sum of all responses within each time interval. Mean values having a common superscript letter had sums that were not significantly different by non-parametric analysis at p> 0.05.

CONCLUSIONS

The crude oils extracted from CELSS candidate oilseed crops differed in their stability to oxidative and hydrolytic rancidity, and in their sensory characteristics and acceptability. Overall, the crude soybean oil prepared here was more stable at 23°C and acceptable in sensory evaluation than were crude peanut and canola oils. However, further testing with modified pretreatments prior to screwpressing, or perhaps another type of extraction procedure, may improve the yield and quality of crude oils from oilseed crops.

REFERENCES

AOAC. *Official Methods of Analysis*, 15th ed. Association of Official Analytical Chemists, Washington, D.C. (1990).

AOCS. *Official Methods and Recommended Practices of the American Oil Chemists' Society*, 4th ed., 2nd printing (additions and revisions through 1993). American Oil Chemists' Society, Champaign, IL (1990).

Bell, J.M., Nutritional Characteristics and Protein Uses of Oilseed Meals, in *Oil Crops of the World, Their Breeding and Utilization* edited by G. Röbbelen, R.K. Downey, and A. Ashni, p. 1942, McGraw- Hill Publ. Co., New York, NY (1989).

Boodram, R., Private communication, BAR North America, Inc., Urbana, IL (1994).

Box, G.E., W.G. Hunter, and J.S. Hunter, *Statistics for Experimenters: An Introduction to Design, Data Analysis, and Model Building,* John Wiley & Sons, New York, NY (1978).

Davies, C.S., S.S. Nielsen, and N.C. Nielsen, Flavor improvement of soybean preparations by genetic removal of lipoxygenase-2, *JAOCS* **64**, 1428-1432 (1987).

Frick, J., S. Nielsen, and C. Mitchell, Yield and seed oil content response of dwarf, rapid-cycling *Brassica* to nitrogen treatments, planting density, and CO_2 enrichment, *J. Amer. Soc. Hort. Sci.* **119**, 1137-1143 (1994).

Hanson, W.D., R.C. Leffel, and R.W. Howell. Genetic analysis of energy production in the soybean, *Crop Sci.* **1**, 121-126 (1961).

Hoff, J.E., J.M. Howe, and C.A. Mitchell, Nutritional and cultural aspects of plant species selection for a regenerative life support system. NASA Contract Report 166324, Ames Research Center, NASA, Moffet Field, California (1982).

Jung, M.Y., and D.B. Min, Effects of α -,γ -, and δ - tocopherols on oxidative stability of soybean oil, *J. Food Sci.* **55**, 1464-1465 (1990).

Jung, M.Y., S.H. Yoon, and D.B. Min, Effects of processing steps on the contents of minor compounds and oxidation of soybean oil, *JAOCS* **66**, 118-120 (1989).

Kahan, G.D. Cooper, A. Papavasiliou, and A. Kramer, Expanded tables for determining significance of differences for ranked data. *Food Technol.* **27**(5), 62, 64, 65, 68, 69 (1973).

Kwon, T.W., H.E. Snyder, and H.G. Brown, Oxidative stability of soybean oil at different stages of refining. *JAOCS.* **61**, 1843 - 1846 (1984).

Labuza, T., Kineticis of lipid oxidation in foods, *CRC Crit. Rev. Food Technol* **2** (3), 355-405 (1971).

Meilgaard, M., G.V. Civille, and B. Carr, *Sensory Evaluation Techniques*, CRC Press, Inc., Boca Raton, FL (1990).

Nelson, J.L., W.B. Wijeratne, S.W. Yeh, T.M. Wei, and L.S. Wei, Dry extrusion as an aid to mechanical expelling of oil from soybeans, *JAOCS* **64**, 1341-1347 (1987).

Niewiadomski, H., Progess in the Technology of Rapeseed Oil for Edible Purposes, in *Chemistry and Industry*, p. 833-888 (1970).

Pascoe, G.A., C.T. Duda, and D.J. Reed, Determination of α-tocopherol and α-tocopherylquinone in small biological samples by high-performance liquid chromatography with electrochemical detection, *J. Chrom.* **414**, 440-448 (1987).

Pandurang Rao, B., S.D. Thirumala Rao, and B.R. Reddy, Rapid methods for determination of free fatty acids contents in fatty oils. *JAOCS* **49**(5), 338-339 (1972).

Salisbury, F.B., and B. Bugbee, Plant productivity in controlled environments, *Hort. Sci.* **23**, 293-299 (1988).

SAS Institute Inc., *SAS/STAT User's Guide*, Version 6, Fourth Edition, Volume 1, 943 pp., Cary, North Carolina, Sas Institute Inc. (1989).

Steel, R.G.D., and J.H Torrie, *Principles and Procedures of Statistics*, 2nd ed. McGraw-Hill Book Co., New York, NY (1980).

Warner, K., Private communication. United States Department of Agriculture, Northern Regional Research Center, Peoria, IL (1994).

Watkins, B.A., B. Henning, and M. Toborek, Dietary Fat and Health, in *Bailey's Industrial Oil: Fat Products Edible Oil a Fat Products: General Applications,* edited by Y.H. Hui, 5th ed., Vol. 1, p.159-214, John Wiley & Sons, Inc., New York, NY (1996).

Paper No. 15180 of Purdue University Agricultural Research Programs. Research supported in part by NASA grant NAGW - 2329.

Pergamon

Adv. Space Res. Vol. 20, No. 10, pp. 1891–1894, 1997

Printed in Great Britain
0273-1177/97 $17.00 + 0.00

PII: S0273-1177(97)00856-9

YIELD COMPARISONS AND UNIQUE CHARACTERISTICS OF THE DWARF WHEAT CULTIVAR 'USU-APOGEE'

B. Bugbee and G. Koerner

Crop Physiology Laboratory, Plants, Soil and Biometeorology Department, Utah State University, Logan, UT 84322-4820, U.S.A.

ABSTRACT

Extremely short, high yielding cultivars of all crop plants are needed to optimize the food production of bioregenerative life support systems in space. In the early 1980's, we examined over a thousand wheat genotypes from the world germplasm collection in search of genotypes with appropriate characteristics for food production in space. Here we report the results of 12 years of hybridization and selection for the perfect wheat cultivar. 'USU-Apogee' is a full-dwarf hard red spring wheat (*Triticum aestivum* L.) cultivar developed for high yields in controlled environments. USU-Apogee was developed by the Utah Agricultural Experiment Station in cooperation with the National Aeronautics and Space Administration and released in April 1996. USU-Apogee is a shorter, higher yielding alternative to 'Yecora Rojo' and 'Veery-10', the short field genotypes previously selected for use in controlled environments. The yield advantage of USU-Apogee is 10 to 30% over these other cultivars, depending on environmental conditions. USU-Apogee (45-50 cm tall, depending on temperature) is 10 to 15 cm shorter than Yecora Rojo and 1 to 4 cm shorter than Veery-10. USU-Apogee was also selected for resistance to the calcium-induced leaf tip chlorosis that occurs in controlled-environments. Breeder seed of USU-Apogee will be maintained by the Crop Physiology Laboratory and seed is available for testing on request.

INTRODUCTION AND PEDIGREE

USU-Apogee was named after the point in an orbit that is the farthest from the Earth. USU-Apogee (Reg. no. CV-840; PI 592742) originated from the cross 'Parula'/'Super Dwarf', both of which were obtained from the International Center for Wheat and Maize Improvement (CIMMYT; Obregon, Mexico) germplasm collection in 1984. Parula has the pedigree: FKN/3/2*FCR//'KenyaAD'/'Gabo 54'/4/Bluebird /'Chanate'; where FKN = 'Frontana'/'Kenya58'//'Newthatch'. Parula was selected for its small leaf size. Super Dwarf has the CIMMYT germplasm number CMH79.481-1Y-8B-2Y-2B-0Y; and the pedigree: *T. sphaerococcum* /2*H-567.71/3/'Era'/'Sonora64'/ /2*Era. Super Dwarf was selected for its short stature (25 cm tall).

SELECTION CRITERIA AND PROCEDURES

Single head selections were made in the F_2 to F_4 generations for short height (less than 50 cm tall), erect tillering habit, reduced tillering, and small leaves. These traits are desirable in high yield conditions (Donald, 1968; 1979). Small leaves are often more photosynthetically efficient than large leaves, and two small leaves may be better than one large leaf (Morgan et al., 1990; LeCain et al., 1989; Bhagsari and Brown, 1986) Mass selections for short height and yield were made in the F_5 to F_8 generations (1988 to 1989). All selections were made in a CO_2-enriched, temperature-controlled greenhouse that had a supplemental photosynthetic photon flux (PPF) of 400 μmol m^{-2} s^{-1} (35 mol m^{-2} d^{-1}) from high pressure sodium lamps. The photoperiod was 24-h (continuous light). The root-zone was a hydroponic soilless medium, watered twice daily with nutrient solution. Continuous cultivation made it possible to evaluate 3 to 4 generations per year. Yields in this environment (about 16 Mg ha^{-1}; 240 bushels per acre) are double the best irrigated field yields when the greenhouse temperature is set at a low temperature (see section below on YIELD STUDIES) and the life cycle is extended to about 100 days (Bugbee and Salisbury, 1988).

Preliminary yield evaluations, in the near-optimal conditions of the CO_2-enriched greenhouse, were begun in the F_7 generation. Mice got into the greenhouse prior to harvest in the F_8 generation and damaged all six replicate plots of USU-Apogee. No other plots were damaged. USU-Apogee had the least leaf tip necrosis, but had considerable variability for plant height, so 67 single heads selected from the F_9 generation were grown as head rows (a row planted from the seed of a single head). Additional selections for yield were made among the best lines in the next six generations (F_{10} to F_{15}). In the F_{16} generation, 100 heads of USU-Apogee were selected and grown as head rows. After roguing off-type and nonuniform rows, the remaining 90 F_{16} lines were harvested and bulked as breeders seed.

RESISTANCE TO CALCIUM-DEFICIENCY INDUCED "TIPBURN"

USU-Apogee is resistant to leaf tip chlorosis that occurs in wheat under rapid growth conditions, particularly in continuous light. This chlorosis (caused by a calcium deficiency) can kill the top 30% of the flag leaf. The chlorosis is severe in Veery-10 and also occurs in Yecora Rojo. Calcium deficiencies, such as tip burn in lettuce (*Lactuca sativa* L.)and blossom end rot in tomatoes (*Lycopersicun esculentum* L.)are common in controlled-environment crop production because Ca has low phloem mobility and is thus not sufficiently translocated to growing meristems. Foliar Ca applications and increased root-zone Ca are not effective because they do not reach the meristematic tissue (Marschner, 1995). USU-Apogee has significant rates of guttation during dark periods and guttation occurs even during the light period when the stomates are partly closed by elevated CO_2. Significant amounts of Ca can be translocated by guttation (Marschner, 1995). The segregating lines with the smallest leaves had the least chlorosis. Tissue analysis by inductively coupled plasma emission spectrophotometry indicated adequate Ca in the top 30% of small leaves (0.4% Ca), but inadequate amounts (0.05% Ca) in large leaves. USU-Apogee has smaller flag leaves (11 to 20 cm long, depending on temperature) than Yecora Rojo and Veery-10 (20 to 30 cm long).

DEVELOPMENTAL CHARACTERISTICS

USU-Apogee has an extremely rapid development rate, which helps reduce leaf size and excessive vegetative growth. Heads emerge 23 days after seedling emergence in continuous light with a constant 25°C temperature. Heads of Yecora Rojo and Veery-10 emerge about 6 days later under these conditions. In field conditions, USU-Apogee heads about 3 days earlier than Yecora Rojo and 6 days earlier than Veery-10.

YIELD STUDIES IN GREENHOUSE, GROWTH CHAMBER, AND FIELD ENVIRONMENTS

We have examined the yield advantage of USU-Apogee in 5 greenhouse studies, 2 field studies, and 2 sets of growth chamber studies (Table 1). Most studies compared USU-Apogee to Veery-10 because these cultivars are similar in height. USU-Apogee out-yielded Veery-10 by an average of 29 ±1% in two greenhouse studies at 23°C (60 day life cycle), but by an average of 16 ±9% in 3 greenhouse studies at 23°C decreasing to 17°C (95-day life cycle). USU-Apogee out yielded Veery-10 by 8% in a growth chamber study under high light (PPF=1500 $\mu mol\ m^{-2}\ s^{-1}$; 20-h photoperiod; 108 $mol\ m^{-2}\ d^{-1}$). Grotenhuis and Bugbee (1997) examined the effects of elevated and super-elevated CO_2 on USU-Apogee and Veery-10. USU-Apogee out yielded Veery-10 by an average of 11% in 12 growth chamber trials, and both cultivars responded similarly to elevated CO_2. The average yield in all controlled environment trails was 0.33 ±0.04 grams of edible seed yield per mole of photosynthetic photons. Side lighting was minimized by guard rows or Mylar screens at the edges of the plot in all trials.

USU-Apogee out yielded Veery-10 by 15±3% in replicated field trials in 1994 and 1995, and out yielded Yecora Rojo by 14% in 1995 (Table 1). The yield of USU-Apogee was 160% of Super Dwarf and 100.1% of Fremont (an adapted semi-dwarf Utah wheat cultivar) in the 1995 field trial. Neither Veery-10 nor Yecora Rojo are specifically adapted to Utah field conditions.

Table 1. Results of yield studies in 3 environments. Yield was measured as g of dry seed per mole of photosynthetic photons and ranged from about 0.3 g mol^{-1} in greenhouse studies to 0.15 g mol^{-1} in field studies. Data are indicated as % yield relative to Veery-10 to facilitate comparisons among the diverse studies. USU lines 1, 10, and 56 are from the same hybrid cross that produced USU-Apogee. The 12 growth chamber studies at PPF=700 are described in detail by Grotenhuis and Bugbee (1997). The greenhouse studies included 4 to 6 replicate plots per genotype. A dashed line indicates that the genotype was not included in the study.

	---- Hydroponic, CO_2 Enriched ----							Utah Field Studies	
	-------- Greenhouse --------					Growth Chamber			
Cultivar Name	**Feb.-May '94**	**Mar.-June '95**	**July-Sept. '95**	**Nov. 95 to Feb.'96**	**Aug.-Dec. '96**	**12studies PPF=700 '92 -'96**	**1 study PPF=1500 '94**	**4 Reps. '94**	**6 Reps. '95**
USU-Apogee	129	101	130	130	116	111	108	118	112
USU-Line 56	128	99	127	--	--	--	98	114	111
USU-Line 1	111	106	115	--	--	--	108	104	96
USU-Line 10	108	104	112	--	--	--	--	98	107
Veery 10	100	100	100	100	100	100	100	100	100
Yecora Rojo	--	--	--	141	101	--	--	--	97
Statistical Significance	0.05	n.s.	0.01	0.01	0.05	--	0.08	0.05	0.05

HARVEST INDEX AND YIELD COMPONENTS

The primary cause of the increased yield of USU-Apogee is increased harvest index, which is 5 to 15% higher than that of Veery-10. Using USU-Apogee, we achieved harvest indexes of 56 and 60% (not including roots) in two greenhouse studies with phasic environmental control (23 °C decreasing to 17 °C after anthesis). Assuming that the root mass was 6% of the total biomass at harvest, the harvest index including roots in these trials would be 50 and 54%. The harvest index of Veery-10 (without roots) was 48 and 49% in these same trials (8 to 11% less than USU-Apogee). The increased harvest index of USU-Apogee is primarily caused by a reduced number of late forming tillers, which are often sterile or low yielding.

In warm environments (constant 23 °C; 60 days from emergence to harvest), heads per m^2 and seeds per head are about 25% higher in USU-Apogee than in Veery-10, and mass per seed is about 25% less. In two studies in a cool environment (23°C, decreasing to 17°C after anthesis; 100 days to harvest), heads per m^2 averaged 20% greater, seeds per head was 5% greater, and mass per seed was 5% less than Veery-10.

BREADMAKING QUALITY

Breadmaking quality was evaluated by the USDA-ARS Western Quality Wheat Laboratory at Pullman, Washington,USA. Milling and baking tests indicated that USU-Apogee has similar quality to Veery-10 and slightly poorer quality than Yecora Rojo. Yecora Rojo is among the best bread wheats grown in the USA.

FUTURE BREEDING EFFORTS

We are continuing to select for even shorter wheat cultivars. We are now conducting a greenhouse yield trial of an advanced breeding line that is 7 cm shorter than USU-Apogee. This line was a re-selection from the F_3 seed that eventually produced USU-Apogee and has the same early maturity, small leaf size, and resistance to leaf tip chlorosis. In March 1996, we planted seed from the F_2 generation from the original Parula/Super Dwarf cross and re-selected for genotypes less than 40 cm tall. We are now evaluating selections in the F_5 generation. These genotypes look exceptionally promising. We hope to obtain homozygous lines with heights of 30 to 40 cm that have green leaf tip color and high yield. These lines will probably have lower yields than USU-Apogee, but they appear to have higher yields than Super Dwarf.

ACKNOWLEDGMENTS

We gratefully acknowledge the conscientious assistance of Steve Johnson, Maki Monje, Dave Kadlec, Noble Mabuchi, and all the other students who helped with planting and data collection.

REFERENCES

Bhagsari, A. and R. Brown. 1986. Leaf photosynthesis and its correlation with leaf area. Crop Sci. 26:127-132.
Bugbee, B. and F. Salisbury. 1988. Exploring the limits of crop productivity. Plant Physiol 88:869-878.
Donald, C. 1968. The breeding of crop ideotypes. Euphytica 17:325-403.
Donald, C. 1979. A barley breeding program based on an ideotype. Jour. Agric. Sci. Camb. 93:261-269.
Grotenhuis, T. and B. Bugbee. 1997. Super-optimal CO_2 reduces seed yield but not vegetative growth in wheat. Crop Sci. 37: 4 (July/August issue, In Press).
LeCain, D., J. Morgan, and G. Zerbi. 1989. Leaf anatomy and gas exchange in nearly isogenic semidwarf and tall winter wheat. Crop Sci. 29:1246-1251.
Marschner, H. 1995. Mineral nutrition of higher plants. Academic Press, NY.
Morgan, J., D. LeCain, and R. Wells. 1990. Semidwarfing genes concentrate photosynthetic machinery and affect leaf gas exchange of wheat. Crop Sci. 30:602-608.

Pergamon

Adv. Space Res. Vol. 20, No. 10, pp. 1895–1899, 1997

Printed in Great Britain
0273-1177/97 $17.00 + 0.00

PII: S0273-1177(97)00857-0

EFFECT OF LAMP TYPE AND TEMPERATURE ON DEVELOPMENT, CARBON PARTITIONING AND YIELD OF SOYBEAN

T. A. O. Dougher and B. Bugbee

Department of Plants, Soils, and Biometeorology, Utah State University, Logan, UT, 84322-4820, U.S.A.

ABSTRACT

Soybeans grown in controlled environments are commonly taller than field-grown plants. In controlled environments, including liquid hydroponics, height of the dwarf cultivar "Hoyt" was reduced from 46 to 33 cm when plants were grown under metal halide lamps compared to high pressure sodium lamps at the same photosynthetic photon flux. Metal halide lamps reduced total biomass 14% but did not significantly reduce seed yield. Neither increasing temperature nor altering the difference between day/night temperature affected plant height. Increasing temperature from 21 to 27°C increased yield 32%. High temperature significantly increased carbon partitioning to stems and increased harvest index.

INTRODUCTION

Short-stature, high yielding cultivars are desirable in controlled environments because volume is often limited. However, soybeans grown in controlled environments are taller than field-grown plants (Downs and Thomas, 1990). Red:far red ratios, specifically phytochrome 660:730 nm, have been implicated as the cause of internode elongation (Pausch *et al.*, 1991), although soybeans may also respond to a balance of red and blue light (Britz and Sager, 1990). Wheeler *et al.* (1991) reported that there was a threshold intensity of blue light (30 $\mu mol \bullet m^{-2} \bullet s^{-1}$) necessary to reduce stem elongation. However elongation is also dependent upon the total photosynthetic photon flux (PPF) from lamps (Tibbitts *et al.*, 1983).

Another factor affecting stem length is temperature, which is easily manipulated in a controlled environment. The reduced plant water potential usually associated with high temperatures can be minimized by growing plants hydroponically at high humidity and elevated CO_2. Elevated CO_2 reduces photorespiration which generally increases with temperature. Thus crops grown in controlled environments should have higher temperature optimums than field grown plants. Optimum field temperatures for soybean have been characterized (Raper and Kramer, 1987), but the results may be affected by reduced plant water potential and increased photorespiration.

Our objective was to study soybean canopy height, carbon partitioning, and yield under high pressure sodium (HPS) versus metal halide (MH) lamps at varied and constant day/night temperatures in a CO_2-enriched, hydroponic, controlled environment.

MATERIALS AND METHODS

Soybean cv. 'Hoyt' canopies were grown in Plexiglas chambers (0.47 x 0.36 x 0.61 m) at a density of 36 plants•m^{-2} (6 plants per chamber). This density was used based on preliminary trials, which indicated that higher densities increased stem elongation and lower densities increased time to canopy closure. An extensive controlled environment screening showed 'Hoyt', a determinate cultivar, to be the shortest and highest yielding. All indeterminate cultivars were unacceptably tall. Chambers were positively pressurized for an open gas exchange system as described by Bugbee (1992). Seeds were germinated in moist diatomaceous earth (Isolite) and transplanted when the hypocotyls had elongated to at least 4 cm (about 6 days). Plants were transferred to an aerated nutrient solution, 21 cm deep, in a 30 L tub. Closed cell foam plugs in a blue Styrofoam lid supported the plants. Nutrient solution was replenished to maintain solution level. Nutrient solution electrical conductivity (140 ± 44μS) and pH (5.6 ± 0.6) were monitored and controlled as necessary. Ammonium sulfate was added as needed to counteract the rise in pH caused by nitrate uptake.

Five day/night temperature regimes were used: 29/25, 26/22, 24/24, 23/19, and 21/21 °C. Root temperatures were kept constant at the average daily temperature of the shoot: 27, 24, 24, 21, and 21 °C, respectively. Shoot air and root-zone temperatures were measured with thermocouples and maintained by computer-controlled heaters. Each set of temperature treatments (5 chambers) was placed under either MH or HPS lamps. All chambers were in a single growth room with light treatments separated by a heavy Mylar sheet. A photosynthetic photon flux of 450 μmol•m^{-2}•s^{-1} was maintained at the top of the canopy. This supplied approximately 40 and 140 μmol•m^{-2}•s^{-1} of blue light in HPS and MH lamps, respectively. Intensity was maintained within four percent by shading each chamber with neutral density filters. Aluminized Mylar around the chamber was maintained at canopy height to minimize the edge effect caused by side lighting. The photoperiod was 12-h. Carbon dioxide concentration was enriched to 1100 μmol•mol^{-1}.

Days to first flower was recorded as appearance of visible flower color. Plants were harvested at physiological maturity as indicated by loss of green color from the pods (Gbikpi and Crookston, 1981). At harvest, canopy height, from stem base to the top of the leaves, was measured *in situ*. Then plants were extended to their full height and measured to the growing tip of the main stem and longest branch. These different length measures were used as a more specific indication of internode elongation. Plants were separated into leaves, stems, pods, and roots, dried at 80°C for 48 hours, and weighed. Seed and pod number was recorded. Yield parameters and carbon partitioning (organ DW / total DW)were calculated from the harvest data. Main effects were tested using the light by temperature interaction error term with SAS ANOVA (SAS Institute, NC). Net canopy photosynthesis (P_{net}) was calculated from the measured change in CO_2 (infrared gas analyzer in differential mode) times air flow rate through the chamber divided by chamber ground area (Bugbee, 1992).

RESULTS

Effect of Lamp Type

MH lamps significantly reduced canopy height but slightly increased relative branch length (Table 1). The main stem of HPS plants were 87% the length of the longest branch while MH main stem was 75% of the longest branch. MH canopy height was taller than the longest branch length because canopy height included petiole lengths. Although not measured, petioles appeared to contribute more to height in MH canopies.

Table 1. Three plant length measures, seed yield, and yield components of soybeans grown under two lamp types. Each parameter is an average of the five chambers with different temperature regimes.

lamp type	canopy height (cm)	main stem length (cm)	longest branch length (cm)	seed yield ($g \cdot m^{-2} \cdot d^{-1}$)	photo-synthetic efficiency† ($g \cdot mol^{-1}$)	total biomass ($g \cdot m^{-2} \cdot d^{-1}$)	pods per m^2	seeds per pod	mass per seed (mg)
HPS	46.4	41.2	47.1	4.99	0.257	13.7	1486	1.91	159
MH	33.2	19.9	26.6	4.62	0.238	12.0	1385	1.85	167
p-value	<0.01	<0.01	<0.01	0.24	0.24	0.04	<0.01	0.10	0.42

†grams of seed per mol of PPF

Plants grown under MH lamps had 14% less biomass compared to plants under HPS lamps (Table 1). Reduced stem mass in MH plants was associated with an increase in harvest index (HI) (Table 2). All other component partitioning was similar. MH lamps also had slightly less seed yield than HPS lamps (Table 1). A difference in pod number and seeds per pod accounted for the trend in seed yield. P_{net} measurements were consistent with the yield differences between lamp types (Figure 1a).

Table 2. Carbon partitioning of soybeans under two lamp types. Measures are a percent of total dry weight. Sum of the five components equals 100%.

lamp type	seed (harvest index)	stem	leaves	pod	root
HPS	36.4	14.7	27.9	12.8	8.2
MH	37.7	12.2	28.6	12.2	9.3
p-value	0.18	<0.01	0.14	0.19	0.18

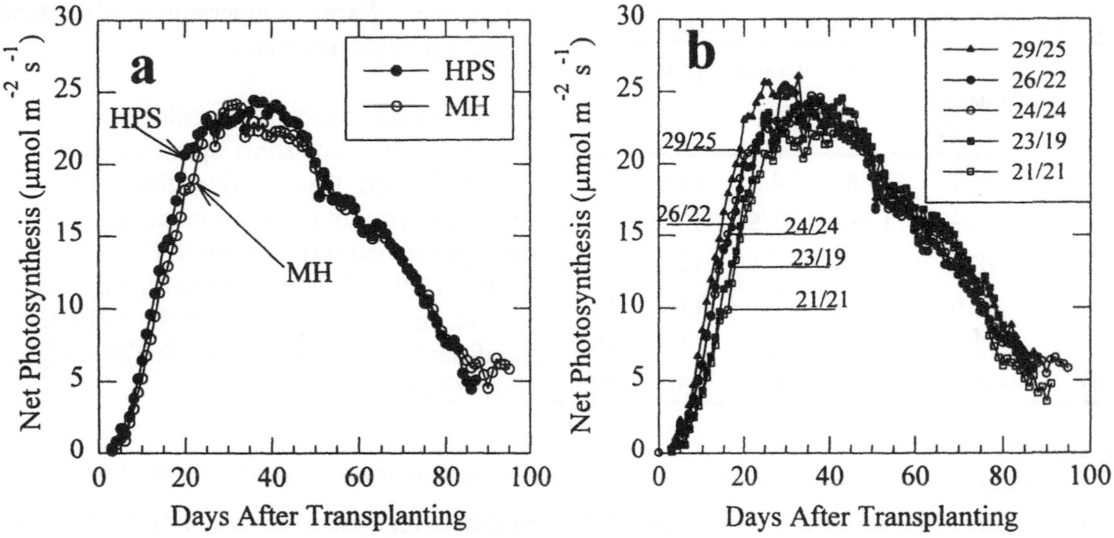

Fig. 1. Net photosynthesis (CO_2 uptake) of soybean canopies. Endpoints are an average day of harvest.
a) Comparison of lamp types. Measurements are an average of the five chambers of different temperatures.
b) Comparison of temperatures. Measurements are an average of the two chambers of differing lamp types.

Effect of Temperature

Higher temperatures increased seed yield via increased number of pods per square meter and seeds per pod (Table 3). We were surprised to find that cooler temperatures reduced the seed fill period. Higher temperatures increased P_{net} early in the life cycle (Figure 1b) but no trend was apparent after 35d. Total biomass (Table 3) and HI (Table 4) tended to decrease with lower temperatures.

Table 3. A comparison of yield and yield components for soybeans grown under five temperature regimes. Each parameter is an average of the two chambers of differing lamp type.

day/night temperature	seed yield ($g•m^{-2}•d^{-1}$)	PE† ($g•mol^{-1}$)	total biomass ($g•m^{-2}•d^{-1}$)	pods per m^2	seeds per pod	mass per seed (mg)	days to first flower	days to harvest	seed fill (days)
29/25	5.46	0.281	13.6	1483	1.96	167	19	87	68
26/22	5.43	0.280	13.1	1550	2.03	153	24	87	63
24/24	5.02	0.258	12.9	1594	1.89	160	27	90	63
23/19	4.13	0.213	12.6	1321	1.79	167	32	90	58
21/21	3.98	0.205	11.9	1230	1.73	169	33	90	57
p-value	0.06	0.06	0.46	<0.01	0.01	0.74	<0.01	0.67	

†PE = photosynthetic efficiency

Table 4. Carbon partitioning of soybeans under five temperature regimes. Data are a percent of total dry weight. Sum of the five components equals 100%.

day/night temperature	seed (harvest index)	stem	leaves	pod	root
29/25	40.0	15.8	24.3	13.0	7.1
26/22	40.9	12.9	25.5	13.8	6.9
24/24	38.9	12.5	27.1	12.9	8.5
23/19	32.6	13.5	32.3	11.2	10.4
21/21	33.0	12.5	32.1	11.5	10.9
p-value	<0.01	0.03	<0.01	0.07	0.06

Percent leaf mass decreased with increasing temperatures probably because of delayed leaf senescence. Warm temperatures also tended to decrease percent root mass.

The day/night temperature scheme did not affect canopy height. For some species Erwin and Heins (1995) showed that altering the difference (DIF) between day/night temperature changed plant height but larger seeded species showed little response to DIF (Erwin, 1991). In our experiment, the canopies at +4 DIF (42 cm @ 26/22°C, 41 cm @ 23/19°C) were not significantly taller than at zero DIF (38 cm @ 24/24°C, 36 cm @ 21/21°C).

DISCUSSION

While short-stature canopies are desired in controlled environments, high yield is also a priority. The mechanism underlying biomass differences with spectral quality, specifically orange bias (HPS) versus a balanced spectrum (MH), is unknown. However, this biomass difference did not significantly affect yield. Because plant height and seed yield commonly are positively correlated (Wells *et al.*, 1993), a slight difference in yield was to be expected. Taller plants under HPS lamps may have had better light interception. Higher P_{net}, longer internodes, larger leaves (data not shown), and more rapid canopy closure (data not shown) suggest that there is better light distribution and capture in the HPS canopy. Increasing plant density under MH lamps might overcome canopy closure differences but this would probably increase stem elongation after canopy closure, which would reduce the height advantage.

A lack of significant effect on plant height indicates temperature can be manipulated to some extent to maximize yield without increasing canopy height. The high temperatures increased yield by increasing pod and seed number. Rapid canopy closure and higher photosynthesis contributed to the differences. High temperatures also hastened development as evidenced by shorter time to final vegetative-stage (data not shown) and decreasing time to first flower. We are currently testing temperatures above 29°C.

Measurement of P_{net} is important for calculating oxygen production for a bioregenerative life support system. Regardless of treatment, there was a broad peak in P_{net} between days 25 and 45. Early life cycle rate of increase in P_{net} was caused by rate of canopy closure and radiation capture. Differences between lamp types and between temperatures were apparent during this part of the life cycle. The decrease in P_{net} was due to senescence and treatment had no effect on the rate of decrease. Therefore we are focusing on environmental changes early in the life cycle to increase canopy closure.

ACKNOWLEDGMENTS

The authors gratefully acknowledge the assistance of Darren Hall, Noble Mabuchi, Shelly Barlow, Derek Knight, and Gus Koerner in collecting this data. We also wish to thank Drs. Ted Tibbitts and Ray Wheeler for their insightful reviews of this manuscript.

REFERENCES

Britz, S.J., and J.C. Sager, Photomorphogenesis and Photoassimilation in Soybean and Sorghum Grown under Broad Spectrum or Blue-Deficient Light Sources, *Plant Physiol.*, 94, 448 (1990).

Bugbee, B., Steady-state Canopy Gas Exchange: System Design and Operation, *HortScience*, 27, 770 (1992).

Downs, R.J., and J.F. Thomas, Morphology and Reproductive Development of Soybeans under Artificial Conditions, *Biotronics*, 19, 19 (1990).

Erwin, J., Temperature Effects on Plant Growth, *Proc. of the Twelfth Ann. Conf. on Hydroponics*, 1 (1991).

Erwin, J., and R.D. Heins, Thermomorphogenic Responses in Stem and Leaf Development, *HortScience*, 30, 940 (1995).

Gbikpi, P.J., and R.K. Crookston, A Whole-plant Indicator of Soybean Physiological Maturity, *Crop Sci.*, 21, 469 (1981).

Pausch, R.C., S.J. Britz, and C.L. Mulchi, Growth and Photosynthesis of Soybean (*Glycine max* (L.) Merr.) in Simulated Vegetation Shade: Influence of the Ratio of Red to Far-Red Radiation, *Plant, Cell and Env.*, 14, 647 (1991).

Raper, D., and Kramer, Stress Physiology, in *Soybeans: Improvement, Production and Uses*, edited by J.R. Wilcox, pp. 590-591, Agronomy Publishers, Madison, WI (1987).

Tibbitts, T.W., D.C. Morgan, and I.J. Warrington, Growth of Lettuce, Spinach, Mustard, and Wheat Plants under Four Combinations of High-pressure Sodium, Metal Halide, and Tungsten Halogen Lamps at Equal PPFD, *J. Amer. Soc. Hort. Sci.*, 108, 622 (1983).

Wells, R., J.W. Burton, and T.C. Kilen, Soybean Growth and Light Interception: Response to Differing Leaf and Stem Morphology, *Crop Sci.*, 33, 520 (1993).

Wheeler, R.M., C.L. Mackowiak, and J.C. Sager, Soybean Stem growth under High-Pressure Sodium with Supplemental Blue Lighting, *Agron. J.*, 83, 903 (1991).

Pergamon

Adv. Space Res. Vol. 20, No. 10, pp. 1901–1904, 1997

Printed in Great Britain
0273-1177/97 $17.00 + 0.00

PII: S0273-1177(97)00858-2

SUPER-OPTIMAL CO_2 REDUCES WHEAT YIELD IN GROWTH CHAMBER AND GREENHOUSE ENVIRONMENTS

T. Grotenhuis, J. Reuveni and B. Bugbee

Crop Physiology Laboratory, Plants, Soils and Biometeorology Department, Utah State University, Logan, UT 84322-4820, U.S.A.

ABSTRACT

Seven growth chamber trials (six replicate trials using 0.035, 0.12, and 0.25 % CO_2 in air and one trial using 0.12, 0.80, and 2.0% CO_2 in air) and three replicate greenhouse trials (0.035, 0.10, 0.18, 0.26, 0.50, and 1.0% CO_2 in air) compare the effects of super-optimal CO_2 on the seed yield, harvest index, and vegetative growth rate of wheat (*Triticum aestivum* L. cvs. USU-Apogee and Veery-10). Plants in the growth chamber trials were grown hydroponically under fluorescent lamps, while the greenhouse trials were grown under sunlight and high pressure sodium lamps and in soilless media. Plants in the greenhouse trials responded similarly to those in the growth chamber trials; maximum yields occurred near 0.10 and 0.12 % CO_2 and decreased significantly thereafter. This research indicates that the toxic effects of elevated CO_2 are not specific to only one environment and has important implications for the design of bio-regenerative life support systems in space, and for the future of terrestrial agriculture.

INTRODUCTION

Carbon dioxide levels in regenerative life support systems in space can exceed 1%. Levels this high are generally not toxic to humans, but may be quite harmful to plants. Several detrimental effects of elevated CO_2 level have been reported. Wheeler *et al.* (1993) studied soybean productivity in 0.05, 0.10, 0.20, and 0.50% CO_2 and found that biomass and seed yield decreased when concentrations exceeded 0.10% CO_2. Macko-wiak and Wheeler (1996) showed that potato growth and yield peaked at 0.10 % CO_2 and declined at higher levels. In our lab, super-optimal CO_2 concentrations (>0.12 %) have been shown to cause significant decreases in seed yield for wheat grown under fluorescent bulbs (20-h photoperiod) and hydroponically in growth chambers. In this closed environment, CO_2 concentrations of 0.25 % resulted in seed yield decreases as high as 22% when compared to those at near-optimal (≈0.12 %) concentrations (Grotenhuis, 1996; Grotenhuis and Bugbee, 1996). Further elevation to 0.8 and 2.0 % CO_2 in the growth chamber resulted in only small additional decreases in yield compared to 0.25 % CO_2. Subsequently, a larger growing system was built in our greenhouse which increased the area of our test plots by a factor of six, allowing even more CO_2 concentrations to be tested. This new system is significantly different from the previous growth chamber setup in that all plants are grown under a combination of sunlight and high pressure sodium bulbs (24-h photoperiod) and in soilless media (1:1, Peatmoss:Perlite). This paper compares the previous study of Grotenhuis and Bugbee (1996) with preliminary yield data from the new greenhouse system.

MATERIALS AND METHODS

Growth Chamber Trials. Seven trials using wheat (*Triticum aestivum* L. cvs. USU-Apogee and Veery-10) were conducted in a controlled-environment chamber, which contained six Plexiglas cylinders (30-cm diameter x 62-cm height). Photosynthetic photon flux (PPF) was provided with cool-white, VHO fluorescent lamps. The photoperiod was 20-h. The PPF was averaged for each trial from the weekly measurements and varied among trials from 328 to 660 μmol m^{-2} s^{-1} (23.6 to 47.5 mol m^{-2} d^{-1}). Temperature was maintained at 22.5/21.5 ± 0.2°C (light/dark) until anthesis, and at 17.5/16.5 ± 0.2°C from anthesis to maturity. The six Plexiglas chambers shared a common, recirculating nutrient solution developed for wheat (Bugbee and Salisbury, 1988). Chamber CO_2 concentration was individually controlled by mixing pure CO_2 with outside air. Air flow into each cylindrical plant growth chamber was maintained at 30 L min^{-1} to provide a rapid air turnover rate (once per minute). Possible contaminants in the CO_2 (Morison and Gifford, 1984) were removed by humidifying the CO_2 and passing it through a column of potassium permanganate. In addition, chamber air was routinely sampled and tested with a gas chromatograph for ethylene concentrations with a sensitivity of 0.1 μmol mol^{-1}. CO_2 concentrations were maintained at 0.035, 0.12, and 0.25 % in six trials, while CO_2 concentrations in one trial were maintained at 0.12, 0.80, and 2.0 %.

Greenhouse Trials. Twelve Plexiglas chambers (36 cm width x 47 cm length x 60 cm height: no bottoms) were placed on two benches (92 cm x 184 cm) at six chambers/bench. Each bench supported a mixture of soilless media (1:1, Peatmoss:Perlite) at a depth of 23 cm onto which wheat seed (*Triticum aestivum* L. cvs. USU-Apogee and Veery-10) had been planted. Irrigation tubes ran the entire length of each bench allowing for adequate irrigation with the standard greenhouse nutrient solution [Peter's Soluble Plant Food 20-10-20 PL (0.25 g/L), Fe-EDDHA (10μM), $CuSO_4$ (0.1 μM), and K_2SiO_3 (10 μM)] two times daily. Direct and diffuse sunlight as well as High Pressure Sodium (HPS) 1000 W lamps provided the PPF, which varied at the top of the canopy between 350-1000 μmol m^{-2} s^{-1} depending on time of day and stage in lifecycle. About half of the PPF came from the supplemental lamps. Temperature was maintained at 24 ± 2°C/20.0 ± 0.5°C day/night. CO_2 treatments were controlled and monitored in the same manner as in the growth chamber trials. Three trials using six CO_2 levels (0.035, 0.10, 0.18, 0.26, 0.50, and 1.0 %) and two replicates each were conducted.

Measurements and Calculations. Total seed mass and total vegetative biomass were measured. From these measured parameters, three indicators of yield were calculated: seed yield (g m^{-2} d^{-1}), harvest index (seed mass / total shoot biomass without roots), and the vegetative growth rate (g m^{-2} d^{-1}; not including seed mass). Root mass was not recorded in the greenhouse trials.

RESULTS AND DISCUSSION

The effect of CO_2 concentration on seed yield, harvest index, and the rate of vegetative growth for the growth chamber and greenhouse trials is shown in Figure 1. Elevated CO_2 altered seed yield, harvest index, and the vegetative growth rate similarly regardless of the growing medium, radiative environment, nutrient solution, or diurnal temperature. The response of seed yield to elevated CO_2 was nearly identical for both closed environments. Increasing CO_2 from 0.035 to 0.10 % in the greenhouse increased seed yield by 29%; and increased CO_2 from 0.035 to 0.12% in the growth chambers increased yield by 16%. There was a linear decrease in seed yield between 0.10 and 0.26 % in both environments. Beyond 0.26 % the decrease in seed yield continued, but only slightly. The greenhouse trials yielded slightly better at all CO_2 levels probably because of the higher daily PPF levels. The vegetative growth rate increased as expected from 0.035 to 0.10 and 0.12 % for both controlled-environments. Beyond these concentrations, vegetative growth rate decreased slightly, but not significantly, indicating a maximum at about 0.12 % CO_2.

Root mass was not separated from the soilless media in the greenhouse studies and was thus not included in total plant mass measurements for the greenhouse trials. However, previous measurements have indicated that the root mass is typically 6% of the total biomass at harvest in wheat. The vegetative growth rate (stems and leaves) and the harvest index (seed mass / total shoot biomass) for these trials is thus lower than in the growth chamber trials. The harvest index in the greenhouse would have been about 3% less if the roots had been included.

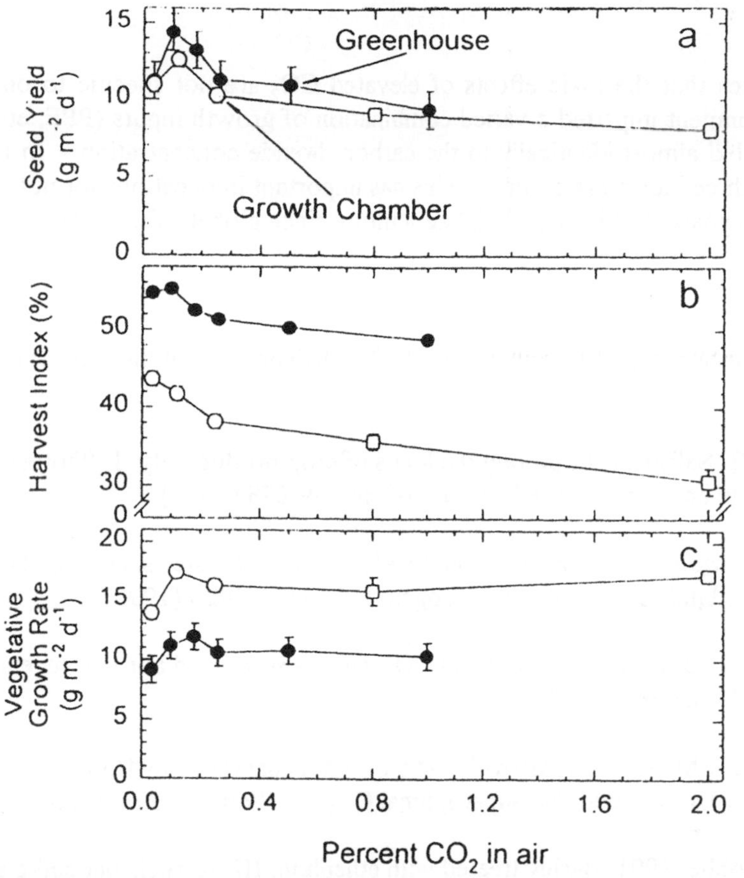

Fig. 1. Effect of elevated CO_2 on seed yield, harvest index, and the vegetative growth rate of wheat grown in six growth chamber (-○-) trials and three greenhouse (-●-) trials. Error bars represent the least significant difference at $\alpha = 0.01$ for the growth chamber (-○-) trials and $LSD_{0.05}$ for the greenhouse (-●-) trials. CO_2 was further elevated to 0.8 and 2.0 % in an additional trial (-□-) conducted in the growth chamber.

The harvest index steadily decreased with increasing CO_2 levels in both environments. A decreased harvest index between ambient and near-optimal CO_2 would normally occur unless the sink capacity increased to exactly match the increase in source capacity. Total biomass did not increase after 1200 μmol mol^{-1} so the decreased harvest index was caused by a decrease in assimilate partitioning to the grain. Wheeler et al. (1993) also observed decreasing harvest index with increasing CO_2 in similar studies with soybean. Measurements of average seed number per head indicate that the reduction in harvest index was caused by reduced seed set.

The detrimental effects of super-optimal CO_2 could be mediated by an increase in ethylene synthesis at high CO_2. CO_2 concentrations above 5% inhibit ethylene formation (Abeles et al., 1992), but elevated CO_2

concentrations below this induce ethylene synthesis. A close correlation between elevated CO_2 and increased ethylene synthesis has been observed in diverse species (see Woodrow and Grodzinski, 1993 and references cited there in). There appears to be a significant relationship between ethylene, pollen formation, and seed set. Ethephon, a commercial product used to generate ethylene, is widely used to reduce plant height and lodging in wheat and barley. Ethephon inhibits pollen formation and can cause full male sterility. Moes and Stobbe (1991) reported that ethephon reduced seeds per head by up to 26% when used to prevent lodging in barley (*Hordeum vulgare* L.).

CONCLUSION

This research indicates that the toxic effects of elevated CO_2 are not specific to only one environment. Although each environment imparted a varied combination of growth inputs (PPF, etc.) the plants in both environments responded almost identically to the carbon dioxide concentration. In fact, the curves were almost parallel for the three indicators of yield. This has important implications for not only bio-regenerative life support systems in space, but for terrestrial agriculture since global CO_2 levels are rising.

REFERENCES

Abeles F.B., P.W. Morgan, and M.E. Saltveit Jr. 1992. Ethylene in plant biology, 2nd ed. Academic Press. London.

Bugbee, B.G., and F.B. Salisbury, Exploring the limits of crop productivity: I. Photosynthetic efficiency of wheat in high irradiance environments, *Plant Physiol.* 88:869-878 (1988).

Grotenhuis, T.P., *Superoptimal CO_2 reduces seed yield in wheat*, Master's Thesis, Plants, Soils, and Biometeorology Dept., Utah State University, Logan, UT 84322-4820 (1996).

Grotenhuis, T.P., and B. Bugbee, Super-optimal CO_2 reduces seed yield but not vegetative growth in wheat, *Crop Science 37: In Press (1997).*

Mackowiak, C.L., and R.M. Wheeler, Growth and stomatal behavior of hydroponically cultured potato (*Solanum tuberosum* L.) at elevated and superoptimal CO_2, *J. Plant Physiol.* (in press:1996).

Moes, J., and E.H. Stobbe. 1991. Barley treated with ethephon: III. Kernels per spike and kernel mass. Agron. J. 83:95-98.

Morison, J.I.L., and R.M. Gifford, Ethylene contamination of CO_2 cylinders: Effects on plant growth in CO_2 enrichment studies. *Plant Physiol.* 75:275-277 (1984).

Wheeler, R.M., C.L. Mackowiak, L.M. Siegriest, and J.C. Sager, Supraoptimal CO_2 effects on growth of soybean [*Glycine max* (L.) Merr.], *J. Plant Physiol.* 142:173-178 (1993).

Woodrow, L., and B. Grodzinski. 1993. Ethylene exchange in *Lycopersicon esculentum* mill. Leaves during short- and long-term exposures to CO2. J. Exp. Bot. 44: 471-480.

Pergamon

Adv. Space Res. Vol. 20, No. 10, pp. 1905–1908, 1997

Printed in Great Britain
0273-1177/97 $17.00 + 0.00

PII: S0273–1177(97)00859–4

CO_2 ENRICHMENT INFLUENCES YIELDS OF 'FLORUNNER,' 'GEORGIA RED' AND 'NEW MEXICO' PEANUT CULTIVARS

D. G. Mortley, Philip A. Loretan, J. H. Hill and J. Seminara

Tuskegee University NASA Center for CELSS, GEORGE Washington Carver Agricultural Experiment Station, Tuskegee University, Tuskegee, AL 36088, U.S.A.

ABSTRACT

Three peanut cultivars, 'Florunner,' 'Georgia Red,' and 'New Mexico,' were grown in reach-in chambers to determine response to CO_2 enrichment. CO_2 treatments were ambient (400 µmol mol^{-1}) and 700 µmol mol^{-1}. Growth chamber conditions included 700 µmol m^{-2} s^{-1} photosynthetic photon flux (PPF), 28/22C, 70% RH, and 12/12 h photoperiod. Growth media consisted of a 1:1 mixture (v/v) of vermiculite and sterilized sand. Six 10 L pots of each cultivar were fertilized three times per week with 250 mL of nutrient solution containing additional Ca (10 mM) and NO_3 (25 mM) and watered well. Beginning 21 days after planting (DAP) and every three weeks thereafter up to 84 days, the second leaf from the growing axis (main stem) was detached to determine CO_2 effect on leaf area, specific leaf area (SLA) and dry weight. Plants were harvested 97 DAP, at which time total leaf area, leaf number, plant and root weights and pod production data were taken. Numbers of pods per plant, pod fresh and dry weights, fibrous root and plant dry weights were higher for all cultivars grown at 700 µmol mol^{-1} than at ambient CO_2. Also, leaf area for all cultivars was larger with CO_2 enrichment than at ambient. SLA tended to decline with time regardless of CO_2 treatment. Percentage of total sound mature kernels (%TSMK) was similar for both treatments. Plants grown at 700 µmol mol^{-1} CO2 had slightly more immature pods and seeds at final harvest.

INTRODUCTION

In tightly closed environments such as are envisioned for space missions, CO_2 partial pressures can reach 500 to 1000 Pa (5000-10,000 µmol mol^{-1}). Since the peanut is a candidate CELSS crop, its response to elevated CO_2 must be determined. Few studies have focused on this. Hardy and Havelka (1977) reported a 60% increase in peanut dry weight and a 76% increase in nitrogen fixation in plants exposed to 1000 to 1500 µmol mol^{-1} CO_2 in open top chambers. In pot studies in semi-closed chambers, peanut plants exposed to 340 and 1000 µmol mol^{-1} CO_2 exhibited higher photosynthetic rates, lower stomatal conductance, and increased plant biomass at 1000 µmol mol^{-1} CO_2 (Chen and Sung, 1990). However, in their studies marketable seed yield did not increase with increased CO_2. In this study, peanuts (Florunner, Georgia Red and New Mexico) were exposed to CO_2 levels of 400 and 700 µmol mol^{-1}, and yield responses were evaluated.

MATERIALS AND METHODS

Six peanut (*Arachis hypogaea* L.) plants each of cvs. Florunner, Georgia Red, and New Mexico were grown in 10 L pots containing a 1:1 mixture (v/v) of vermiculite and sterilized sand, fertilized three times per week with 250 mL of nutrient solution containing additional Ca (10 mM) and NO_3 (25 mM) and watered well. All studies were conducted in reach-in plant growth chambers with fluorescent/incandescent lighting maintained at a PPF of 700 µmol $m^{-2}s^{-1}$ with a 12 h photoperiod and matching 28/22 thermoperiod. Relative humidity was maintained constant near 70%. CO_2 concentrations were maintained at approximately 400 (ambient) and 700 µmol mol^{-1} with an infrared gas analyzer. Beginning 21 DAP and every three weeks thereafter up to 84 days, the second leaf from the growing axis (main stem) from each of six plants was detached to determine the CO_2 effect on leaf area, SLA, and dry weight. Plants were harvested at 97 DAP. The experiment was repeated and total leaf area, leaf number, plant and root weights, and pod production data were taken.

RESULTS AND DISCUSSION

Table 1 shows that the number and weight of mature seeds were higher for plants exposed to 700 µmol mol^{-1} than for plants grown at ambient CO_2. Plant dry weight increased with CO_2 enrichment and was highest for Florunner at 700 µmol mol^{-1} with a similar trend observed for all three cultivars for root dry weight. There was a trend for immature/unmarketable seed yield to increase with enrichment, apparently because enrichment during seed fill may have maximized competition between developing seeds and gynophores causing inhibition of seed growth (Chen and Sung, 1990). The difference in percentage of total sound mature kernels (%TSMK)—the percentage of mature seeds with no damage from mold, insects, discoloration or premature sprouting—between ambient and enriched CO_2 environments was highest in 'Florunner,' indicating differences in cultivar responses.

Table 1. Yield of Three Peanut Cultivars in Response to CO_2 Enrichment

Treatment	Plant dry wt. (g)	Fibrous roots dry wt. (g)	Mature seeds no./plt.	Mature seeds wt. (g)	Imm. seeds no.	Imm. seeds wt. (g)	TSMK (%)
Florunner							
Amb.	20.6	18.6	24.6	11.9	11.8	0.91	67.6
700	33.9	22.0	39.4	20.3	13.9	0.93	73.9
Georgia Red							
Amb.	13.0	7.1	25.7	11.4	25.2	1.59	50.4
700	20.6	12.4	37.1	20.2	31.6	1.83	54.0
New Mexico							
Amb.	12.3	5.5	27.7	10.2	13.0	1.12	68.1
700	23.4	8.1	44.2	19.0	19.3	1.76	69.6
Amb.	15.3	10.4	26.0	11.2	16.4	1.20	61.3
700	26.0	14.2	40.2	19.8	21.6	1.53	65.0

The number of pods per plant and pod fresh and dry weights were higher for the three cultivars with increased enrichment (Table 2). Pod yield was similar across cultivars showing that enrichment had a stronger effect than cultivar. With CO_2 enrichment, seed dry weight per plant increased 73%, 71% and 100% for Florunner, Georgia Red and New Mexico, respectively.

Table 2. Peanut Pod Development in Response to CO_2 Enrichment

Treatment	Pod No.	Pod fresh wt.	Pod dry wt.	Seed dry wt.
Florunner				
Amb.	24.4	34.4	14.5	12.8
700	31.2	55.1	27.0	22.1
Georgia Red				
Amb.	35.6	49.0	15.5	13.0
700	46.5	73.8	27.6	22.2
New Mexico				
Amb.	23.5	37.0	14.4	10.4
700	34.3	63.8	26.2	20.8

The number of leaves per plant and total leaflet area were increased for all cultivars by CO_2 enrichment (Table 3) with Florunner producing the greatest number of leaves and leaflet area per plant. The specific leaf area (SLA) indicates that leaves were thinner for Florunner and thicker for New Mexico with CO_2 enrichment, but elevated CO_2 did not affect the leaf thickness of Georgia Red, and this response could be related to differences in cultivar—although leaf size is also influenced by the lighting environment. Generally, large thin leaves are more desirable than small thick leaves because they have more dry matter per unit area (Tibbitts et al., 1994).

Table 3. Peanut Leaf Growth in Response to CO_2 Enrichment

Treatment	No./plt.	Total leaflet area (cm^2)	SLA (m^2/kg)
Florunner			
Amb.	133	2132	24.3
700	204	3850	30.0
Georgia Red			
Amb.	66	1207	28.2
700	118	2769	29.6
New Mexico			
Amb.	42	766	30.2
700	92	2124	23.0

CONCLUSIONS

Results from this study showed that growth of seeds, pods, leaves and roots in three peanut cultivars were enhanced by increasing CO_2 from 400 to 700 µmol mol^{-1}. Specific leaf area seemed to vary with cultivar. CO_2 enrichment resulted in a greater percentage of total sound mature kernels in only one of the three cultivars—Florunner.

ACKNOWLEDGMENTS

This research was supported by funds from the U.S. National Aeronautics and Space Administration (Grant No. NAGW-2940) and USDA/CSREES (Grant No. ALX-SP-1) and is Contribution No. 266 of the George Washington Carver Agricultural Experiment Station, Tuskegee University.

REFERENCES

Chen, J.J., and J. M. Sung. 1990. Gas exchange rate and yield responses of Virginia-type peanut to carbon dioxide enrichment. Crop·Sci. 30:1085-1089.

Hardy, R.W.F., and U.D. Havelka. 1977. Possible routes to increase the conversion of solar energy to food and feed by grain legumes and cereal grains (Crop Production): Carbon dioxide and N fixation, foliar fertilization and assimilate partitioning. Pp. 299-322 in Biological Solar Energy Conversions (A. Matsui, Ed.). Academic Press, New York.

Tibbitts, T.W., W. Cao, and R. W. Wheeler. 1994. Growth of potatoes for CELSS. NASA Cont. Rep. 177646, Ames Research Center, Moffitt Field, California.

BIOLOGICAL EFFECTS OF CLOSURE AND RECYCLING IN A CELSS

Proceedings of the F4.8 Symposium of COSPAR Scientific Commission F which was held during the Thirty-first COSPAR Scientific Assembly, Birmingham, U.K., 14–21 July 1996

Edited by

T. W. TIBBITTS

Department of Horticulture, University of Wisconsin, 1575 Linden Drive, Madison WI 53706-1590, U.S.A.

Pergamon

Adv. Space Res. Vol. 20, No. 10, p. 1911, 1997

Printed in Great Britain
0273-1177/97 $17.00 + 0.00

PII: S0273-1177(97)00624-8

PREFACE

Closure in CELSS causes the accumulation of volatile compounds and populations of microorganisms in the nutrient solution that must be understood, and possibly controlled. The papers in this session detailed the large number of volatile organic compounds that have been found in closed systems and indicated which were of biological origin. Papers emphasised the importance of ethylene emission by plants and the influence of wind velocity on emission rates of this gas. Other papers were directed toward understanding the microorganism complex that accumulates in recirculating nutrient systems, how this might be monitored and the impact that the microorganisms have on plants.

Thanks to N. S. Pechurkin for serving as Deputy Organizer and Richard Strayer as a program committee member. A special thanks to the several individuals that provided reviews of one or more of the papers for this session.

T. W. Tibbitts

Pergamon

Adv. Space Res. Vol. 20, No. 10, pp. 1913–1922, 1997
Published by Elsevier Science Ltd on behalf of COSPAR
Printed in Great Britain
0273-1177/97 $17.00 + 0.00

PII: S0273–1177(97)00625–X

ACCUMULATION AND EFFECT OF VOLATILE ORGANIC COMPOUNDS IN CLOSED LIFE SUPPORT SYSTEMS

G. W. Stutte* and R. M. Wheeler**

** Dynamac Corporation, Mail Code DYN-3, Kennedy Space Center FL 32899 U.S.A.*
***Biomedical Operations and Research Office, Kennedy Space Center, FL 32899 U.S.A.*

ABSTRACT

Bioregenerative life support systems (BLSS) being considered for long duration space missions will operate with limited resupply and utilize biological systems to revitalize the atmosphere, purify water, and produce food. The presence of man-made materials, plant and microbial communities, and human activities will result in the production of volatile organic compounds (VOCs). A database of VOC production from potential BLSS crops is being developed by the Breadboard Project at Kennedy Space Center. Most research to date has focused on the development of air revitalization systems that minimize the concentration of atmospheric contaminants in a closed environment. Similar approaches are being pursued in the design of atmospheric revitalization systems in bioregenerative life support systems. In a BLSS one must consider the effect of VOC concentration on the performance of plants being used for water and atmospheric purification processes. In addition to phytotoxic responses, the impact of removing biogenic compounds from the atmosphere on BLSS function needs to be assessed. This paper provides a synopsis of criteria for setting exposure limits, gives an overview of existing information, and discusses production of biogenic compounds from plants grown in the Biomass Production Chamber at Kennedy Space Center.

Published by Elsevier Science Ltd on behalf of COSPAR

INTRODUCTION

Bioregenerative life support systems (BLSS) for long duration space missions are being designed to operate with limited resupply, utilizing biological systems to revitalize the atmosphere, purify water, and produce food. The diversity of material used in construction of such a facility, the incorporation of plant and microbial processes for life support, and daily human activities suggest that volatile organic compounds will be produced and accumulate in the atmosphere. NASA has traditionally taken an industrial hygiene approach to the monitoring and control of organic constituents in the atmosphere and has established exposure limits and monitoring protocols (NASA, 1990; NASA, 1991).

As NASA develops atmospheric control strategies for long-duration space missions, the role of volatile compounds on human physiology, human psychology, and plant physiology will need to be understood. Crew health and safety has been, and will continue to be, the overriding criterion when establishing standards for components used during space flight (NASA, 1991). Although standards have been established to maintain crew health and safety for long-duration space missions, no analogous standards have been established for non-human biological components of a BLSS. It is generally accepted that volatiles play a vital role in the normal growth and development of plants (Abeles et al., 1992; Sharkey et al., 1991), but the threshold levels of phytotoxicity have not been established for most known biogenic compounds.

HISTORICAL OVERVIEW

Atmospheric monitoring and sampling of the atmosphere from shuttle missions has identified over 250 organic compounds in the cabin atmosphere of the Space Shuttle (Buoni et al., 1989). Most of these compounds are found in only trace amounts and are well below their spacecraft maximum allowable concentration (SMAC) limits (NASA, 1990). In addition, a number of unknown compounds have been detected at or near the detection limit of the GC/MS. Although indicative of the types of volatile compounds present, these data are of limited value for designing a BLSS since atmospheric revitalization systems are used to remove all organic compounds from the atmosphere. These data indicate that air revitalization systems are generally effective, but do not allow an assessment of potential toxicity or interactions between humans and plants in a BLSS.

Volatiles reported by Zlotopolsk'ii and Smolenskaja (1996) in the atmosphere of closed chambers used for plant growth included acetone, ethanol, methanol, toluene, acetaldehyde, ethylacetate, methyethylketone, and cyclohexane. Charron et al. (1996) reported on the production of hexenal, hexenol, and hexenyl-acetate from lettuce grown in a closed chamber. Wheeler et al. (1996) detailed the developmental production of ethylene by soybean, lettuce, wheat, and potato in the BPC.

One of the first long-duration human-rated bioregenerative testbed studies using higher plants was the BIOS-3 study conducted in 1977 by the Department of Biophysics, L.V. Kirenskii Institute of Physics, in Krasnoyorsk, Russia. (Terskov et al., 1981; Gitelson and Okladnikov, 1994). This study documented VOCs in a human habitat in an atmosphere being regenerated continuously for four months by higher plants. The primary volatile constituents observed in the human habitat are summarized in Table 1.

Table 1: Estimates of the concentration of classes of volatile organic compounds in the living compartment of the "BIOS-3" Experiment during a 4-month test conducted in 1977[z].

Substances	*With Catalytic Furnace*	*Without Catalytic Furnace*
Aldehydes, (mg m^{-3})	0.56	0.51
Alcohols, (mg m^{-3})	2.5	3.6[y]
Mercaptans, (mg m^{-3})	0	0
Acetic Acid, (mg m^{-3})	1.15	0.92
Carbon Monoxide, (mg m^{-3})	6.8	0
Readily oxidizable, total, (mg O_2 m^{-3})	5.0	7.8[y]
Difficult to oxidize, total, (mg O_2 m^{-3})	34.3	30.1

[z]Data derived from Terskov et al., 1981. NASA TM-76452
[y]Increased concentrations are associated with use of ethanol for microbial monitoring.

The volatiles detected during BIOS-3 were primarily associated with man-made materials and not of biological origin. During the four-month test, the classes of compounds reported fluctuated around a low, consistent concentration (Terskov et al., 1981; Gitelson and Okladnikov, 1994). This observation is interesting since the facility had no physical or chemical atmospheric regeneration system incorporated during the initial stages of the experiments, suggesting an interaction with the bioregenerative component

(Gitelson and Okladnikov, 1994). Incorporation of a catalytic furnace during the final stages of the experiment reduced the concentration of CO, but had limited effect on the other compounds (Terskov et al., 1981).

More recently, trace contaminants of the atmosphere were monitored twice a week during the Early Human Testing Initiative: Phase I Experiment at NASA's Johnson Space Center (NASA, 1996; Edeen et al., 1996). A number of compounds were reported to be present in all samples at trace concentrations. The background compounds included solvents, Freon 113, siloxanes and silenes. Tetrahydrofuran and ethylbenzene were detected as transients in the chamber. Although a number of VOCs were detected, all were at concentrations < 0.1 part per million with the exception of methanal and isopropyl alcohol. The only plant produced volatile reported was ethylene, which reached concentrations of 0.9 mg m^{-3}, which is high enough to affect growth and development of plants (Abeles et al., 1992), but does not pose any health risk to the crew (NASA, 1990).

CLASSIFICATION OF VOCS

Within a closed environment, volatile organic compounds can be classified based on whether the compounds are anthrogenic or biogenic. Anthrogenic compounds originate from man-made sources, such as construction materials, solvents, or physical/chemical processes. Biogenic compounds originate from biological systems including humans, animals, plants and microbes. This categorization is useful in distinguishing between volatiles that can be partially controlled through pre-treatment (off-gassing of materials) and those expected to be produced by the plant or human subsystems. In addition to identifying the source of the volatiles, this classification provides a means of estimating the relative impact of the physical and biological systems on total atmospheric VOC load.

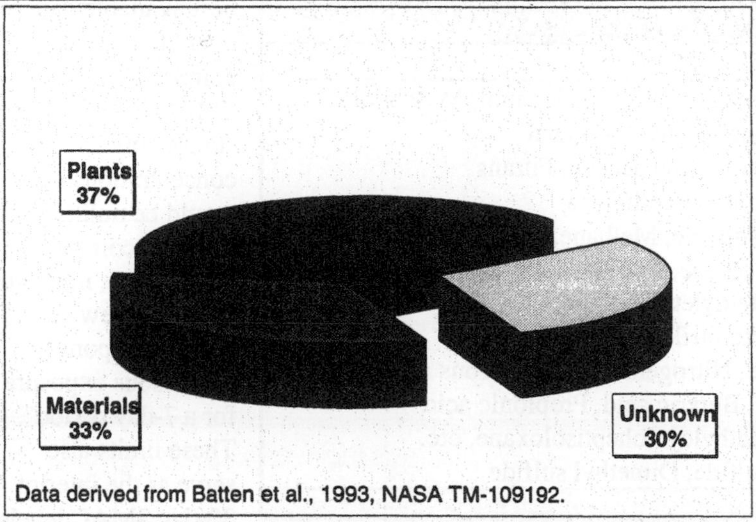

Figure 1: Relative percentage of volatile organic compounds detected in the Biomass Production Chamber headspace (based on total number of compounds) during an 84-day experiment with wheat cv. Yecoro Rojo.

In one study of wheat grown in the Biomass Production Chamber (BPC) at KSC, a total of 89 VOCs were identified from 94 samples of air analyzed during the growout (Batten et al., 1993). The results revealed that approximately equal numbers of compounds were derived from biogenic and anthrogenic sources. (Figure 1). Analysis of the atmosphere of all experiments in the BPC have resulted in similar distributions of compounds (unpublished results). However, the total mass of VOC exposure was primarily due to compounds off-gassing from materials. (Figure 2). These results have been confirmed with other crops grown in the BPC. Materials accounted for 87% of total VOC mass in a tomato study and 92% of the total mass in a soybean study (Stutte, 1996). Siloxanes and siloxones have been the primary anthrogenic VOC constituents in the atmosphere of the BPC in all cases.

The data reported from many experiments are based on a functional classification of the compounds. This type of classification assumes that compounds with similar chemical structures will behave similarly within the environment.

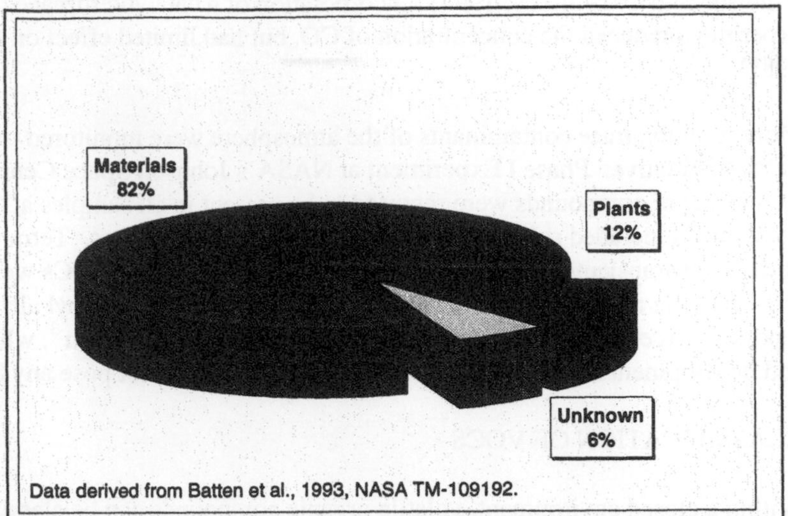

Figure 2: Relative percentage of volatile organic compounds, by source, detected in the Biomass Production Chamber headspace during an 84-day experiment with wheat cv. Yecoroa Rojo.

Leban and Wagner (1989) listed 219 potential atmospheric contaminants used to develop trace contaminant load models for space station. This classification was used by Wydenven and Golub (1990) to describe the waste streams in a typical space habitat and to establish the spacecraft maximum allowable concentrations (SMAC) for volatile compounds (NASA, 1991).

Limits of human exposure for many of the compounds listed in the functional classification discribed in Table 2 have been established for industrial and space habitat applications. The SMAC limits have been established using the best available data for over 200 volatile compounds. The SMAC limits are "those atmospheric concentrations that would be unlikely to cause discomfort, impairment, illness or injury to crew members upon continuous exposure for a 7-day period". These limits also serve as the interim 30-day SMAC levels for extended orbiter missions. For compounds without SMACs, or where insufficient data are available to establish a SMAC, a default value of 0.1 mg m^{-3} has been assigned. The sources of toxicological limits are the American Conference of Governmental Industrial Hygienists (ACGIH) and U.S. Occupation Safety and Health Administration (OSHA). OSHA has the power to enforce maintenance of exposure limits in workplaces in the USA. The ACGIH has no enforcement power, but their recommendations for threshold limit values (TLV) for chemical substances are widely followed.

Table 2: Functional classification of potential VOCs in the atmosphere of space habitats with characteristic examples of the class[z]

Category	*Examples*
Alcohols	Butanol, Isopropyl alcohol, Ethanol, Methanol
Aldehydes	Acetaldehyde, Butanal, Ethanal
Esters	Butyl Acetate, Ethyl acetate, Furans
Halocarbons	Freons, Tetrachloroethylene
Hydrocarbons, Aliphatic	Ethylene, Isoprene, Methane,
Hydrocarbons, Aromatic	Benzene, Toluene, Xylene
Ketones	Acetone, Methylethylketone
Organic Nitrogens	Indole, Acetonitrile
Nitrogen Oxides	Nitric oxide, Nitrogen dioxide, Nitrous oxide
Organic Acids	Acetic acid, Butyric acid, Propionic acid
Silanes and Siloxanes	Tetradecamethylcycloheptisoloxane, etc.
Sulfides	Carbon disulfide, Dimethyl sulfide

[z]**Data excerpted from NASA, 1990**

In a closed environment with several known sources of VOCs, it is probable that a mixture of compounds will be present. The SMAC assumes that compounds with a similar toxicological category will have an additive effect. If the ratio of the concentration of detected compounds to their SMAC is < 1.0, then the toxicological exposure within this category is considered acceptable.

Space station trace contaminant control is being designed around threshold values (T-values) and SMACs as outlined in NASA-NHB 8060.1C (1991). The exposure limit is established so that the cumulative T-value for all compounds, without separation into toxicological categories is less than 0.5. This assumes that no compound exceeds its SMAC threshold. JSC 20584 establishes exposures limits based on total concentration of a compound within a toxicological category. This means that a compound can be "counted" more than once in the exposure limit in JSC 20584 if it has multiple toxicological classifications (e.g. an irritant and aphixicant).

VOCS IN THE BIOMASS PRODUCTION CHAMBER AT KSC

In order to establish the relative abundance and potential biological impact of VOCs within the biomass production component of a BLSS, a database of VOC production by candidate crops is being compiled by the Breadboard Project at Kennedy Space Center. Data on VOC concentration in the BPC throughout development has been reported for wheat (Batten et al., 1993; Batten et al., 1995), lettuce (Batten et al., 1996), potato (Batten et al., 1996), soybean (Stutte, 1996) and tomato (Stutte, 1996).

Table 3. Primary biogenic compounds detected from wheat cv. *Yecoro Rojo* in the atmosphere of the BPC at different stages of crop development.

Vegetative	*Anthesis*	*Grain Fill*	*Senescence*
2-Ethyl-1-hexanol	Benzaldehyde Nonanal	2-Methyl furan 3-Methyl furan	Tetramethyl urea Tetramethyl thiourea

Data derived from Batten et al., 1995, Phytochemistry 39: 1351-1357

In one study with wheat, the effect of crop development on the production of biogenic volatiles was studied (Batten et al., 1995). Air samples were collected on a weekly basis approximately 2 hours after the lamps came on. The experimental conditions used to quantify VOCs in the Biomass Production Chamber utilized modifications of standard methodologies. The samples (16-liter) were collected in passivated stainless steel canisters (SUMMA® polished stainless steel). The canisters and pumping system were checked for cleanliness and background levels by filling the canisters with ultra-pure nitrogen and analyzing for any contaminants. The air was analyzed according to a modification of a standard method (EPA method TO14). The analyses were carried out by concentrating 1.1-liter samples onto a multibed graphitized carbon adsorbent trap. The adsorbent trap was interfaced through a cryofocusing unit to a gas chromatograph-mass spectrometer (GC-MS) equipped with a 30 m x 0.596 mm DB™ 5.625 bonded phase column and a jet separator. During this study, using these methods, 23 compounds were determined to be of biogenic origin. Of those compounds, eight were found to differ significantly in apparent concentration throughout the development cycle of the plant. The compounds that were characteristic of each stage of development are noted in Table 3. Batten et al. (1995) suggested that these compounds may provide a readily monitored, non-destructive developmental marker for assessing crop development and physiology in a closed environment.

Table 4: Relative abundance of presumed biogenic volatile organic compounds detected in the Biomass Production Chamber at Kennedy Space Center over the growth cycle of crops being evaluated for bioregenerative life support systems.

Compound [z]	*Soybean*	*Wheat*	*Potato*	*Lettuce*	*Tomato*
Benzaldehyde	+++[y]	++	+	+	++
Butanal				+	
Butanol, 1-		+			
Butanone, 2-	+++	+		+	
Carbon disulfide	+	+	+	+	+
Carene					++
Cymene, p-	++				
Decanol		+			
Dimethyl sulfide			+	+	
Ethylene	+	+	+	+	+++
Ethyl-1-hexanol, 2-		+	++	+	+
Ethylfuran, 2-	++				+
Fenchene, α- and cyclo-					++
Furan		+		+	
Heptadecane					+
Heptanal	+(+++)[w]	+			+++
Hexadecanol, 1	+(+++)[w]				
Hexanal	+(+++)[w]	+			+++
Hexen-1-ol acetate, 2-	+(+++)[w]				+++
Isoprene				+	++
Limonene	+				++
Methyl-2-propamine, 2-			++		
Methylfuran, 2-		+		+	++
Methylfuran, 3-		+		+	
Nonanal	+(+++)[w]	+		+	+++
Ocimene	+(+++)[w]				+
Pentylfuran, 2-		+			
Phenol		+			
Pinene, α-	++			+	+++
Pinene, β-	++			+	++
Propanone, 2-			+		
Propenenitrile, 2-			+		
Terpinene, α-	+(+++)[w]				+++
Terpinene, γ-	+				++
Tetrahydrofuran	++				+++
Tetramethylthiorea		+	+		
Tetramethylurea		+	++	+	
Thiobismethane		+	+	+	
Thiourea		++			
Thujone, α-					++

[z] Compounds were detected and identified by either GS/MS standards or spectral matches with Wiley Spectral Library on at least two dates during a growout.
[y] Relative abundance where: + = <10 $\mu g\ m^{-3}$; ++ = 10-100 $\mu g\ m^{-3}$; +++ = > 100 $\mu g\ m^{-3}$.
[w] Relative abundance following horticultural practice of removing senescent leaves.

In over six years of monitoring, a number of different compounds have been identified in the atmosphere of the BPC which may be of biogenic origin. The criteria for determining whether a compound was biogenic included direct determination of the compound from plant tissue (Bertis et al., 1993), detection in BPC only when plants were present (Batten et al., 1995), or previous citation in the literature (Duke, 1992; Buttery et al., 1993). It is difficult to rigorously quantify the concentration of a particular compound because losses associated with surface adsorption, sample transfer, and GC/MS analysis are only partially accounted for with the analytical methods used for the analysis. However, these data do provide a series of "snapshots" of the concentration of compounds, and suggest what compounds may be associated with particular developmental events or horticultural practices. A summary of the compounds tentatively identified as being biogenic that have been detected from crops grown in the BPC is presented in Table 4. This table also displays the relative abundance of the compounds. It should be noted that these VOCs are at trace concentrations: the (+++) abundance of >100 $\mu g\ m^{-3}$ is approximately 10 ppb for a compound with a molecular weight of 250 at 20° C.

IMPACT OF VOCS ON BLSS FUNCTION

The biological functions of most VOCs at low concentrations are not known. Terpenes were detected from all species, with tomato having relatively large concentrations during flowering and ripening periods. Terpenes are compounds that are derived from isoprene skeletons. Of the known terpenes, only pinene and isoprene, the building block compounds, have been studied in any detail. Species appear to produce either one or the other in a significant amount. Isoprene is a characteristic compound of deciduous forests and pinene of coniferous forests. High concentrations of terpenes were detected in soybean following the horticultural practice of removing of senescent leaves from the chamber. It is hypothesized that the process of crushing the leaves released the volatile terpenes. This suggests that modifing horticultural practices may reduce the overall VOC load.

Only a few of the compounds identified in the BPC, benzaldehyde, carbon disulfide, and 2-ethyl-1-hexanol, were detected in atmosphere of all the crops. The other compounds were detected in only a few of the crops. Although the biological functions of these compounds are not known, terpenes have been reported to have antifungal activity by several authors (Sharkey et al., 1992 and references therein). In addition, isoprene may have some role in the development of high temperature tolerance for some species (Sharkey and Singsaas, 1995). Several of the other compounds detected also have been reported to have ecological functions related to pest mitigation. Among the compounds reported to have antifungal properties are aldehydes (Aharoni and Stadelbacher, 1973), nonanol (Vaughn, 1995), and hexanal (Song et al., 1996). Undoubtedly, the biological role of these compounds will become clearer with increased research activity.

As previously noted, over 80% of the total concentration of volatiles have been from non-biological sources for every experiment conducted in the BPC, with the primary contaminants being siloxanes, which are the outgassing products of sealants used to maintain chamber integrity. Activated carbon filters (Heath and Manukian, 1992) have been used to reduced the concentration of these anthrogenic compounds. The GC/MS analyse have indicated that concentrations of alkanes, furans, and aromatic hydrocarbons are significantly reduced in the BPC with passive filtering through the activated charcoal filters, but these filters were not very effective for removing halocarbons. In addition, the filters were ineffective at reducing ethylene concentrations in the BPC. However, the activated charcoal filters were effective at reducing the concentration of sulfides, resulting in an enhancement of the olfactory quality of the atmosphere (Stutte et al., 1995).

Although NASA's primary approach has been the removal of volatiles from closed atmospheres, there is mounting evidence for the age-old belief that contact with plants fosters psychological well-being and helps reduce physiological stress (Ulrich and Parsons, 1992). The physiological benefits to stress reduction appear to be achieved when the individual perceives that plants are present, even when direct physical or visual contact is not maintained (Heerwagen and Orians, 1986). The strong psychological benefit to having plants in the closed environments has been recognized by Russian scientists (Nechitailo and Mashinsky, 1993) who have routinely maintained plants on the MIR space station.

The roles of volatiles in inducing positive psychological and physiological responses is not known. However, it is known that the olfactory system is especially sensitive and can trigger both positive and negative psychological responses. For example, the aroma of fresh cut grass is typically pleasing, and flavor is enhanced when terpenes are added to "bland" food products. As a consequence, the strategy to remove all volatiles from the atmosphere should be approached cautiously, since their biological activity may have positive effects on human health as well as the growth and development of plants.

CONCLUSIONS

The most abundant VOCs in the closed environment are of man-made origin and threshhold exposure limits for humans have been established. There is significantly less known of the effects of VOCs on the performance of plants within a BLSS. The limited available information suggests that biogenic compounds will constitute a minor fraction of the total VOC component. Of the individual biogenic VOCs, the biological activities are largely unknown, with the notable of exception of ethylene. There may be potential positive benefits to maintaining low concentrations of biogenic compounds in the atmosphere to obtain normal growth and development practices and to provide some protection against potential pathogens. Each species appears to have a unique VOC production profile, which changes during the development cycle of the plant. Additional work is required to determine the production rates and impact of anthrogenic and biogenic compounds on BLSS performance.

ACKNOWLEDGMENTS

Portions of this research were conducted under the auspices of Biomedical Operations and Research Office, John F. Kennedy Space Center, FL, USA under NASA contract NAS10-12180 by the Dynamac Corporation. We wish to thank Barbara V. Peterson for her assistance in the collection and analysis of gas samples.

REFERENCES

Abeles, F.B., P.W. Morgan, M. E. Saltveit, Jr. *Ethylene in Plant Biology*, 2nd edition. Academic Press, New York (1992).

Aharoni, Y. and G.J. Stadelbacher. The Toxicity of Acetaldehyde Vapor to Postharvest Pathogens of Fruits and Vegetables. *Phytopathology* **63**: 544-545. (1973)

Batten, J.H., B.V. Peterson, E. Berdis, and R.M. Wheeler. Biomass Production Chamber Analysis of Wheat Study (BWT931). *NASA Technical Memorandum* **109192** (1993).

Batten, J.H., G.W. Stutte, and R.M. Wheeler. Effect of Crop Development on Biogenic Emissions From Plant Populations Grown in Closed Plant Growth Chambers. *Phytochemistry.* **39**, pp. 1351-1357 (1995).

Batten, J.H., G.W. Stutte, and R.M. Wheeler. Volatile Organic Compounds Detected The Atmosphere Of NASA's Biomass Production Chamber. *Adv. Space Res.* **18 (4/5)**: pp 189-192 (1996).

Berdis, E., B.V. Peterson, N.C. Yorio, J. Batten and R.M. Wheeler. Development of a Sparging Technique for Volatile Emissions from Potato (*Solanum tuberosum*). *NASA Technical Memorandum* **109199** (1993).

Buoni, C., R. Coutant, R. Barnes, and L. Slivon. Space Station Atmospheric Monitoring Systems. *Proceedings of the 37tsh International Astronautical Congress, Innsbuck, Austria.* **Paper A87-15845.** (1989).

Buttery, R. G. and L. C. Ling. Volatile Components of Tomato Fruit and Plant Parts: Relationship and Biogenesis. in *Bioactive Volatile Compounds From Plants.* edited by R. Teranishi, R.G. Buttery and H. Sugisawa. pp 23-34. American Chemical Society. Washington, D.C., (1993).

Charron, C.S., D.J. Cantliffe, R.M. Wheeler, A. Manukian, and R.R. Heath. Photosynthetic Photon Flux, Photoperiod, and Temperature Effects on Emissions of (Z)-3-Hexenal, (Z)-3-Hexenol, and (Z)-3-Hexenyl Acetate from Lettuce. *J. Amer. Soc. Hort. Sci.* **121**: 488-494. (1996).

Duke, J. A. *Handbook Of Phytochemical Constituents of GRAS Herbs and Other Economic Plants.* CRC Press, Boca Raton, FL pp. 272-277. (1992).

Edeen, M.A., J.S. Dominick, D.J. Barta and N.J.C. Packham. Control of Air Revitalization Using Plants: Results of the Early Human Testing Inititative Phase I Test. *26th Int. Conf. on Environmenta. Systems,* Monterey, CA **SAE Technical Paper 961522** (1996).

EPA. The Determination of Volatile Organic Compounds (VOCs) in Ambient Air Using SUMMA ® Passivated Canister Sampling and Gas Chromatographic Analysis. *EPA Method TO14* (1987).

Gitelson, J.I, and Y. N. Okladnikiv. Man as a Component of a Closed Ecological Life Support System. *Life Support and Biospherics.* **Vol 1(2)**: 73-81 (1994).

Heath, R.R. and A. Manukian. Development and Evaluation of Systems to Collect Volatile Semiochemicals from Insects and Plants Using a Charcoal-infused Medium for Air Purification. *J. Chemical Ecology* **18:** 1209-1226 (1992).

Heerwagen, J.H. and G. Orians. Adaptations to Windowlessness: A Study of the Use of Visual Decor in Windowed and Windowless Offices. *Environment and Behaviour* **18:** 623-639 (1986).

Leban, M.I., and P.A. Wagner. "Space Station Freedom Gaseous Trace Contaminant Load Model Development" *Proceedings of 19th Intern.Conference on Environmental Systems,* San Diego, CA **Paper No. 891513** (1989).

NASA, Office of Safety and Mission Quality. *Flammability, Odor, Offgassing, and Compatibility Requirements and Test Procedures for Materials in Environments that Support Compustion.* **NHB 8060.1C** (1991).

NASA, *Early Human Testing Initiative Phase I: Final Report.* **JSC-33636** (1996).

NASA, *Spacecraft Maximum Allowable Concentrations for Airborne Contaminants.* NASA Technical Report **JSC-20584** (1990).

Nechitailo, G.S., and A.L. Mashinsky. *Space Biology: Studies on Orbital Stations*, Translated by N. Lyubimov. Mir Publishers, Moscow (1993).

Sharkey, T.D. and E.L. Singsaas. Why Plants Emit Isoprene. *Nature* **374**: 769 (1995).

Sharkey, T.D., E.A. Holland, H.A. Mooney. *Trace Gas Emissions by Plants.* Academic Press, New York (1991).

Song, J., R. Leepipattananit, W. Deng, R.M. Beaudry. Hexanal Vapor is a Natural, Metabolizable Fungicide: Inhibition of Fungal Activity and Enhancement of Aroma Biosynthesis in Apple. *J. Amer. Soc. Hort. Sci.* **121:** 937-942 (1996).

Stutte, G.W. and B.V. Peterson. 1996. Biogenic Volatile Organic Compound Production by Soybean and Tomato. *Proc. Plant Growth Regul. Soc. Amer.* Calgary, Alberta, Canada. **23:**295-300 (1996).

Stutte, G.W., R.M. Wheeler and B.V. Peterson. Effectiveness of Activated Charcoal Filters for Atmospheric Purification of Closed Environments. *Proc. American Institute of Aeronautics and Astronautics Conf. on Life Sciences and Space Medicine* Houston, TX, pp. 107-108 (1995).

Terskov, I.A., I.I. Gitelson, G.G. Kovrov et al., *Closed System: Man-Higher Plants (Four Month Experiment). NASA TM 76452* (1981).

Ulrich, R.S. and R. Parsons. Influences of Passive Experiences With Plants on Individual Well-Being and Health. In, *The Role of Horticulture in Human Well-Being and Social Development*. Edited by Relf, D. pg 93-105,Timber Press, Portland, OR (1992).

Vaughn, S.F. Natural Products from Plants that Influence Microorganisms. *Proc. Plant Growth Regulator Society of America.* **22**: 16-22 (1995).

Wheeler, R.M., B.V. Peterson, J.C. Sager, and W.M. Knott. Ethylene Production by Plants in a Closed Environment. *Adv. in Space Res* **18 (4/5)**: 193-196. (1996)

Wydeven, T. and M.A. Golub. Generation Rates and Chemical Compositions of Waste Streams in a Typical Crewed Space Habitat. **NASA TM 102799** (1990).

Zlotopolsk'ii and T.S. Smolenskaja. *Gas Filter for Space Orangery [Greenhouse]*),. 26th International Conference on Environmental Systems, Monterey , CA **SAE Technical Paper 961411** (1996).

Pergamon

Adv. Space Res. Vol. 20, No. 10, pp. 1923–1926, 1997

Printed in Great Britain
0273-1177/97 $17.00 + 0.00

PII: S0273–1177(97)00626–1

EFFECT OF WIND VELOCITY ON ETHYLENE RELEASE FROM LETTUCE PLANTS

A. Tani and M. Kiyota

Collage of Agriculture, Osaka Prefecture University, Sakai, Osaka, 593, Japan

ABSTRACT

Effect of wind velocity on ethylene release rate of intact lettuce plant was investigated. Lettuce plants were grown at wind velocities of 0.1, 0.4, 0.8, and 1.4 m s^{-1} for 25 to 33 days and then used for ethylene measurement. When ethylene release rate of the plants grown at a wind velocity of 0.1m s^{-1} was measured at wind velocities of 0.2, 0.6 and 1.0m s^{-1}, the rate was not affected by wind velocity. This result indicates that ethylene diffusion from lettuce leaf to atmosphere is not affected by boundary layer conditions. When ethylene release rate of the plants grown at wind velocities of 0.1, 0.4, 0.8 and 1.4 m s^{-1} was measured at the same wind velocity as growing conditions, the rate was scarcely increased by high velocity of wind. A strong wind (4.0 m s^{-1}), which induced wounding damage in small areas of the leaves, had no measurable effect on a ethylene release of the whole plant.

INTRODUCTION

The trace gas, ethylene, is released from plants and can accumulate in closed cultivation facilities in CELSS (Wheeler et al., 1996 ; Tani et al., 1996a). In order to predict ethylene concentration in a closed plant cultivation facility and in order to determine the capacity of an ethylene removal system, it is necessary to measure the ethylene release rate of growing plants and to establish the relationship between ethylene release rates and environmental conditions, including physical stress.

In the present study, we focused on wind velocity, because wind is likely to be unevenly distributed over plant cultivation beds, especially in large-scale facilities. It is well known that increasing wind velocity increases leaf boundary layer conductance to CO_2 and water vapor. Therefore, photosynthesis is promoted by the proper velocity of wind. But excess velocity of wind leads to stomata closure, decrease in photosynthetic rate, and morphological changes of plants. The wind also can lead to physical stress to leaves through vibration, mutual touching and wounding (Mitchell et al., 1975). These factors may stimulate ethylene release from plants (Yang and Hoffman, 1984).

We used lettuce as a plant material and investigated the effect of wind velocity on ethylene release from growing lettuce.

MATERIALS AND METHODS

Lettuce (*Lactuca sativa* L. cv. Okayama) plants were grown on rockwool cubes (10×10×10 cm) under continuous lighting with cool white fluorescent lamps (three band type, Matsushita Co. ltd.). The photosynthetic photon flux (PPF) was 150 μmol $m^{-2}s^{-1}$. Air temperature and relative humidity were 24-26 ℃ and 60-70 %, respectively. The wind velocity applied to plants was 0.1, 0.4, 0.7, and 1.4 m s^{-1}. The plants were grown under these conditions for 25-33 days and then used for ethylene measurement.

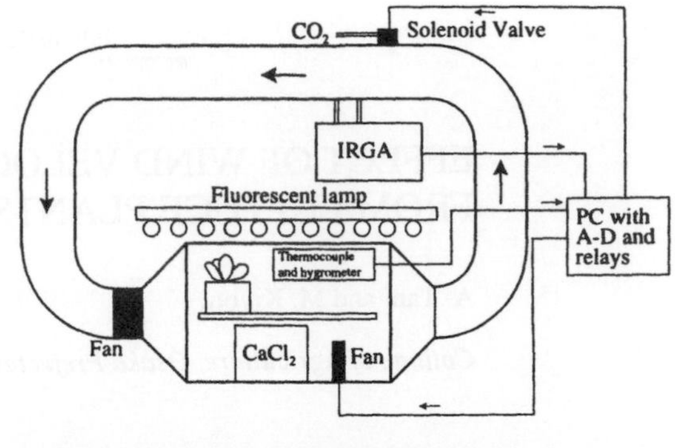

Fig. 1. Schematic diagram of a closed type chamber used for ethylene measurement

Figure 1 shows the schematic diagram of the closed chamber used for ethylene measurement. The chamber was made from polyvinyl-chloride sheets. The growing area and total volume of the closed chamber were 2300 cm^2 and 176 L. In this chamber, it was possible to control environmental factors such as CO_2 concentration, air temperature, relative humidity, and wind velocity. Wind velocity was controlled by controlling the input voltage to the fan incorporated in the chamber. The number of air exchanges per hour within the chamber was 0.016 h^{-1}. Rates of ethylene release, photosynthesis and transpiration of lettuce plants were obtained by measuring the concentrations of ethylene, CO_2 and water vapor, respectively, in the closed chamber in which three lettuce plants were grown for 3 to 4 hours. Photon flux, temperature and humidity during each experiment in the closed chamber were similar to the growing conditions.

Ethylene release rate is calculated as :

$$\Delta R_E = \Delta C_E + \Delta L_E + \Delta Ad_{Wall} + \Delta Ad_{CaCl2}$$

where ΔR_E is ethylene release rate of intact plant (mol s^{-1}), ΔC_E is ethylene increase rate in chamber (mol s^{-1}), ΔL_E is ethylene leakage rate (mol s^{-1}), ΔAd_{Wall} is ethylene adsorption rate by inside wall (mol s^{-1}) and ΔAd_{CaCl2} is ethylene absorption rate by $CaCl_2$ (mol s^{-1}) which was used to absorb water vapor and control humidity inside the chamber. The sum of ΔL_E, ΔAd_{Wall} and ΔAd_{CaCl2} was less than 12% of ΔC_E in all measurements (Tani et al., 1996b).

The ethylene was collected on an absorbent (Porapack Q) at a liquid oxygen temperature, in order to concentrate the small amounts of ethylene, before being introduced into a gas chromatograph equipped with FID (Tani et al., 1996b).

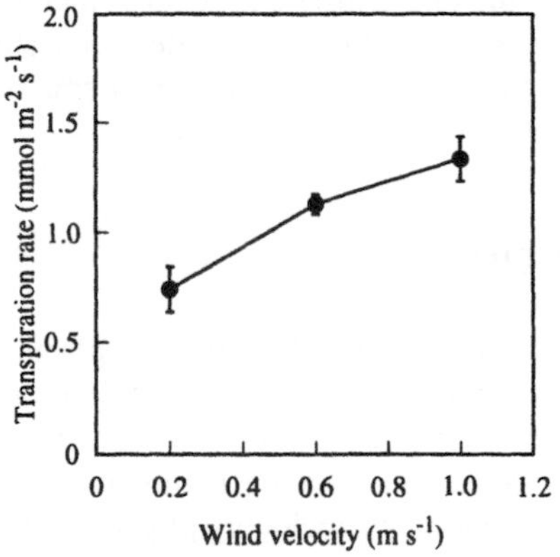

Fig. 2. Effect of wind velocity during measurement on transpiration rate of intact lettuce plants. The plants were grown at wind velocity of 0.1 m s^{-1} for 25 to 33 days and then used for measurement.

RESULTS AND DISCUSSIONS

In the first measurements, we used the plants grown at low velocity of wind and measured ethylene release rate of lettuce plants at different wind

Table 1 Stomatal conductance (mol H_2O $m^{-2}s^{-1}$) of expanded lettuce leaves grown at different wind velocities.

Expanded leaf	Wind velocity ($m\ s^{-1}$) 0.1	0.4	1.4
Young	0.19±0.03	0.11±0.04	0.08±0.04
Old	0.08±0.02	0.06±0.02	0.03±0.01

velocities. The purpose of the measurement was to evaluate the contribution of leaf boundary layer conductance to ethylene diffusion from leaf to atmosphere. Figure 2 shows the relationship between transpiration rate of lettuce and wind velocity. Transpiration rate increased with wind velocity, which suggests that leaf conductance, which is boundary layer conductance plus stomatal conductance, was greater at higher wind velocity. Figure 3 shows the relationship between ethylene release rate and wind velocity. Ethylene release rate was not affected by wind velocity. This result indicates that ethylene diffusion from leaf to atmosphere is not affected by leaf boundary layer conditions.

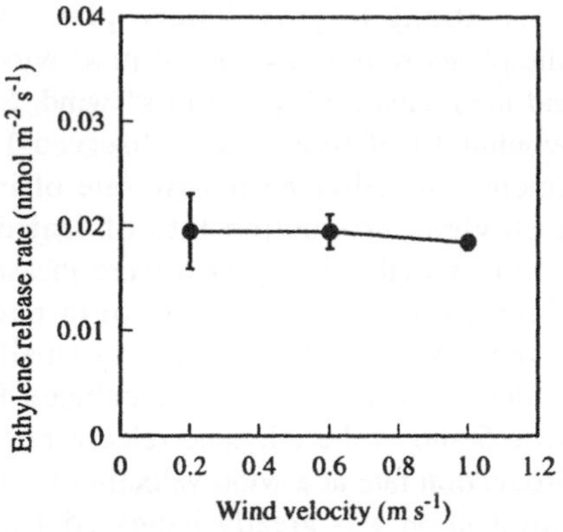

Fig. 3. Effect of wind velocity during measurement on ethylene release of intact lettuce plants. The plants were grown at a wind velocity of 0.1 $m\ s^{-1}$ for 25 to 33 days and then used for measurement.

In the second measurements, plants grown at different wind velocities were studied. Plants were subjected to wind velocities during measurements that were the same as the wind velocities under which plants were grown. Figure 4 shows the relationship between ethylene release rate and wind velocity. The ethylene release rates at wind velocities of 0.4, 0.7 and 1.4 $m\ s^{-1}$ were greater by 20% to 30% than that at velocity of 0.1 $m\ s^{-1}$. At wind velocities of 0.7 and 1.4 $m\ s^{-1}$, leaves of lettuce vibrated and touched each other. However, ethylene release rate was scarcely affected by long term treatment of wind, and was almost constant at any wind velocity in the range from 0.4 $m\ s^{-1}$ to 1.4 $m\ s^{-1}$.

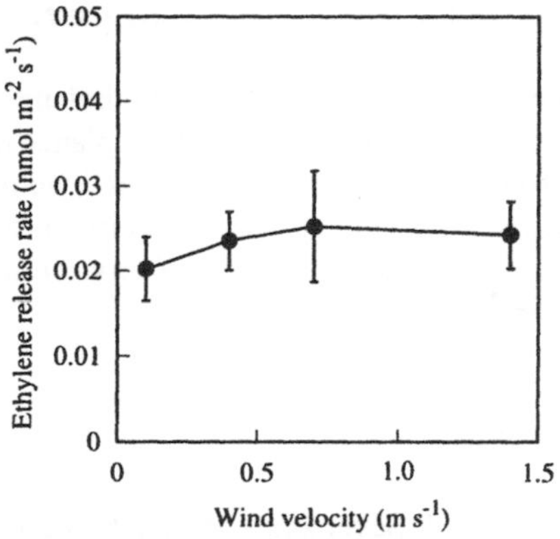

Fig. 4. Effect of wind velocity during growth on ethylene release of intact lettuce plants. The plants were grown at wind velocities of 0.1, 0.4, 0.8 and 1.4 $m\ s^{-1}$ for 25 to 33 days and measurement was made at same wind velocities.

Table 1 shows stomatal conductance to water vapor of expanded leaves of lettuce grown at wind velocities of 0.1, 0.4 and 1.4 $m\ s^{-1}$. Stomatal conductance to water vapor was measured with LI-COR LI6400 photosynthetic meter. The boundary layer conductance was 1.42 mol $m^{-2}s^{-1}$. In plants grown at a wind velocity of 0.1 $m\ s^{-1}$, the highest conductance was observed both in young leaves and in old leaves, followed by 0.4 $m\ s^{-1}$ and finally 1.4 $m\ s^{-1}$, which implies that stomatal opening was small in the plant exposed to high wind velocity for long term. The plants grown at wind velocities of 0.4 to 1.4 $m\ s^{-1}$ had higher ethylene release rates than plants grown at a wind velocity of 0.1 $m\ s^{-1}$, in spite of the lower stomatal conductance. This fact seems to result from the increase of ethylene production activity in leaf tissues at high wind velocity. But, the factors relating ethylene release and stomatal opening are unclear.

In order to investigate the effect of strong wind on ethylene release from lettuce, lettuce plants were subjected to 4.0 m s^{-1} wind for only a few days. Because this wind was too strong to grow lettuce plants from seedling, the plants were grown at 1.4 m s^{-1} wind for 30 days and then subjected to 4.0 m s^{-1} wind. Small areas of wounded leaf tissue were observed 1 day after treatment. The ethylene release rate of intact plant and ethylene production rate of leaf discs excised from this visible injury area were measured. The ethylene production rate was obtained by measuring the ethylene concentration in small test tubes in which six leaf discs were incubated for 3 hours. Figure 5 shows the ethylene release rate and ethylene production rate at a wind velocity of 4 m s^{-1}. Ethylene production was greatly increased 1 and 2 days after treatment with 4 m s^{-1} wind.

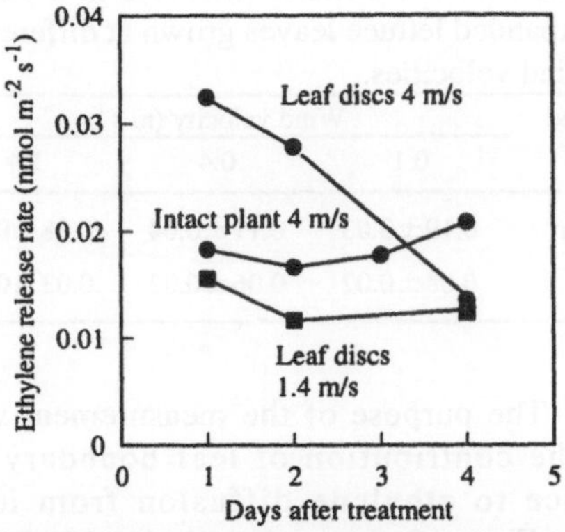

Fig. 5 Effect of wind velocity on ethylene release rate of leaf discs and of intact lettuce plants. The plants were grown at a wind velocity of 1.4 m s^{-1} for 30 days and then subjected to 4.0 m s^{-1} for 1 to 4 days.

But ethylene release rate of the plants did not increase, because the area of injury on the leaves was small (approximately only 1-2 % of total leaf area). This result indicates that the increase in ethylene production rate in leaf discs and excised tissues is not always in agreement with the ethylene release rate of intact plants when environmental stress is imposed on plants. Thus ethylene release rate was not greatly increased by 4.0 m s^{-1} of wind, that led to wounding of the leaves.

From these results, it became clear that ethylene diffusion from lettuce leaves to atmosphere is not affected by boundary layer conditions. When lettuce plants were subjected to high velocity of wind for long term, which induced leaf vibration, mutual touching and wounding of lettuce plants, ethylene release from growing lettuce was only scarcely affected by wind velocity. In this experiment, we used lettuce only. Further experimentation is needed to investigate the effects of environmental stress, including wind, on ethylene release from other candidate plants for CELSS.

REFERENCES

Mitchell, C. A., C. J. Severson, J. A. Wott, and P. A. Hammer. Seismomorphogenic regulation of plant growth. J. Amer. Soc. Hort. Sci., 100, 161 (1975).

Tani, A., M. Kiyota, I, Aiga, K. Nitta, Y. Tako, A. Ashida, K. Otubo, T. Saito, Measurements of trace contaminants in closed-type plant cultivation chambers, Adv. Space Res., 18(4/5) 181 (1996a).

Tani, A., M. Kiyota, and I. Aiga, Studies on trace contaminant accumulated in closed system -Characteristics of ethylene evolution from intact lettuce plant-, Environ. Cont. Biol., 34, 29 (1996b). (In Japanese with English summary)

Wheeler, R. M., B. V. Peterson, J. C. Sager, and W. M. Knott, Ethylene production by plants in a closed environment, Adv. Space Res., 18(4/5), 193 (1996).

Yang, S. F., and N. E. Hoffman, Ethylene biosynthesis and its regulation in higher plants, Ann. Rev. Plant. Physiol., 35, 155 (1984).

Pergamon

Adv. Space Res. Vol. 20, No. 10, pp. 1927–1930, 1997

Printed in Great Britain
0273-1177/97 $17.00 + 0.00

PII: S0273-1177(97)00627-3

IMPAIRED GROWTH OF PLANTS CULTIVATED IN A CLOSED SYSTEM: POSSIBLE REASONS

J. I. Gitelson, L. S. Tirranen, E. V. Borodina, V. Ye. Rygalov

Institute of Biophysics (Russian Academy of Sciences, Siberian Branch), Krasnoyarsk 660036, Russia

ABSTRACT

Plants in experiments on "man-higher plants" closed ecosystem (CES) have been demonstrated to have inhibited growth and reduced productivity due to three basic factors: prolonged usage of a permanent nutrient solution introduction into the nutrient medium of intra-system gray water, and closure of the system. Gray water was detrimental to plants the longer the nutrient solution was used. However, higher plant growth was mostly affected by the gaseous composition of the CES atmosphere, through accumulation of volatile substances.

BACKGROUND

The problem of "soil fatigue" (inhibited growth and development of plants, reduced harvest and deterioration of its quality when cultivated for a long time in one place) has long been known, since De Candolle (1813). Suppressed growth and reduced harvest of hydroponic monocultures using recirculating nutrient solution and the same substrate has been noted by many authors (Muller, 1971; Kalmykova, 1974;Lisovsky, 1979).

Closure of a biological life support system with higher plants involves long-term employment of the same nutrient solution and substrate. Impaired condition and reduction in productivity of plants growing in a closed ecological system (CES) can reduce, or terminate, the necessary functions of the higher plants such as gas, water, and food regeneration, and with the absence of other photoautotroph components, can make further existence of the ecosystem impossible.

This work makes an attempt to investigate the causes of inhibited growth and reduced productivity of plants in CES. The prolonged use of nutrient solution, the addition of gray water, and the gaseous composition of the atmosphere were assumed to be the possible reasons for inhibited growth, and studied in different experiments.

POLYCULTURE

Plants in experiments without system closure, but with the same solution did not appear to have changes in growth and condition for the first two months. However, by the end of the fourth month of using the same solution, the productivity of radish (wet mass of roots) dropped by 30% and wheat (wet mass of seeds) by almost 50%. Long use of the same solution by vegetables in polyculture affected the harvest less than that with wheat in monoculture.

PERMANENT SOLUTIONS

Since the prolonged use of solution built up organic substances, owing to root emissions, root residue, and microorganism metabolites, numbers of microorganisms in the medium increased 50 to 300%. It may be deduced that accumulated metabolites, of plant and microbial origin, had an effect on productivity of plants and growth of seedlings. Some authors (Ivanov, 1973) have suggested that metabolism products, mostly physiologically active substances, react with roots and affect their biochemical processes. Excesses of physiologically active substances may inhibit growth of plants, sometimes resulting in their death, especially at the plantlet phase.

GRAY WATER ADDITIONS

Long-term experiments with CES, particularly wheat growing on a nutrient solution with gray water, demonstrated inhibited growth of plants in the first months. By the end of the fourth month stems and roots had ceased to elongate and the seed harvest dropped to zero (Table 1). The impaired condition of plants with additions of sanitary and hygiene water to the nutrient solution seems to be due to the accumulation in the medium of substances toxic to plant growth; products of plant metabolism, microorganisms, and supply of organic substances with the sanitary and hygiene water. The continued supply of the intrasystem sanitary and hygiene water caused the build up organic substances, increasing numbers of all groups of microorganisms, and alterations in the microbial complex of the higher plants component ecology that had formed prior to closure. Especially drastic was the increase of bacteria; the colon bacillus group, proteus group, lactic acid, and yeasts, indicating contamination of the medium. An equilibrium between supply of organic substances in sanitary and hygiene water and species composition of microorganisms utilizing these substances, did not occur. Thus, organic substances did not mineralize completely and were not completely assimilated by the microflora.

Table 1. Wheat grain yield of successive crops grown in solution provided with varying quantities of gray water for varying periods of time.

Planting	Growth (days)			Treatment with gray water		Yield of grain (kg/m2)
	Before closure	After closure	Total	Rate (1/day)	Period (days)	
A	63	0	63	None		1.182
B	51	12	63	None		1.109
C	42	21	63	None		1.283
D	33	30	63	None		1.292
E	24	39	63	24-36	16	1.033
F	15	48	63	16-36	16	0.807
G	6	57	63	7-36	16	0.742
H	0	63	63	1-36	16	0.739
I	0	63	63	1-33	16	0.803
J	0	63	63	1-24	16	0.824
K	0	63	63	1-16	16	
				31-36	32	0.079
L	0	63	63	1-8	16	
				23-36	32	0.000
M	0	63	63	15-36	32	0.000
N	0	63	63	5-36	32	0.000

MICROBIAL INTERACTIONS

To identify the microorganisms whose metabolites were affecting growth of plants, we studied the effect of cultural fluids of the dominating species of bacteria isolated from different part of wheat and from the nutrient solution. Out of 85 strains under investigation, inhibiting capability was found in 20% of the strains (*Pseudomonasfluorescens, Xanthomonas glycines, Erwinia sp., Pseudomonas denitrificans* (not all strains), *Flavobacterium fucatum, Micrococcus luteus, Bacillus subtilis* and some others. Stimulating abilities were found in 3% of the strains (*Nocardia restricta, Pseudomonas putrefaciens, Pseudomonas ambigua)*. Some strains had varying (positive or negative) effects on growth of seedlings depending on dilution of the cultural fluid. High dilution stimulated growth, whereas low dilution inhibited growth.

In addition, we pricked wheat leaf blades to study the ability of plant leaves to form necrotic zones and the ability to produce vitamins of "B" group. We used strains dominant in CES experiment with wheat grown on the nutrient solution with sanitary and hygiene water. Out of 90 freshly isolated bacterial strains checked for necrotic zone formation, two (*Erwinia sp.*, close to *Erwinia carotovora*) caused death of wheat seedlings, three strains (the same species as above and two strains of *Pseudomonas putida*) formed necrotic patches at pricking sites. In eight strains responses were questionable, and the remaining strains were negative.

When checking the vitamin-forming ability of 94 bacterial strains, many microbes were found to produce one or two vitamins, mostly biotin, pantothenic or nicotinic acids and rarely, five or six. Among the cultures, that produced three or four vitamin-inhibitors, stimulation of growth of seedlings occurred more frequently. The strains that caused complete death of seedlings, by applying bacterial suspension at prick sites, did not produce vitamins of the "B" group. Strain No. 534b was a strong inhibitor of seedling growth. Wheat seeds soaked in this bacterial cultural fluid diluted 50, 100 and 200 times did not germinate. Attempts to find the cause of the toxic effect of this bacteria strain have not met with success.

ATMOSPHERIC CLOSURE

The commonly found reduction of plant productivity in CES experiments cannot be accounted for solely by the effect of long use of solution and supply of sanitary and hygiene water. Other factors, such as gas composition of the system atmosphere might also be contributing to this reduction. Productivity in CES experiments compared to that in outside open systems have demonstrated that the crop yield in experiments without closing the system, higher than in CES experiments. Productivity of radish, for example, by the end of a four month-period of utilizing the solution with plants grown outside CES was 4 kg/m^2 of wet edible biomass, while in CES at the same period it was not more than 2.2 kg/m^2, and two months after the system was opened in the same experiment it increased to 4 kg/m^2. The yield of wheat seeds in CES experiments was also lower.

Seedling growth, in terms of wet mass and the length of stems and roots, on solutions continuously recycled in CES and grown outside, did not show effects of system closure. The effect of solution on seedling growth was practically identical. The effect of solutions outside was checked under laboratory conditions where such factors as anaerobic medium and gas composition of the system atmosphere were not taken into account, therefore this test is not satisfactory enough to make conclusions about the effect of system closure.

Volatile products of both plants and of microorganisms (Stotzky, 1976; Sprecher, 1983; Wilkins and Scholl, 1989)) and of other components of the system, such as man and the technological equipment mounted inside the closed ecosystem, affect the gas composition of the system atmosphere and therefore the condition of plants. Experiments showed that volatile metabolites of microorganisms can build up in the habitat, dissolve in the water from the atmosphere, and retain their biological activity for many days (Kovrov and Tirranen, 1982). It is common knowledge that volatile products produced by microorganisms may be toxic to plants (Kholodny, 1957). This statement is supported by our observations. Experiments demonstrated inhibiting effects of some volatile metabolites produced by microorganisms on the growth of wheat seedlings. Growth of seedling roots was inhibited more than of stems (Table 2).

Table 2. Growth of wheat seedlings following inoculation of nutrient solutions with different bacterial strains

	Length (% of untreated)	
Bacterial Strain	Roots	Stems
19	28.8	51.5
14	32.5	57.4
18	36.3	65.9
36a	43.6	56.9
26a	44.8	86.7
6	46.8	72.3
22	51.9	94.6
30a	52.7	81.3
43	53.7	86.8
44	73.8	72.8

SUMMARY

Thus our studies showed that inhibition of plant growth and reduction of their productivity in a closed system are due to the effect of a number of factors. Among them are sanitary and hygiene water use, and closure of the system atmosphere. These factors had the greatest effect on condition and yields of the cultures.

We think that polyculture is advantageous over a monoculture; the microbial complex of polyculture, is more stable, the species composition is more diverse, it functions with more reliability, and the crop yields are larger.

Up-to-date automatic methods of research combined with traditional techniques can produce more objective information about the mechanism and dynamics of microflora, and the ecological consequences of altered condition for the higher plants in CES. The question of which is the cause and which is the effect, for the changes in the microflora and plant growth, has not been resolved and invites further research efforts.

REFERENCES

Lisovsky, G.M. ed. Closed System: Man-Higher Plants, Novosibirsk, "Nauka", 159 p. (1979).

DeCandolle A.P., Theorie Elemantaire de la Botanique. Paris, 1813 (cited by Grummer G. Allelopathy, "Inostrannaya Literatura", 1957)

Ivanov V.P., Plant Emissions and their Significance in the Life of Phytocenoses, Moscow: Nauka, 296 p. (1973)

Kalmykova N.A., Microorganisms as a Factor of Soil Toxicity under Permanent Tilling of Agricultural Plants, Autoreferat of Candidate's Dissertation, Leningrad (1974)

Kholodny N.G., On Physiological Effect of Volatile Organic Substances on Plants, Selected Works, Kiev: Izd-vo AN USSR, pp. 338-340 (1957)

Kovrov B.G., Tirranen L.S. Accumulation and Retention of Volatile Organic Substances Emitted by Bacteria, Mikrobiologiya, vyp. 6, pp. 905-909 (1982)

Muller H., Preising F., Unterglasge Musenbau (Tomaten, Salatgurken, Einlegegurken) Unterglasgemusenbau, Bd. 1, S. 108-164 (1971)

Sprecher E., Hanssen H-P., Distribution and Strain-Dependent Formation of Volatile Metabolites in the Genus Ceratocystis, Ant. Van Leeuwenhoek, V. 49, No. 4-5 (1983)

Stotzky G., Schenk S. Volatile Organic Compounds and Microorganisms, CRC Crit. Revs Microbiol., V. 4, No. 4, pp. 333-382 (1976)

Wilkins C.K., Scholl S., Volatile Metabolites of some Barley Storage Molds, Int. J. Food Microbiol., V. 8, No. 1, pp. 11-17 (1989)

Pergamon

Adv. Space Res. Vol. 20, No. 10, pp. 1931–1937, 1997

Printed in Great Britain
0273-1177/97 $17.00 + 0.00

PII: S0273-1177(97)00628-5

DENSITY AND COMPOSITION OF MICROORGANISMS DURING LONG-TERM (418 DAY) GROWTH OF POTATO USING BIOLOGICALLY RECLAIMED NUTRIENTS FROM INEDIBLE PLANT BIOMASS

J. L. Garland*, K. L. Cook*, M. Johnson**, R. Sumner** and N. Fields**

**Dynamac Corporation, Mail Code DYN-3, Kennedy Space Center, FL 32899*
***The Bionetics Corporation, Mail Code BIO-4, Kennedy Space Center, FL 32899*

ABSTRACT

A study evaluating alternative methods for long term operation of biomass production systems was recently completed at the Kennedy Space Center (KSC). The 418-day study evaluated repeated batch versus mixed-aged production of potato grown on either standard 1/2-strength Hoagland's nutrient solution or solutions including nutrients recycled from inedible plant material. The long term effects of closure and recycling on microbial dynamics were evaluated by monitoring the microbial communities associated with various habitats within the plant growth system (i.e., plant roots, nutrient solution, biofilms within the hydroponic systems, atmosphere, and atmospheric condensate). Plate count methods were used to enumerate and characterize microorganisms. Microscopic staining methods were used to estimate total cell densities. The primary finding was that the density and composition of microbial communities associated with controlled environmental plant growth systems are stable during long term operation. Continuous production resulted in slightly greater stability. Nutrient recycling, despite the addition of soluble organic material from the waste processing system, did not significantly increase microbial density in any of the habitats.

INTRODUCTION

The dynamics of microbial communities is an important component of the overall stability of Advanced Life Support systems given the potential costs of human or plant disease. Previous studies have indicated that although significant numbers of microorganisms are present in prototype biomass production systems (i.e., up to 10^{11} bacterial cells g dw root^{-1}), both the density and composition of microbial communities become relatively stable (Strayer 1991, Strayer 1994, Garland 1994). Recycling nutrients, via the effluent from aerobic bioreactors processing inedible plant biomass, also was shown to have little effect on microbial communities (Mackowiak et al. 1996). Bioreactor effluents appear to contain recalcitrant dissolved organic material that does not stimulate microbial activity. Labile dissolved organic material in the inedible biomass, which can increase microbial respiration and denitrification in the root zone, are degraded within the bioreactor prior to addition to the plant growth system (Mackowiak et al. in press).

Previous assessments of stability were based on batch plant growth tests lasting no longer than 100 days. Microbial stability during longer operation of plant production systems has not been evaluated. One concern is excessive microbial growth (i.e., biofouling) over time, particularly when recycling nutrients using bioreactor effluents. We examined the density and composition of microorganisms in all major habitats (root zone or rhizosphere, nutrient solution, biofilms on hardware surfaces within the nutrient delivery systems, atmospheric condensate, and air) during a 418-day test of potato production in a large, prototype biomass production system to assess the question of long term stability. The study evaluated the effects of plant production mode (repeated batch growouts vs. mixed-age production resulting from staggered planting) and nutrient recycling procedures (control inorganic-salt based hydroponic medium vs. recycled bioreactor effluent).

METHODS

Potato (*Solanum tuberosum* L. cv. Norland) was hydroponically grown in the large controlled environmental chamber called the Biomass Production Chamber (BPC) at the Kennedy Space Center (KSC). The chamber contains two compartments with separate atmospheres. Each compartment contains two nutrient delivery systems (or levels) consisting of sixteen 0.25 m^2 plant growing trays. Plants were grown in either a batch mode (all sixteen trays planted simultaneously) or mixed-age mode (4 trays harvested and replanted every 26 days). All plants were grown under the following environmental conditions: 12-h light / 12-h dark photoperiod, 20/16°C (light/dark) air temperature, 70% relative humidity, and average photosynthetic photon flux (PPF) of 814 $\mu mol\ m^{-2}\ s^{-1}$. The 418-day study consisted of 4 repeated batch growouts of 104 days in the upper compartment and sixteen 26-day harvests in the lower compartment. The first three harvests from the mixed-age treatment were of immature plants (26, 52, and 78-day old plants). A single compartment was grown under batch or mixed-aged mode. One level from each production mode was supplied with either a control nutrient solution (1/2-strength Hoagland's) or recycled solution (bioreactor effluent). Solutions were used continuously, without replacement, for the entire 418 days. For the recycled solutions, nutrients were regenerated from inedible biomass using microbial degradation within an aerobic, continuously stirred tank bioreactor (CSTR). Details on the design and operation of this reactor are presented elsewhere (Finger and Alazraki 1995, Strayer and Cook 1995). Effluent was filtered through a 0.22 µm pore size filter, and analyzed for chemical composition using colorimetric methods for N and P and inductively coupled plasma (ICP) spectrometry for other nutrients. The effluent was amended with reagent-grade salts so that relative concentration of all individual nutrients were similar to those found in the control solution. The bioreactor effluent supplied approximately 50% of the total nutrient requirement of the plant, although the percentage recycled varied among the individual nutrients (Mackowiak et al., this volume). Amended bioreactor effluent or inorganic salt-based stock solution was automatically added to the nutrient delivery systems when conductivity fell below 1.2 $dS\ m^{-1}$. Weekly checks of nutrient concentrations were performed using colorimetric methods and ICP as discussed above.

Rhizosphere (root) samples (a 1 cm x 1 cm section of root mat) were taken every 26 and 52 days from mixed-age and repeated batch treatments, respectively. The repeated batch treatment was sampled less intensively to minimize root damage to immature plants. Root samples were removed from four separate trays. Nutrient solution samples were removed every 14 days from the nutrient tanks from each level. Polyvinyl chloride (PVC) sampling coupons (1 cm x 2 cm) were placed within each nutrient delivery system at the beginning of the experiment and 3 coupons were removed every 100 days. Condensate from each compartment was collected every 14 days from either the collector directly below the cooling coils or a larger holding tank where condensate is stored prior to being recycled to the nutrient delivery tanks. Air samples were collected every 14 days from each compartment and from outside the chamber.

Suspensions of microorganism from root samples and PVC chips were made by shaking the material in filtered sterilized nutrient solution (either control or recycled depending on sample origin) and glass beads for 2 minutes. Total microbial cell density in both the liquid samples (nutrient solution and condensate) and the suspensions (rhizosphere and PVC surfaces) was determined by staining with Acridine Orange (AO) using a modification of the method of Hobbie *et al.* (1977). AO stained cells were filtered onto 0.22 µm black polycarbonate filters and counted using epifluorescence microscopy. To obtain an estimate of viable cell density, the same samples was serially diluted and spread plated onto R2A media. The plates were incubated for 48 hours and then numbers of colony forming units (CFU) were determined. Every 100 days, 25 isolates were randomly selected from each environment (rhizosphere, nutrient solution, condensate, and PVC surfaces) and identified using the Vitek system (BioMerieux, Hazelwood, MO). Sampling of air was accomplished using a Matson-Garvin slit-to-agar sampler connected to gas sample ports located in the air return ducts. Air was sampled at a rate of 0.028 m^3 of air per minute for 30 minutes. Tryptic Soy Agar (TSA) was used to enumerate airborne bacteria, and Inhibitory Mold Agar (IMA) was used to enumerate airborne fungi.

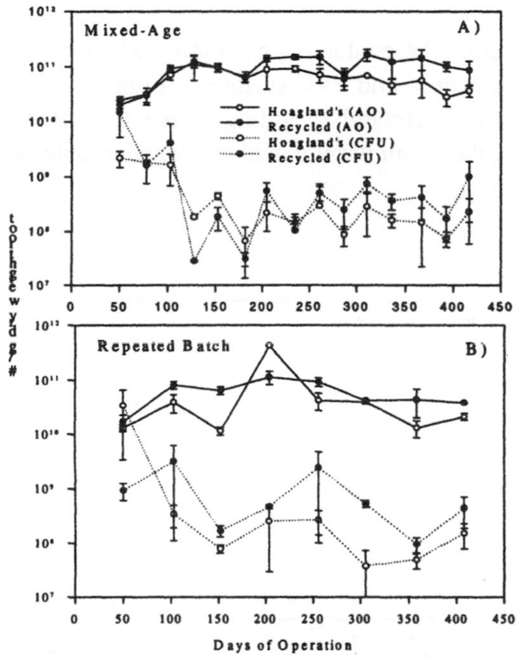

Fig. 1. Bacterial density in the rhizosphere over time in A) mixed-age or B) repeated batch production.

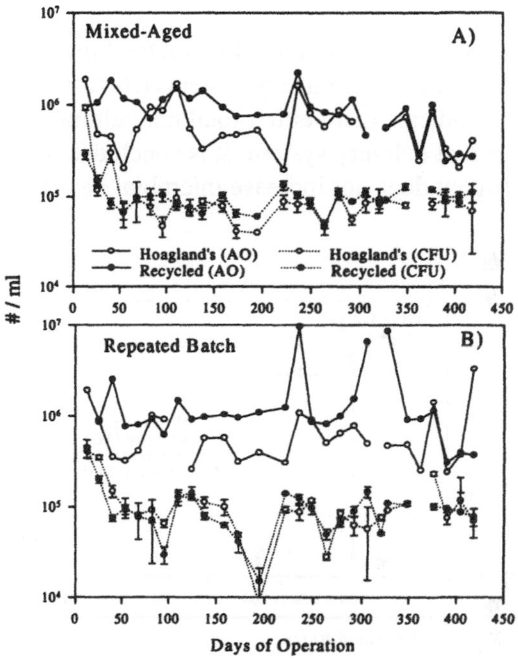

Fig. 2. Bacterial density in the nutrient solution over time in A) mixed-age or B) repeated batch production.

RESULTS AND DISCUSSION

The dynamics of microbial cell density in the nutrient delivery systems (rhizosphere, nutrient solution, and PVC surfaces) followed two major trends (Figures 1-3). First, total density (AO counts) remained relatively constant over the course of the experiment, with the exception of a slight increase during the first 100 days in the rhizosphere as plants matured. Second, culturable cell density (R2A plate counts) decreased during the first one hundred days. The first finding indicates that total microbial cell density reaches a steady state within the nutrient delivery systems, and that excessive biofouling does not appear to be a problem. The second observation indicates that the percentage of culturable cells decreases during the initial period of operation of the plant growth systems, reaching levels less than 10% in the nutrient solution (Figure 2) and less than 1% in the rhizosphere and on PVC surfaces (Figures 1 and 3). The reason for the decrease in culturability is unclear. It is possible that it is related to increased complexity in the communities with time and the concomitant inability of the organisms to grow in isolation from one another. The lower culturability in samples associated with root and PVC surfaces supports this hypothesis since biofilm communities are extremely interactive and many organisms within the biofilm require products of other organisms for growth (Characklis *et al.*, 1990). The higher % culturability in the condensate samples (Figure 4) also supports this hypothesis since these samples were dominated by only a few types of organisms (Table 5, and see below for further discussion of microbial characterization).

Only bacterial cell densities in samples from the nutrient delivery system and condensate are reported since fungi were not observed in appreciable numbers within these samples. This agrees with previous studies that found that fungal densities were several orders of magnitude less than bacterial densities (Strayer 1991).

The effects of nutrient recycling on total microbial density could be evaluated in samples taken from within the nutrient delivery systems (rhizosphere, nutrient solution, and PVC surfaces) (Figures 1-3). Results indicate that neither total nor culturable cell density was affected by adding bioreactor effluent to the nutrient delivery system. It is concluded that addition of the recalcitrant dissolved organic material in the effluent does not increase microbial density.

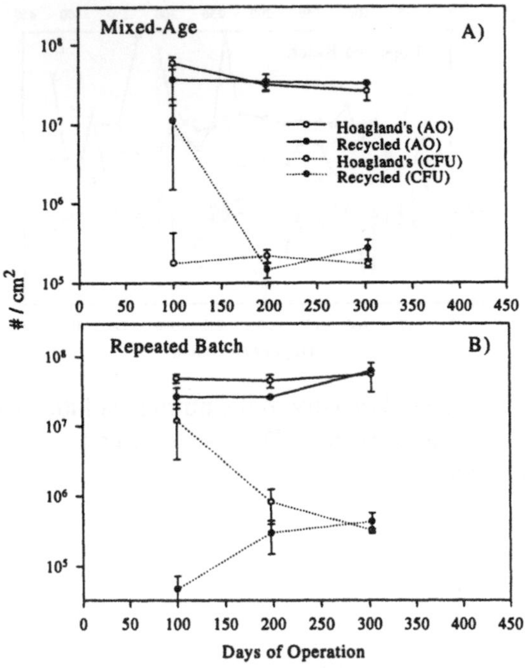

Fig. 3. Bacterial cell density on PVC surfaces within nutrient delivery systems over time in A) mixed-age or B) repeated batch production.

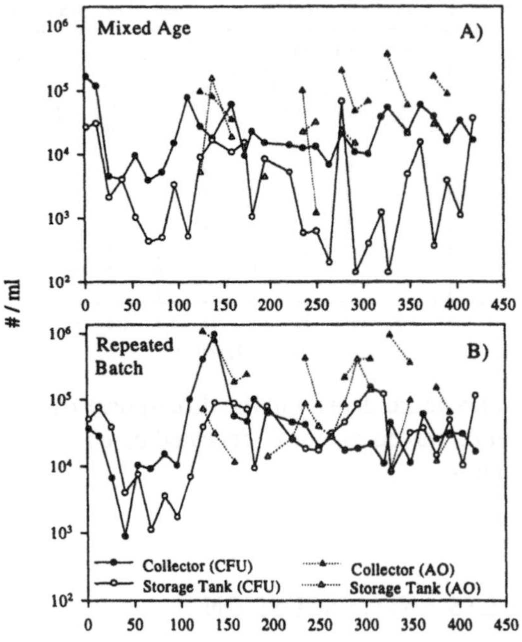

Fig. 4) Bacterial density in the condensate recovery system over time in A) mixed-age or B) repeated batch production.

The effects of mixed-age versus repeated batch production could be evaluated for samples taken from the nutrient delivery systems and from the atmosphere since the production modes were distinctive between the two compartments within the BPC. Neither total nor culturable cell density was consistently different in samples from the nutrient delivery system between repeated batch and mixed-age production (Figures. 1-3), although variation in cell density appeared to be less in mixed-age production for samples from the rhizosphere and nutrient solution. Cell density in the condensate collectors was approximately 10^4 ml^{-1} from both chamber compartments, but cell density was slightly lower in the condensate storage tank from the mixed-age continuous treatment (Figure. 4). The differences in storage tanks between levels may be related to differences in the condensate recovery systems between compartments rather than treatment affects since cell densities in samples of the atmospheric condensate in the collectors were similar.

Bacterial and fungal counts in air samples from both the repeated batch and mixed-age compartments were less than 20 and 200 colony forming units (CFU) m^{-3}, respectively, during the course of the experiment (Figure. 5). Cell densities in air from within the BPC were consistently lower then in air from outside the chamber. Current guidelines for air quality on spacecraft suggest operational levels of 500 total CFU m^{-3}, although a widely sanctioned standard is not available (NASA-STD-3000). General guidelines for indoor air quality suggest that levels of saprophytic airborne contaminants (microbial and fungal) less than 2000 CFU m^{-3} do not constitute a contamination problem (Burge 1987). Given that the levels in the BPC never exceeded 500 CFU m^{-3}, and averaged less than 100 CFU m^{-3}, it appears that microbial contamination of the air from biomass production systems (at least those containing potato) is not a human health problem.

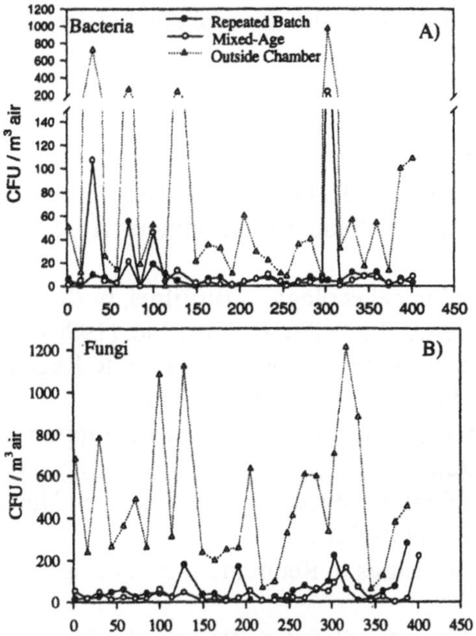

Figure 5. Density of culturable A) bacteria and B) fungi in the atmosphere inside and outside the Biomass Production Chamber.

Table 1. Microbial Characterization[a]

Organism	Percentage of Total Isolates
Rhizosphere	
Pseudomonas sp.	50%
Vibrio sp.	14%
Nonfermenting Gram Negative Bacillus[b]	5%
Flavobacterium sp.	5%
Ochrobactrum anthropii	5%
No identification	7%
Other	14%
Nutrient Solution	
Pseudomonas sp.	45%
Agrobacterum tumefaciens	13%
Nonfermenting Gram Negative Bacillus	10%
Ochrobactrum anthropi	6%
No identification	4%
Other	22%
Condensate	
Pseudomonas sp.	72%
Flavobacterium sp.	8%
No identification	18%
Other	2%
Surfaces	
Nonfermenting Gram Negative Bacillus	47%
Pseudomonas sp.	21%
Flavobacterium sp.	7%
No identification	11%
Other	14%

[a]Listing of predominate isolates
[b]No genus or species name given

A general listing of the major bacteria present in different habitats are presented in Table 1. As previously reported (Strayer 1991, Strayer 1994), organisms from the genus *Pseudomonas* were the most common isolates in rhizosphere, nutrient solution, and condensate collection systems of the BPC . *P. paucimobilis, P. vesicularis,* and *P. pickettii* were the most commonly isolated species. *Flavobacterium sp., Ochrobactrum anthropi,* and *Agrobacterium tumefaciens* made up a lesser portion of the isolates but were found in appreciable quantities in one or more of the habitats sampled. Nonfermenting gram negative bacillus (NFGNB) was the most common characterization given by Vitek to isolates from surface samples. This category (NFGNB) encompasses a broad range of microbial genera (including *Pseudomonas, Moraxella, Flavobacterium,* etc.), but does not identify the isolate to any one of these genera. The high percentage of unidentified or unspecified isolates points to the need for a change in the method of microbial identification. Current identification systems were developed for medical, rather then environmental, research. It may be more beneficial for the purposes of monitoring to base isolate characterization on morphology and biochemical properties rather than nomenclature. Even with improved characterization techniques, the utility of evaluating isolates is limited given the low % culturability in our system. Quantifying the densities of a limited number of specific microorganisms of interest (i.e., plant pathogen or human pathogens) using non-cultural methods (e.g., fluorescent antibodies or gene probes) may be more valuable.

In conclusion, we found that the density and composition of microbial populations within plant chambers growing potatoes reach a steady-state which is maintained throughout long-term cultivation of the crop. The use of processed plant material for nutrient replenishment had little or no effect on microbial cell density and did not result in biofouling. Continuous cultivation of potato appeared to result in a more steady-state environment than batch production, but there was no significant difference in total or viable cell density based on production mode. The decrease in culturability in the rhizosphere, nutrient solution and on PVC surfaces merits further examination.

REFERENCES

Burge, H., M. Chatigny, J. Feeley, K. Kreiss, P. Morey, J. Otten, and K. Petersen, Bioasrosols: Guidelines for Assessment and Sampling of Saprophytic Bioaerosols in the Indoor Environment, *Applied Industrial Hygiene*, **2**, 10-16 (1987).

Characklis, W.C., G.A. McFeters, and K.C. Marshall, Physiological ecology in Biofilms, in *Biofilms*, edited by W.C. Characklis and K.C. Marshall, pp. 341-394, John Wiley & Sons, Inc. (1990).

Finger, B.W. and R.F. Strayer, Development of an Intermediate-scale Aerobic Bioreactor to Regenerate Nutrients from Inedible Crop Residues, SAE Technical Paper Series 941501 (1994).

Garland, J.L., Coupling Plant Growth and Waste Recycling Systems in a Controlled Life Support System (CELSS), NASA Technical Memorandum 107544 (1992).

Garland, J.L., C.L. Mackowiak, and J.C. Sager, Hydroponic Crop Production Using Recycled Nutrients from Inedible Crop Residues, SAE technical paper 932173 (1993).

Garland, J.L., The Structure and Function of Microbial Communities in Recirculating Hydroponic Systems, *Adv. Space Res.*, **14**, 383-386 (1994).

Hobbie, J.E., R.J. Daley, and S. Jasper, Use of Nucleopore filters for Counting Bacteria by Fluorescence Microscopy, *Appl. Environ. Microbiol.*, **33**, 1225-1228 (1977).

Mackowiak, C.L., J.L. Garland, and J.C. Sager, Recycling crop residues for use in recirculating hydroponic crop production, *Acta Horticulturae* (in press)

Mackowiak, C.L., R.M. Wheeler, G.W. Stutte, N.C. Yorio, and J.C. Sager. Use of biologically reclaimed minerals for continuous potato production in a CELSS, *Adv. Space Res.* (this volume)

Mackowiak, C.L., J.L. Garland, R.F. Strayer, B.W. Finger and R.M. Wheeler, Comparison of Aerobically-Treated and Untreated Crop Residue as a Source of Recycled Nutrients in a Recirculating Hydroponic System, *Adv. Space Res.*, **18**, 281-287 (1996).

NASA STD 3000. Man-Systems Integration Standards. Revision B. (1995)

Strayer, R.F., Microbiological Characterization of the Biomass Production of Chamber During Hydroponic Growth of Crops at the Controlled Ecological Life Support System (CELSS) Breadboard Facility, SAE technical paper 911427 (1991).

Strayer, R.F., Dynamics of microorganism populations in recirculating nutrient solutions. *Adv. Space Res.* **14**, 357-366 (1994)

Pergamon

Adv. Space Res. Vol. 20, No. 10, pp. 1939–1943, 1997

Printed in Great Britain
0273-1177/97 $17.00 + 0.00

PII: S0273-1177(97)00629-7

PLANTS–RHIZOSPHERIC ORGANISMS INTERACTION IN A MANMADE SYSTEM WITH AND WITHOUT BIOGENOUS ELEMENT LIMITATION

L. A. Somova*, N. S. Pechurkin*, V. I. Polonsky*, T. I. Pisman* A. B. Sarangova*, M. Andre** and G. M. Sadovskaya***

**Institute of Biophysics (Russian Academy of Sciences, Siberian Branch), Krasnoyarsk 6600036, Russia*
***CEA-Sciences du Vivant, DEVM, CEA/Cadarache, 13108 Saint-Paul-lez-Durance, Saint-Paul-Lez-Durance Cedex France*
****Computing Center Russian (Academy of Sciences, Siberian Branch), Krasnoyarsk 6600036, Russia*

ABSTRACT

The effect has been studied of inoculation of seeds of wheat with two species of rhizospheric microorganisms, - *Pseudomonas fluorescens* and *Pseudomonas putida* - on young plant growth with complete and with nitrogen deficit mineral nutrition.

With complete mineral medium, plants grown from seeds inoculated with bacteria of *Pseudomonas* genus (experiment plants) have been found to have better growth over plants not inoculated with these bacteria (control plants). The experiment plants had increased transpiration and their biomass had higher organic nitrogen content.

With nitrogen deficit medium, the plants inoculated with bacteria and those without them, have not revealed changes in growth. Neither case demonstrated competition of microorganisms with plants for nitrogen sources.

INTRODUCTION

Plants are known to emit through their roots a variety of diverse organic compounds. Substances produced by plants are a source of nutrition for rhizospheric microorganisms. The latter are known to have favorable effects on plants, stimulating photosynthesis and growth processes [Frommel et al., 1991]. This may be due to several reasons: 1) vitamins and phytohormones released by microorganisms [Polonsky, 1995]; 2) production of active substances of antibiotic type inhibiting development of pathogenous fungi [Skilyagina, 1973]; 3) conversion of mineral elements, difficult for assimilation by roots, into easily assimilated forms. The performance of the entire range of these stimulation mechanisms for plants occurs under more or less optimal external conditions. Under the effect of limiting factors, deficient mineral nutrition elements in particular, there can arise a situation when the rhizospheric microorganisms actually compete with the plant roots for the mineral nutrition elements. This may substantially reduce the stimulating effect of rhizospheric microorganisms or retard growth processes in plants.

It is obvious that this effect will be enhanced when a plants nutrition changes from seed to autotroph type. Work on interrelations of rhizospheric microorganisms with plants under deficiency of mineral elements in the environment are few [Diaz et al., 1993; Gifford, 1995; Polyanskya et al., 1994].

This work is concerned with effect of two species of rhizospheric microorganisms of *Pseudomonas* genus: *Pseudomonas fluorescens* and *Pseudomonas putida* on young wheat plants growth under complete mineral nutrition and under deficient nitrogen in man-made sterile soil.

MATERIALS AND METHODS

Krasnoyarskaya 83 summer wheat seedlings were grown and two strains of rhizospheric bacteria: *Pseudomonas fluorescens* and *Pseudomonas putida* were kindly provided by Prof. Panikov (Institute of Microbiology of the Russian Academy of Sciences) and D.E. Polonskaya - head of microbiological laboratory of Krasnoyarsk Agrarian University. Preliminary experiments were conducted to estimate the effect of rhizospheric microorganisms on wheat seed germination. Washed wheat seeds were inoculated with different amounts of bacteria (106, 107, 108, 109 cells in 5 ml of water) and placed in Petri dishes for 5 days. Then, length of roots and green seedlings was measured in all replicates of experiment and control plants. As compared to control, at inoculate concentrations within the range of 10^6- 10^7 cells, the length of roots and shoot of germinating seeds was doubled. With 10^9 cells, seed germination was inhibited compared to control. These preliminary experiments established the amount of bacteria or inoculate size for the main experiment.

Wheat seeds (average mass of 40 mg) were sterilized for 5 minutes with ethanol and 3% solution of hydrogen peroxide at the ratio of 1:1, inoculated with turbidity-standardized bacterial suspension of 5 ml, germinated for two days and sown in 500 ml vegetation vessels in a sterile mode. The vessel bottom was covered with 2-3 cm thick layer of expanded clay aggregate for drainage. Sterile river sand, 7 cm thick, was used for substrate. Seedlings sown on the sand surface were covered with 2-3 cm of sterile expanded clay. To irrigate the plants, each vessel was equipped with a funnel with a cotton-gauze stopper. The plants were irrigated with sterileKnoppe nutrient solution containing a complex of microelements or with identical solution without nitrogen.

Table 1 Data on wheat growth with and without bacterial inoculation.

Bacterial Innoculation	Leaf Size				Plant Biomass, Dry Weight(mg)				Bacterial Mass (Percent of Root Mass)
	Leaf Blade Length (mm)		Leaf Blade Area (cm^2)						
	5th leaf	6th leaf	5th leaf	6th leaf	Shoots	Roots	Whole plant	Shoot/ Root Ratio	
Complete Nutrient Solution									
None (Control)	87.5	21.0	3.1	0.8	88	14	102	6.2	-
Pseudomonas putida	137.5	88.0	5.3	3.8	123	22	145	5.5	0.15
Pseudomonas fluorescens	129.0	78.0	4.6	3.5	118	20	138	5.9	0.15
Nitrogen Deficient Solution									
None (Control)	41.5	16.0	0.9	0.30	70	23	93	3	-
Pseudomonas putida	34.0	14.0	0.78	0.24	71	24	95	2.9	3.6
Pseudomonas fluorencens	38.5	13.5	0.86	0.29	76	22	98	3.4	1.5

The vegetation vessels carrying 12-17 plants each (total number of vessels in the experiment was 12 with 2 vessels per treatment) were in a room with air temperature 19-21 C. Table 1 presents the experiment treatments. The plants were illuminated round-the-clock by fluorescent lamps located at both sides of the plants. Illuminance was about 7 kilolux. Experiment time from sowing to harvesting was 23 days. To standardize experiment conditions, every two days the position of the vessels were randomly interchanged.

In the course of experiment, plant vessels were weighed every day and the decrease of the mass compensated by addition of the Knoppe medium (with or without nitrogen). Physical evaporation of moisture from the substrate surface was determined by weighing control vegetation vessels without plants. Length of wheat leaf blades was measured every two or three days. At the end of experiment, length and width of the green part of leaves was measured, biomass of shoots and roots was determined. Basic forms of nitrogen in the substrate and shoot of the plants' biomass was determined. At the end of experiment microflora of rhizosphere and roots' rhizoplane was counted.

To count the rhisosphere microorganisms, the roots were washed in a glass with a measured amount of phosphate buffer with pH 5.8, and the solution was serially diluted. The roots were then homogenized, serially diluted and inoculated on meat-peptone agar (MPA) and on mineral medium with glucose for*Pseudomonas*. The sowings were incubated in a thermostat at the temperature of 28-30 C and the formed colonies were counted. The number of microorganism cells were calculated on a root dry biomass basis. The mass of microorganisms as a per cent of root mass was also calculated. To convert bacterial numbers into mass values, the dry mass of one bacterial cell was taken to be $2x10^{-14}$ g.

It should be noted that sterilization of seeds was not complete. Control sowings showed that this seed surface was sterile, but inside (after caryopsis was ground) there remained a certain small amount of microorganisms, later considered background. The background amount of microorganisms in controls was 3 orders less than the initial value without sterilization (about 10^3 cells per caryopsis). Results are represented by mean arithmetic values. When counting the microorganisms, the relative measurement error was 6-8% on the average. Plant variation coefficient was 9% for dimension and 5% for biomass.

RESULTS AND DISCUSSION

Growth and Development of Plants and Microorganisms on Complete Mineral Medium

Table 1 presents measurements made at the end of experiment: length and area of leaf blade (5th and 6th leaves), dry weight of biomass of plants and leaf blade and separately - the shoot and roots, nitrogen content in the shoot of plants; shoot to root ratio: and weight of bacteria as per cent of root weight.

The leaves increased in length exponentially during the first week. This was followed by linear growth both in control and in inoculated plants of the experiment. By the end of experiment, the length and area of leaf blades of the 5th and 6th leaves of inoculated plants were found to differ from control plants by 40 to 300%. The total area of all leaves of plants, however, did not yield significant difference between control and inoculated plants. Table 1 demonstrates that the bacteria-inoculated plants increased their shoot biomass by 35-40% as compared to control plants. The difference between the two inoculated treatments was insignificant (10%). The biomass of roots in control plants was less than in inoculated plants.

When grown on complete mineral medium, the total nitrogen content in the shoot biomass was identical in both control and inoculated plants, yet the amount of organic nitrogen of inoculated plants was more than double, i.e. inoculated plants were better in transforming the nitrogen of nitrates into the organic matter of the plant. The control plants had more nitrate nitrogen in the leaves. Appropriate selection of rhizospheric microorganisms can, probably, reduce nitrate content in the plant biomass.

Compared to the initial amount of bacterial strains inoculated to the seeds, the bacteria in the radical part of the plants in inoculated vessels increased by 2-3 orders by the end of experiment In different treatments of the experiment the inoculation of both *Pseudomonas putida* and *Pseudomonas fluorescens* was 1×10^9. The biomass of bacteria was 0.15% of the root mass.

Growth of Plants on Nitrogen-Deficient Medium

The exponential growth of leaves in the treatments with and without nitrogen in the first week of experiment actually did not differ from growth in nature. Later, because of nitrogen deficiency it reached a plateau. As compared to growth on complete medium, the expansion of the 5th and 6th leaves revealed practically no difference between inoculated and control plants, while the linear growth of leaves was found to be retarded. The biomass of both shoot and roots in inoculated plants did not differ from control. The biomass of plants increased mainly owing to internal supply of nitrogen in the seed. The table demonstrates that rhizospheric bacterial strains used in the experiment did not retard growth of plants on nitrogen-deficient medium. Neither the microorganisms nor the plants were found to compete for nitrogen. The plants' growth is retarded by deficient nitrogen but not by the microorganisms.

In control plants, with only the so-called "bacterial background", microorganisms increased by 3-4 orders during the experiment. The inoculated treatments growing with nitrogen deficiency turned out to have 10-20 times more microorganisms based on the per cent of the dry weight of roots than on the complete mineral medium. Seemingly, microbial growth in the root zone was limited by the rate of supply of organic substances from root exudates in both cases. There is little likelihood that the growth of microorganisms was limited by nitrogen. Most probably, they received ample quantity of nitrogen from root exudates. In the case of nitrogen limitation, the shoot/root ratio was practically double that for plants grown on complete mineral medium (6.2 vs. 3). This suggests a higher intensity of processes operating in the soil, more intense flow of nutrients to plant roots, including exudates utilized by the microorganisms. Besides, it cannot be ruled out that these microorganisms could fix atmospheric nitrogen. The ability of the two strains under investigation to fix nitrogen was shown by the authors under laboratory conditions.

It is assumed that stimulation of the plants' growth processes by microorganisms was brought about by vitamins and phytohormones supplied to the plant through roots. These seem to promote: 1) higher synthesis of enzymes required to enhance photosynthetic and respiratory activity of plants; 2) direct hormonal regulation of division and extension of cells in plants; 3) regulation of stomatal aperture by antagonistic hormones of the abscicic acid group.

The first assumption is favored by higher organic nitrogen content and lower nitrate concentration in the plants biomass. The second, by larger size of the upper leaves. The third assumption suggestion is indirectly supported by high level of transpiration per leaf surface unit.

There is previous research supporting the positive role of kinetin, cytokinins and their analogs in enhancing transpiration intensity by elevated stomatal conductivity [Shmatko, 1995]. Such a stimulation of the growth processes in plants by rhizospheric microorganisms is evolutionally justified, since basic "food" for these microorganisms is supplied by exudates of plant roots. This process is known to be positively correlated both with outflow of assimilates from the leaves and roots and with photosynthesis intensity.

Thus, this study has revealed the contribution of microorganisms in the development of a simple man-made system. The research carried out with a man-made system is an essential stage in investigation of CO2 concentration on behavior of biospheric systems.

REFERENCES:

Frommel, M.I., J. Nowak, G. Lazarovits. Growth Enhancement and Development Modifications of In Vitro Grown Potato (*Solanum tuberosum* ssp. tuberosum) as Affected by a Non-Fluorescent *Pseudomonas* sp. Plant Physiology, 96 (3), pp. 928-936 (1991).

Diaz, S., J.P. Grime, P. Harris, J.E. McPherson. Evidence of Feedback Mechanism Limiting Plant Response to Elevated Carbon Dioxide, Nature, V. 364, pp. 616-617 (1993).

Gifford, R. V., Interaction of Carbon Dioxide with Growth-Limiting Environmental Factors in Vegetation Productivity: Implications for the Global Carbon Cycle. In Advances in Bioclimatology, 4(1), pp. 39-43(1995).

Polonskaya, J.E., The Influence of Epiphytic Bacteria on Growth and Nitrogen Absorption of Pine Seedling Roots (Sterile Conditions and Field Test). Abstracts of Papers of the 14th Long Ashton International Symposium, Bristol, England, pp. 13-15 September (1995).

Polyanskaya, L. M., M.H. Orazova, A.A. Sveshnikova, D.G. Zvyagintsev. Effect of Nitrogen on Barley Radical Zone Colonization by Microorganisms, Mikrobiologiya, 63(2), pp. 308-313(1994).

Skilyagina, T.S. On the Nature of Antagonistic Effect of Pseudomonas aeruginosa (Schroeter) Migula on Phytopathogenous Fungi, in: Microflora of Plants and Harvest. Nauka Publishing Co., Novosibirsk, Russia, pp. 235-250 (1973).

Shmatko, I.G., O.O. Stasik, A.B. Kononchuk, I.A. Grigoryuk. Peculiarities of Interaction of Water Exchange and CO2-Gas Exchange under the effect of Biologically Active substances on Winter Wheat. Physiology and Biochemistry of Agricultural Crops, 27(3), pp. 135-140(1995).

Pergamon

Adv. Space Res. Vol. 20, No. 10, pp. 1945–1948, 1997

Printed in Great Britain
0273-1177/97 $17.00 + 0.00

PII: S0273–1177(97)00630–3

CATALASE ACTIVITY AS A POTENTIAL INDICATOR OF THE REDUCER COMPONENT OF SMALL CLOSED ECOSYSTEMS

A. B. Sarangova, L. A. Somova and T. I. Pisman

Institute of Biophysics (Siberian Branch, Russian Academy of Sciences)Krasnoyarsk 660036, Russia

ABSTRACT

Dynamics of catalase activity has been shown to reflect the growth curve of microorganisms in batch cultivation (celluloselythic bacteria *Bacillus acidocaldarius* and bacteria of the associated microflora *Chlorella vulgaris*). Gas and substrate closure of the three component ecosystems with spatially separated components "producer-consumer-reducer" (*Chl. vulgaris-Paramecium caudatum-B. acidocaldarius*, two bacterial strains isolated from the associated microflora *Chl. vulgaris*) demonstrated that the functioning of the reducer component can be estimated by the catalase activity of microorganisms of this component.

INTRODUCTION

Microbiological processes degrading organic contaminants are in their nature a totality of enzymatic reactions. Control of a man-made ecosystem involves knowledge of classical laws of growth and enzymatic kinetics of microorganisms (Varfolomeyev and Kalyuzhny, 1990). Monitoring of the bacterial reducer component in small closed ecosystems is difficult. However, this can be done by monitoring the physiological condition of the bacteria on the basis of their enzymatic activity. The condition necessary to determine this activity is to choose an enzyme specific for the substrate and perform the integral function for the exchange processes. Catalase is such an enzyme, specific with respect to hydrogen peroxide, whose formation in the cell is related to the nutrient substrate concentration in the medium and the rate of its consumption by bacteria.

MATERIALS AND METHODS

Celluloselythic bacteria *Bacillus acidocaldarius* and two strains isolated from the associated microflora of chlorella, *Paramecium caudatum* and *Chlorella vulgaris,* were utilized.

Bacteria were batch cultivated on a mineral medium (M.S. Yegorov - ed., 1986) with two different sources of carbon: sodium salt of carboxymethylcellulose (Na-CMC) and dead chlorella. For the three component closed "producer-consumer-reducer" system balanced mineral media for cultivation of chlorella, and protozoa (Pisman et al., 1995), and mineral media with Na-CMC were used at the ratio of 3:2:1. This was revealed as an optimal ratio for all the components' growth in the course of the preliminary experiments. Chlorella cell numbers were determined by using the Goryaev chamber (counting techniques), those of paramecia - by direct counting (Pisman et al., 1995). The nephelometry method was applied in estimating the bacterial biomass (Gerhard, 1984). Microorganism catalase activity was determined by a modified permanganometric technique (Sarangova and Somova, 1994).

RESULTS AND DISCUSSION

Experimental evidence indicates that the growth of celluloselythic bacteria on mineral medium with Na-CMC follows the classical law of batch growth (Fig. 1A, curves 1; 2). Experimental curves show that dynamics of catalase activity replicates the growth curve of these bacteria (Fig. 1A, curve 3) .

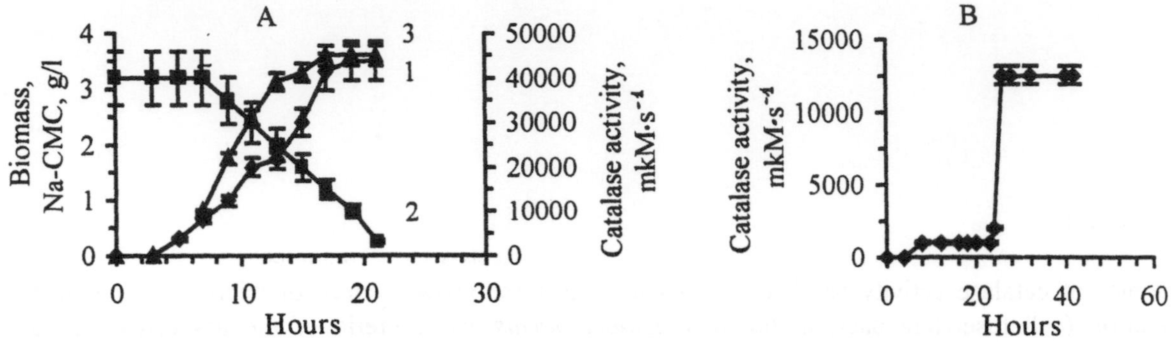

Fig. 1. Batch growth and activity of catalase of *Bacillus acidocaldarius* on Na-CMC (A) and on dead chlorella (B). A: 1 - biomass, g/l; 2 - Na-CMC, g/l; 3 - catalase activity, mkM·s^{-1}

When celluloselythic bacteria were grown on a native substrate - dead chlorella - it was not easy to evaluate its biomass nephelometrically, nonetheless, determination of the catalase activity made it possible to monitor the bacterial growth on dead chlorella (Fig. 1B). The experimental curves demonstrate that bacteria grew, even though slowly. The lag phase was 24 hours, followed by exponential growth phase of 2 hours and then by a prolonged steady-state phase (for 18 hours).

The dynamics of growth and activity of catalase with two different bacterial strains in association with chlorella in batch cultivation on mineral medium with Na-CMC and on dead chlorella was similar to growth and activity of celluloselythic bacteria. Thus, it has been shown that catalase activity appears to be related to the growth rate of a bacterial culture or to the rate of substrate consumption (Sarangova A.V., Somova L.A., Pechurkin N.A., Pisman T.I. Mikrobiologiya (in press)).

The experiments conducted on catalase activity of celluloselythic bacteria and bacterial strains associated with chlorella have formed the basis to study the catalase activity of the reducer component in a gas- and substrate-closed three component "producer-consumer-reducer" system. For the producer component we used microalgae *Chl. vulgaris*, for consumer - *Paramecium caudatum*, for reducer - a community of celluloselythic bacteria *B. acidocaldarius* and two bacterial strains isolated from the associated microflora of chlorella. Fig. 2. shows the chart of flows and interactions of the closed three component system under study.

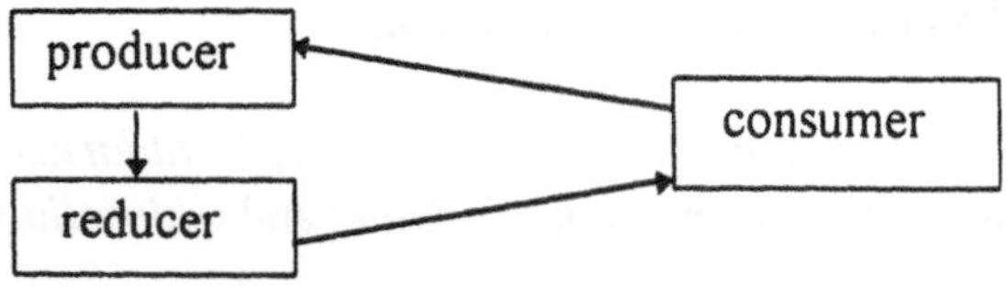

Fig. 2. Chart of flows and interaction of the gas- and substrate-closed three-component "producer-consumer-reducer" system.

Results of this experiment are shown in Fig. 3. Experimental observations demonstrate that the life-time of the system was about 18 days. Initial abundance of the inoculated producer was $5 \cdot 10^6$ cell/ml, of reducer was $5 \cdot 10^7$ cells/ml, of consumer was 20 cells/ml. For 5 days chlorella maintained constant numbers (curve 1), while the protozoa reduced to 4 cells/ml (curve 2). From the experimental observations of the reducer component it can be seen that the reducer's catalase activity decreased (curve 4) and thus it can be assumed that the bacterial biomass and protozoa also decreased. On the third day of the experiment, the solution flow was stopped for one day, increasing the catalase activity 3 times. In parallel with increasing catalase activity, (and, consequently, with increasing bacterial biomass) the protozoa increased to 25 cells/ml. This decrease is likely related to the inhibiting effect of chlorella metabolites (Sirenko L.A. and Kozitskaya V.N. (1988)).

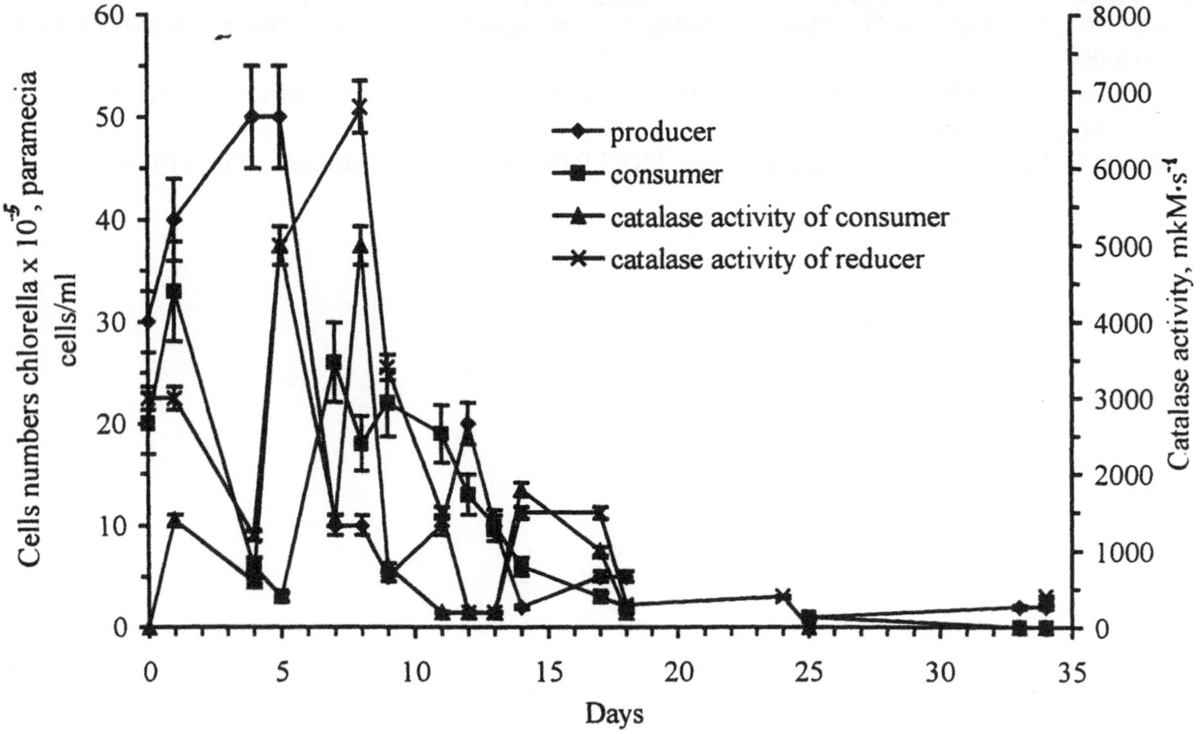

Fig. 3. Dynamics of components in a three-component gas- and substrate-closed three-component "producer-consumer-reducer" system.

It is worth noting that increase of protozoa was accompanied by increasing catalase activity in the fermenter of the consumer (curve 3). Catalase activity in the medium sharply decreased with increasing numbers of protozoa. Increasing enzyme activity in the medium implying presence of bacteria from the fermenter (the reducer) and their utilization by the protozoa for food. After 8 days of the experimental system, the numbers of protozoa drastically reduced. By the 18th day the abundance of protozoa was 2 cells/ml; abundance of bacteria in the reducer component was 10^4 cells/ml, and abundance of chlorella was 10^5 cells/ml. Still, the system did not cease to operate for a succession occured in which paramecia were replaced by *Litonotus* (small flagellates) whose domination is specific for systems with low organic substrate concentration.

REFERENCES:

Gerhardt, P.H. (ed). Manual of Methods for General Bacteriology, Moscow: Mir, (1984)
Pisman T.I., Somova L.A., Sarangova A.B. et al., Experimental Model of Gas-Closed "Autotroph-Heterotroph" Ecosystem, *Mikrobiologiya*, 64, (5), 657-660, (1995)
Sarangova A.B., Somova L.A., Catalase Activity of Microorganisms in Small Ecosystems, *AMSE Transactions*
Scientific Siberian", A. AMSE Press, 14, 125-139, (1994)
Sarangova A.B., Somova L.A., Pechurkin N.S., Pisman T.I. Catalase Activity as an Indicator of Physiological State of Microorganisms, Mikrobiologiya (in press)
Sirenko L.A., Kozitskaya V.N., Biological Substances of Algae and Quality of Water, Naukova Dumka, Kiev, (1988)
Varfolomeyev S.D., Kalyuzhny S.V., Biotechnology Kinetic Fundamentals of Microbiological Processes, Vysshaya Shkola, Moscow, 296, (1990)
Yegorov N.S. Metabolism of Microorganisms. MGU Publishing House, Moscow, 256, (1986)

Pergamon

Adv. Space Res. Vol. 20, No. 10, pp. 1949–1958, 1997
Published by Elsevier Science Ltd on behalf of COSPAR
Printed in Great Britain
0273-1177/97 $17.00 + 0.00

PII: S0273-1177(97)00860-0

SIGNIFICANCE OF RHIZOSPHERE MICROORGANISMS IN RECLAIMING WATER IN A CELSS

C. Greene*, D. L. Bubenheim** and K. Wignarajah*

**Lockheed Martin Engineering and Science Corporation, Ames Research Center, Space Technology Division, Regenerative Life Support Branch, Moffett Field CA 94035*

ABSTRACT

Plant-microbe interactions, such as those of the rhizosphere, may be ideally suited for recycling water in a Controlled Ecological Life Support System (CELSS). The primary contaminant of waste hygiene water will be surfactants or soaps. We identified changes in the microbial ecology in the rhizosphere of hydroponically grown lettuce during exposure to surfactant. Six week old lettuce plants were transferred into a chamber with a recirculating hydroponic system. Microbial density and population composition were determined for the nutrient solution prior to introduction of plants and then again with plants prior to surfactant addition. The surfactant Igepon was added to the recirculating nutrient solution to a final concentration of 1.0 g L^{-1}. Bacteria density and species diversity of the solution were monitored over a 72-h period following introduction of Igepon. Nine distinct bacterial types were identified in the rhisosphere; three species accounted for 87% of the normal rhizosphere population. Microbial cell number increased in the presence of Igepon, however species diversity declined. At the point when Igepon was degraded from solution, diversity was reduced to only two species. Igepon was found to be degraded directly by only one species found in the rhizosphere. Since surfactants are degraded from the waste hygiene water within 24 h, the potential for using rhizosphere bacteria as a waste processor in a CELSS is promising.

Published by Elsevier Science Ltd on behalf of COSPAR

INTRODUCTION

There is a rapidly growing technology now available in which microorganisms are used to degrade waste produced by humans and human activity. Municipal landfill operations, where the input waste contains a mixture of consumer products including toxic or hazardous materials, are developing acceptable management strategies that enhance the speed and efficiency of waste processing through microbial activity (Brummeler, et al, 1991; Irvine, et al, 1992). The practical use of microorganisms in the treatment of wastewater and groundwater reservoirs contaminated by complex organic pollutants has long been considered (Benjes, 1992; Vernick and Walker, 1991; Livingston, 1993a, 1993b). Recently, investigations have shown that uptake of heavy metal cations by microorganisms may be useful in reducing the levels or accumulation of heavy metals in the environment (Niu et al, 1993; Mattuschka and Straube, 1993). Since microorganisms demonstrate the ability to breakdown complex mixtures of waste and environmental pollutants, their use for recycling wastes has become a practical potential in the development of regenerative life support systems for long-term space habitation.

The need to recycle and reuse materials aboard long-term space missions is necessitated by the high cost of carrying cargo in earth-to-orbit transport (Moses et al, 1989). Plants play an important role in recycling in NASA's Controlled Ecological Life Support Systems (CELSS). In addition to supplying fresh food, plants convert carbon dioxide in the atmosphere to oxygen through photosynthesis and purify water via transpiration.

Reclaiming water from waste streams offers considerable cost savings. Water contributes approximately 90% by mass of the life sustaining provisions in a human space habitat. Approximately half, of which, will be used for personal hygiene and dish washing. The primary contaminant of the used hygiene water will be the cleansing agents or soaps used to carry out these functions. Studies conducted to determine soap use in space habitats showed that the hygiene waste water stream is likely to contain 1 g L^{-1} Igepon, the surfactant used in cleansing agents and soaps selected for space (Verostko et al, 1989). Since this raw hygiene water waste stream may likely require processing in a regenerative life support system we selected the value of 1 g L^{-1} for our screening investigations. This concentration exceeds the acute toxicity threshold for lettuce (Greene, 1994; Bubenheim et al, 1997).

The rhizosphere, or area surrounding plant roots, is a region of enhanced microbial activity because roots release a variety of compounds (Atlas and Bartha, 1981). The rhizosphere consists of microbial communities that may be attached to root cells, embedded in the root mucilage, or not physically connected to the root at all (Lugtenberg and de Weger, 1992). The turbulance and rapid flow in a hydroponic system should ensure that microbes are uniformly distributed throughout the solution. Analysis of the nutrient solution should be representative of the microrganisms derived from the rhisosphere. In previous work, naturally occurring rhizosphere communities have demonstrated the ability to mineralize the organic matter of inedible plant material (Garland, 1992). In addition, a bioreactor containing microorganisms isolated from the roots of hydoponically grown lettuce has been successful at degrading Igepon in a CELSS liquid waste stream alleviating all phytotoxicity effect of the solution (Wisniewski and Bubenheim, 1993). An analysis of the microbial ecology during surfactant degradation, however, has not been completed. In an attempt to understand the system and its dynamics as a whole, we identified changes in the microbial ecology associated with the rhizosphere of hydroponically grown lettuce during surfactant exposure.

METHODS AND MATERIALS

Six week old lettuce (*Lactuca sativa* cv. Waldmann's Green) plants grown hydroponically in a greenhouse were transferred to a 0.52 m^3 stainless steel plant chamber. A 40 L recirculating hydroponic system was isolated from the shoot zone by a sheet of ABS plastic through which the lettuce plants were planted in 1 cm holes. The stems were wrapped in open-cell foam plugs to support the plants and maintain an effective barrier between the atmosphere and hydroponic system. Sixteen lettuce plants were used in each experiment with a total projected canopy area of 0.157 m^2. Light was provided with a 1 kW high pressure sodium lamp and the PPF was maintained at 800 $\mu mol\ m^{-2}\ s^{-1}$. Temperature was held constant at 23°C. The chamber atmosphere was maintained at 350 $\mu mol\ mol^{-1}$ CO_2 and 21% O_2 with a constant dew point of 10°C (50% relative humidity).

Microbial cell density and species diversity were determined for nutrient solution samples prior to transfer of the lettuce plants to the chamber.and then again 0.5 h following introduction of the plants. These analysis provide a measure of microbial activity associated with the plants and establishes a microbial baseline prior to introduction of the surfactant, Igepon. Igepon was added to the recirculating nutrient solution to a final concentration of 1.0 g L^{-1}. Ten ml samples of nutrient solution were collected from the chamber at regular intervals over a 72-h period following Igepon addition. Bacterial cell density was determined by measuring absorbance at 590 nm. Surfactant concentration was measured using a Surfactant Electrode (Orion model 93-42). A serial dilution of each solution sample was completed in triplicate. Each sub-sample of the dilution was added to Difco R2A media and poured into petri dishes and allowed to solidify. The pour plate method was utilized to achieve the best distibution of colonies to facilitate identification of unique colonies (Collins et al, 1991). Following a 13-day incubation period at room temperature, individual colonies were characterized. Basic morphological characteristics (shape, color, texture, etc.) were used to describe and differentiate different bacterial types.

The study was repeated three times. Between each replicate run, the chamber was cleaned with a 10% bleach solution. A fourth run of the study, in which no surfactant was added to the nutrient solution, served as the control.

Gram stain analysis was completed on pure cultures of the distinguishable bacterial types. Using aseptic techniques, an inoculating loop full of the pure cultures were inoculated into individual flasks containing 100 ml of a 1 g L^{-1} Igepon solution and incubated at 27°C with constant agitation. The surfactant electrode was used to monitor Igepon degredation over the 24-h incubation period. Cell density was determined at the time of inoculation and at the end of the incubation period. The one species capable of degrading Igepon was negatively stained with a solution containing 1% phosphotungstic acid and 50 $\mu g\ ml^{-1}$ Bacitracin and viewed using a Zeiss Electron Microscope (EM 109). Photographs were taken using Kodak Technical Pan film.

RESULTS AND DISCUSSION

The rhizosphere of lettuce was the only source of bacterial inoculum in the chamber. Colonies did not grow on agar plates from samples collected before lettuce was placed in the chamber, thus it was assumed that the microbes found in the nutrient solution represents the rhizosphere microbes. Nine morphologically distinct bacterial types were identified in the nutrient solution of the hydroponically grown lettuce. The majority of the colonies formed in culture were small, circular in shape with entire edges and appeared smooth without surface features (Table 1). Color was the major variant between the different colonies. Opaque white, yellow, orange, pink, green, and clear colonies were isolated from the solution after lettuce was added to the chamber.

Table 1. Colony morphologies of the nine bacterial types isolated from the rhizosphere of hydroponically grown lettuce. Colonies are described using basic morphological characteristics following a 13-day incubation period. Color was the major variant between the bacterial types.

Character	Species								
	1	2	3	4	5	6	7	8	9
color	white	yellow	light yellow	green	flourescent yellow	orange	pink	clear	brownish yellow
shape	irregular	punctiform	circular-rhizoid	circular	circular	circular	circular	irregular	circular
surface elevation	flat	undetectable	flat	raised	raised	raised	raised	raised	raised
edge	erose	entire	filamentous	entire	entire	entire	entire	entire	entire
topography	rough	smooth	smooth	smooth	smooth	smooth	smooth	smooth	smooth
optical character	opaque, glistening	translucent, glistening	translucent, glistening	opaque	opaque, glistening	opaque, glistening	opaque, glistening	transparent, glistening	opaque, dull

It has been reported that the rhizosphere typically contains a higher proportion of gram negative, nonsporulating, rod shaped bacteria, and a low proportion of gram positive, non-spore forming rods, cocci, and pleomorphic forms (Atlas and Bartha, 1981). The bacterial community associated with root surfaces of lettuce in recirculating nutrient delivery systems of the Biomass Production Chamber at Kennedy Space Center was composed of only gram negative organisms with the majority being rod shaped (Strayer, 1994). This did not hold true, however, in our study (Table 2). Only two species were bacili, or rod shaped, while the majority of species were cocci and diplococci in shape. Additionally, the rhizosphere was composed of almost an equal percent of gram positive and gram negative members.

Table 2. Gram stain reaction and cellular morphology of the nine bacterial types isolated from the rhizosphere of hydroponically grown lettuce.

	Species								
Character	1	2	3	4	5	6	7	8	9
gram reaction	-	+	-	-	-	+	+	-	+
cell shape	cocci	diplococci	diplococci	cocci aggregates	diplococci	bacilli	diplococci	cocci	bacilli

Three species accounted for 87% of the rhizosphere microbial population identified (Table 3, numbered 1, 2 and 3). Their counts increased to represent 100% of the population soon after Igepon was added to the nutrient solution. Six of the original nine species disappeared from solution within six hours of adding Igepon (Figure 1). Once Igepon was completely degraded from the solution the microbial diversity within the nutrient solution was reduced to only two bacterial species (numbers 1 and 2) of the original population.

Table 3. Percentage of total microbial population represented by individual species and surfactant degradation capability of these rhizosphere bacteria.

	Species								
Character	1	2	3	4	5	6	7	8	9
% of total population	48.0	15.4	15.0	7.6	4.0	3.5	3.0	3.0	0.5
degrades Igepon	yes	no	no	no	no	no	no	no	no

A typical bacterial growth curve was realized during the 72 h sampling period following the addition of Igepon, although the microbial population in our study grew without a distinctive lag period (Figure 2). This observation is consistent with that of Knaebel and Vestal (1992) and Federle and Ventullo (1990) who quantified surfactant mineralization by soil (rhizosphere) microorganisms. Typically, during the lag phase, the bacteria begin preparing for reproduction by synthesizing DNA and the enzymes needed for cell division. When a bacterium is inoculated into a new culture medium, the lag phase reflects the time needed to manufacture the enzymes required to utilize that new and specific substrate for growth (Altas, 1988). In this study, the bacterial cultures were physically transferred to the chamber on the plant roots but the substrate remained the same.

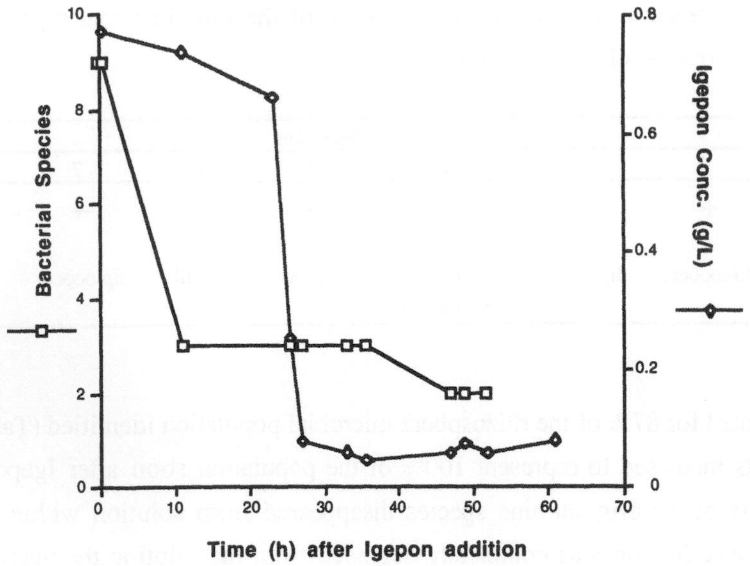

Fig. 1. Changes in rhizosphere species diversity during surfactant degradation. The rhizosphere of hydroponically grown lettuce was reduced from 9 to 3 species within 6 hours of a 1.0 g L^{-1} Igepon exposure.

The sudden increase in cell number without a significant degree of surfactant degradation suggests that the bacteria are actively utilizing resources in the plant nutrient solution for growth and reproduction. The microbes may be metabolizing organic root exudate (primarily sugars, amino acids, and organic acids) and inorganic ions leaked from the roots damaged by Igepon. Tearing and shedding of epidermal cells has been demonstrated in roots exposed to 1.0 g L^{-1} Igepon solutions (Greene, 1994).

The bacterial enzyme used for Igepon degradation is likely specific for surfactants. Since Igepon degradation did not occur immediately, the lag phase in degradation may reflect the time needed by the bacteria to manufacture the enzymes required to utilize the surfactant as a carbon source. Igepon was degraded during the exponential growth phase of the rhizosphere bacteria (Figure 2). During this time the population doubled every 2.3 h. During a 24-h exponential growth period, the number of bacterial species was reduced by one third. Surfactants can be toxic to microorganisms, this give detergents their sanitizing properties. It is possible that 1.0 g L^{-1} of Igepon may be above their toxicity threshold or the three dominant species may have simply out competed the others for the available substrate, the surfactant. Alternatively, the three dominant species may have produced compounds which inhibited the growth of the other six species. Although the specific interactions between the species is not known at this time, it appears that approximately 90% of the Igepon present in solution was degraded by three bacterial species. Once Igepon was depleted, a reduction in total cell counts was observed.

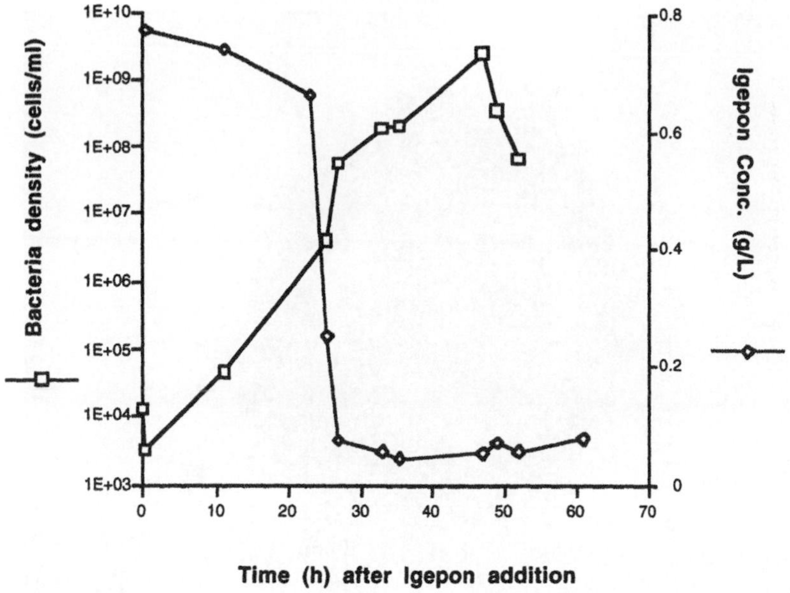

Fig. 2. Surfactant degradation by rhizosphere bacteria. A 10,000X increase in total counts of bacteria corresponds with a 90% decrease in Igepon concentration. This suggests that rhizosphere bacteria utilize the carbon source provided by Igepon to grow and reproduce.

The nine rhizosphere bacterial species were individually isolated in culture in order to determine which are capable of degrading surfactants in solution. Only one of the nine species was capable of completely degrading Igepon (numbered 1) on it's own. The species numbered 2 and 3 were present during Igepon degradation, yet their inability to individually degrade Igepon suggests that they utilize the by products of degradation or root leakage for growth and reproduction.

A number of species of *Pseudomonas* were reported to be present on the rhizosphere of hydroponically grown lettuce and suggested to be responsible for surfactant degradation (Wisniewski and Bubenheim, 1993). However, we eliminated *Pseudomonas* as a putative microorganism capable of degrading Igepon in our study. Electron microscopy analysis of isolates revealed the absence of any flagellum, a characteristic feature of all *Pseudomanas* species (Figure 3). We are currently taking measures to determine the identity of the organism degrading Igepon.

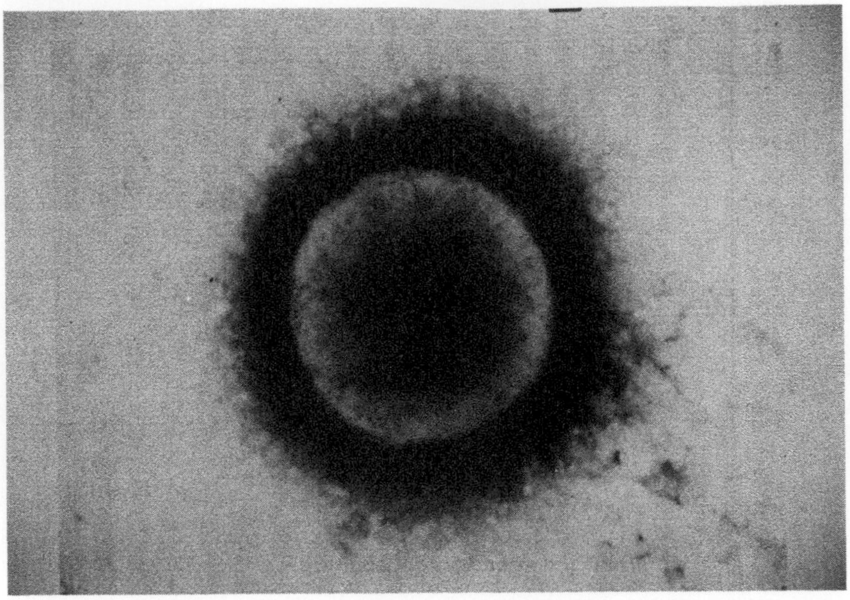

Fig. 3. Transmission electron micrograph of rhizosphere microorganism capable of degrading Igepon in solution. Bar = 0.1 µm. The microorganism is a gram negative cocci lacking fllagellum or pilli.

SUMMARY

This study was undertaken in an attempt to understand the microbial ecology of the rhizosphere of hydroponically grown lettuce when challenged with surfactants. We identified dramatic changes in the population density and composition during acute surfactant exposure. Additionally, we isolated the bacterium capable of degrading the surfactant in a CELSS hygiene waste water stream. Microbial and enzymatic processes have been suggested for waste water reclamation in space (Petrie and Nacheff-Benedict, 1991; Miller, 1991) as they can operate under ambient conditions, requires a minimal input of energy, and typically produces by-products that are not toxic or hazardous. The results of this study support the idea of using rhizosphere bacteria for bioremediation processes. Since surfactants are degraded from the hygiene water within 24 h, the potential for using rhizosphere bacteria as a waste processor in a CELSS is promising.

REFERENCES

Atlas, R. M. and R. Bartha, *Microbial ecology: Fundamentals and applications,* Addison-Wesley Publishing Co., London, (1981).

Altas, R. M., *Microbiology fundamentals and applications,* Macmillan Publishing Co., New York, (1988).

Benjes, H. H., Jr. *Handbook of biological wastewater treatment,* Garland STPM Press, New York, (1980).

Brummeler, E., H.C.J. M. Horbach and I. W. Koster, Dry anaerobic batch digestion of the organic fraction of municipal solid waste, *J. Chem. Tech. Biotechnol.* 50, 191-209 (1991).

Bubenheim, D. L., K. Wignarajah, W. Berry, and T. Wydeven, Phytotoxicity effects of gray water due to surfactants, *J. American Soc. for Hort. Sci.* in press (1997).

Collins, C.H., P.M. Lyne, and J.M Grange. Cultrual Methods (Chapter 5) In Collin's and Lyne's Microbial Methods, Butterwoth-Heinemann Publication, Oxford, UK (1991).

Federle, T. W. and R. M. Ventullo, Mineralization of surfactants by the microbiota of submerged plant detritus, *Applied and Environ. Microbiol.*, 56, 333 - 339, (1990).

Garland, J. L., Coupling plant growth and waste recycling systems in a Controlled Life Support System (CELSS), NASA Tech. Memo. 107544, Natl. Aero. and Space Admin., John F. Kenedy Sapce Center, FL (1992)

Greene, C.M., Dose response analysis of surfactant toxicity in hydroponically grown lettuce seedlings, M.S. Thesis, San Jose State University, San Jose, CA (1994).

Irvine, R. L., R. Chozick and J. P. Earley, Use of biologically based periodic processes in waste treatment, *American Chemical Society* , 471-474 (1992).

Knaebel, D.B. and J.R. Vestal, Effects of intact rhizosphere microbial communities on the mineralization of surfactants in surface soils, *Can. J. Microbiol.* 38, 643 - 653 (1992).

Livingston, A. G., A novel membrane bioreactor for detoxifying industrial wastewater: I. Biodegradation of phenol in a synthetically concocted wastewater, *Biotechnol. and Bioengin.*, 41, 915-926 (1993a).

Livingston, A. G., A novel membrane bioreactor for detoxifying industrial wastewater: II. Biodegradation of 3-Chloronitrobenzene in an industrially produced wastewater, *Biotechnol. and Bioengin.*, 41, 927-936 (1993b).

Lugtenberg, B. J. J. and L.A. de Weger, Plant root colononization by *Pseudomonas* spp., in: *Pseudomonas molecualr biology and biotechnology,* American Society for Microbiology, Washington, D. C., (1992)

Mattuschka, B. and G. Straube, Biosorption of metals by a waste biomass, *J. Chem. Tech. Biotechnol.* 58, 57-63 (1993).

Miller, G. P., R. J. Portier, D. P. Dickey and H. L. Sleeper, Using biological reactors to remove trace hydrocabon contaminants from recycled water, SAE Tech. Paper Series 911504, The Engin. Soc. for Adv. Mobility Land Sea Air and Space, Warrendale, PA (July 1991).

Moses, W.M., T.D.Rogers, H. Chowdhury and R. Cusick, Performance characterization of water recovery and water quality from chemical/organic waste products, in: *Intersociety Conf. on Environ. Systems*, San Diego, CA (1989).

Niu, H., X. S. Xu, J. H. Wang and B. Volesky, Removal of lead from aqueous solutions by Penicillium biomass, *Biotechnol. and Bioengin.*, 42, 785-787 (1993).

Petrie, G. E. and M. S. Nacheff-Benedict, Development of immobilized cell bioreactor technology for waster reclamation in a regenerative life support system, SAE Tech. Paper Series 911503, The Engin. Soc. for Adv. Mobility Land Sea Air and Space, Warrendale, PA (1991).

Strayer, R. F. Dynamics of microorganism population in recirculating nutrient solutions, *Life Sciences and Space Research* XXV (3), 14, (11)357-(11)366, (1994).

Vernick, A.S. and E. C. Walker, *Handbook of Wastewater Treatment Processes*, Marcel Dekker, Inc., New York, 1981.

Verostko, C.E., C.E. Garcia, R. Sauer, RP.. Reysa, A.T. Linton, and T. Elms. Test results on reuse of reclaimed shower water - a summary. SAE Technical paper # 891442, 19th ICES Conference, San Diego, CA (1989).

Wisniewski, R. and D. Bubenheim, Aerobic biological degradation of surfactants in waste water, *AIAA 93-4152* (1993).

Pergamon

Adv. Space Res. Vol. 20, No. 10, pp. 1959–1965, 1997

Printed in Great Britain
0273-1177/97 $17.00 + 0.00

PII: S0273-1177(97)00631-5

PECULIARITIES OF BIOLOGICAL PROCESSES UNDER CONDITIONS OF MICROGRAVITY

G. S. Nechitailo* and A. L. Mashinsky**

*Scientific-Technical Center "Ecology and Space", Timura Frunse 34-7, 119021, Moscow, Russia
**Institute of Biomedical Problems, Choroshevskoe shosse 76a, 123007, Moscow, Russia

ABSTRACT

The results of experiments aboard spacecraft demonstrated the dependence of the pattern of biological processes on microgravity and on the ability of biological objects to adapt themselves to new environmental conditions. This is of fundamental importance for solving theoretical and practical problems of space biology, for elaborating the theory of organism's behavior in weightlessness, and for elucidating the global mechanisms of the action of microgravity on living systems.

INTRODUCTION

In the course of evolution, terrestrial biological objects have been exposed to numerous permanent impacts, such as atmospheric pressure, atmospheric gases, gravitational, electric, and magnetic fields, etc. Gravity is the factor that is least vulnerable to change; on the other hand, it may be most crucial, although the mechanism of its action on the growth and development of biological organisms is uncertain. Space flight is characterized by weightlessness or microgravity, which is one of a few factors that can not be simulated in a terrestrial laboratory for a long time or in parabolic flight (over 1 min). Therefore, its role for biological objects should be thoroughly studied. Weightlessness should be defined as conditions in which gravitational and other external forces influencing a body induce no acceleration of the body normally caused by these forces. In real space flight, however, there is always a wide spectrum of small residual accelerations that are defined as "microgravity" although they are typically in the milligravity range (i.e., $10^{3}g_n$) or less.

There are many known effects related to the action of microgravity on the organism. This has stimulated extensive fundamental and applied research dealing with the mechanism of microgravitational action on the organism, cellular, subcellular and molecular levels. An attempt was made to relate the effects of gravity on the organism to peculiarities of moisture transfer and the pattern of chemical reactions (Sadyhov et. al. 1988).

CONDITIONS OF MICROGRAVITY ABOARD SPACECRAFT

Microgravity acting on bodies under space flight conditions is estimated by taking into account drag from Earth's very thin atmosphere, the orbital trajectory, the size, weight and optic characteristics of the spacecraft surface as well as dynamic operations required by the flight program, and other factors (Mashinsky *et al.*, 1976).

Sunlight exerts a pressure on an orbital craft producing an accelerational force of 10^{-9} g_n. The gravitational gradient is 1×10^{-7} g_n to 3×10^{-6} g_n during 0.5 -10 m removal of the biological object from the spacecraft centroid under conditions of constant or inertial orientation. Perturbations from functional units of the orientation system are 10^{-7} g_n to 10^{-9} g_n, and spatial turns and maneuvers produce microgravity of about 2×10^{-1} g_n to 2×10^{-2} g_n lasting several seconds to dozen of seconds. As the station joins a transport vehicle, microgravity may exceed 5×10^{-1} g_n (Table 1).

Table 1. Microgravity Induced by Various Perturbating Factors at the Salyut Stations

Perturbating factor in space	Maximum predictable microgravity at a 3-m distance from the centroid (g_n)	Pattern of action	Estimated vector of microgravity
Aerodynamic braking effects of low-pressure gases	1×10^{-5} - 1×10^{-7}	Permanently operative during the flight	In all directions
Light pressure	10^{-9}	Same	Same
Gravitational gradient	3×10^{-6} - 1×10^{-7}	Permanent or periodically varying with orbits (depending on programs)	x,y,z
Function of the orientation systems	1×10^{-7} - 1×10^{-9}	Permanent in different directions	x,y,z
Spacial turns and maneuvers	2×10^{-1} - 2×10^{-2}	Occasional	x,y,z
Docking	5×10^{-1} - 2×10^{-2}	Same	x
Movements of the crew*	4×10^{-3} - 2×10^{-4} (occasionally: 1×10^{-2} - 8×10^{-3})	Daily during the flight with variable incidence	y,z
Daily exercise of the crew*	9×10^{-2} - 1×10^{-3}	Daily with alternating vectors, a few hours	y,z

* Microgravity induced by the exercise of crew is given with respect to vibrations of the station

Most of the gravitational perturbations are produced by movements of the crew, especially during physical exercises with the use of exercise equipment. Accelerations may increase from 1×10^{-3} g_n to 4×10^{-3} g_n and even to 8×10^{-3} g_n $\div$ 1×10^{-2} g_n. Acceleration vectors vary, but the greatest microgravity occurs in the X and Y axes. Microgravitational perturbations aboard a spacecraft should be taken into account when dealing with a particular research problem in the course of biological and biotechnological investigations.

MOISTURE TRANSFER UNDER CONDITIONS OF MICROGRAVITY

Moisture transfer in plants and in the root zone is a gravitationally-dependent process affecting the growth and development of plants under the action of microgravity. Water transport in plants occurs through xylem conductive elements. The water regimen in plants is best characterized by the water potential value:

$\psi = p + \rho gh + \pi$ where ψ= water potential of plant tissue; p= hydraulic pressure in a water solution; π= osmotic potential; ρ= water density; g = gravity; h = vertical distance.

The conditions of gas-liquid support of plants during space flight differ from those on Earth. This has been demonstrated by us in direct experiments aboard the orbital stations Salyut and Mir in which the distribution of moisture was determined in a soil substitute with one-dimensional capillary soaking provided by a pulsed water supply (Podolsky et al., 1991).

The results obtained in several space flights indicate that there is excessive moisture near the source of water supply as compared to the ground-based control. Changes in the maximum moisture level and the formation of localized mildly resorbing moisture volumes are likely to occur. It seems likely that, under conditions of microgravity, dissipative processes related to the inertial and frictional forces, that counteract the forces of capillary absorption, take on greater significance. The existence of such a situation with moisture transfer and moisture distribution in a soil substitute may result both in moisture deficiency and moisture excesses and, as a result, produce retarded growth of plants.

PATTERNS OF CHEMICAL REACTIONS IN WEIGHTLESSNESS

It has been theoretically inferred that weightlessness acts on intercellular processes by redistributing cell particles. Effects of mass forces may occur only in cell particles that are larger than 10 nm. We have undertaken studies that demonstrated a gravitational dependence of chemical reactions associated with phase transitions. Gelatin and agar biopolymers and polyacrilamide gels (PAAG) were selected as model systems to assess the effects of space- flight factors on the pattern of chemical reactions.

Studies of the effects of weightlessness on the model systems of gelatin and agar solutions were performed aboard the satellite Kosmos-613, the spacecraft Soyuz-20, and the orbital station Salyut-6. They showed specific changes in their properties. Aging gelatin solutions displayed an increase in characteristic viscosity (η) as compared to the control (Table 2), and a correlation was found between this characteristic of agar and its concentration. The studies were performed using an Ubelloide viscosimeter.

Table 2. Effects of Space Flight Conditions on Characteristic Viscosity (η) of Gelatin and Agar (dL/g)

Varient	Gelatin concentration %	(η)	Agar concentration (%)	(η)
Control	0.84	0.50	010	7.0
Space	0.84	0.75	0.10	5.0
Control	1.90	0.33	0.15	3.0
Space	1.90	0.52	0.15	3.0

Comparison of the tests using gelatin and agar, the biopolymers of different classes, suggests that hydrodynamic gel properties were differently affected by weightlessness. The increase of η in gelatin was reflective of a growing size of molecules with an increase in their asymmetry. On the other hand, the decrease of η in the agar solutions was related to modifications in the shape of clusters and their lesser symmetry. Therefore, it was shown that space flight conditions result in modified properties and shapes of biopolymer molecules (Sadykov *et al.*, 1988). However, the taxotropic properties of gelatin and agar suggest that the observed changes might be related to dynamic impacts of the flight (microgravity, acceleration) during landing. This was a motivation for a series of studies concerning the properties of products of the reaction of irreversible gel synthesis. The experiments were conducted aboard the orbital station Salyut-7 using specially designed devices for photoionization of the reaction of acrylamide and bismethylene acrylamide copolymerization with the resultant gel (PAAG) synthesis. The gels synthesized under weightlessness conditions were used for electrophoretic separation of complex protein mixtures: blood plasma proteins, nuclear proteins (histones) and various forms of cytochrome C. Polyacrylamide gels obtained from the space flight showed considerable changes in their electrophoretic characteristics.

Electrophoresis was performed in a thermostatic device for vertical electrophoresis (LKB-2001). Density measurements were made with an Ultrascan densirometer (LKB).

Changes in physico-chemical properties of gel were also revealed: in NMR spectra, patterns of small-angle X-ray scattering, etc. Thus, chemical systems involving phase transitions in their reactions were found to be acceleration- sensitive under conditions of weightlessness.

We have performed experiments to explore the feasibility of the synthesis of biologically active nucleotides in outer space where ultraviolet radiation of the sun, ionizing radiation and temperature act together as a source of energy for reactions. Vacuum-sealed ampoules were loaded with dry mixtures of chemical substances: adenine-ribose, adenine-deoxyribose, thymine-deoxyribose.

As a result, these studies have demonstrated that the exposure of dry mixtures of the above-mentioned chemical substances to the outer space results in the synthesis of nucleoside-like substances. The molecular mass of these new compounds is greater than that of natural nucleosides. The synthesized products differ from natural nucleosides by their molecular mass and the maximum of absorption in the UV-spectrum (Nechitailo and Mashinsky, 1993).

The evidence obtained suggests a possible impairment of the pattern of chemical reactions in the absence of accelerational force and the possibility of obtaining new reaction products different from terrestrial products.

GROWTH AND DEVELOPMENT OF HIGHER PLANTS UNDER SPACE FLIGHT CONDITION

An extended program of biological investigations with higher plants has been carried out aboard orbital stations. The program is aimed at solving the problems of plant growth and development in space flight.

Various growth-chamber devices, as well as methods and approaches reducing the effect of unfavorable space flight factors, have been devised for orbital stations. These methods involve a complex of measures for treating seeds with antimutagens and with different physical factors (artificial gravity, magnetic fields, gradient application of electric and electrochemical potentials). Investigations with these devices and methods have made it possible to provide answers about morphogenetic features, spatial orientation metabolism and productivity of plants, under space flight conditions (Nechitailo and Mashinsky, 1993).

To carry out the experiments in space, the following equipment has been developed:

- biogravistat used in Space experiment with damping of gravitational effects on the seedlings; further, biogravistats 1 and 2 were made where the artificial gravity was created through the centrifuge; (Laurinavichyus *et al.*, 1984)
- device for cultivation of bulbous plants (Vason) used for experiments with orchids, tulips, and several with onion;
- for prolonged cultivation of higher plants, a family of Oasis devices was developed in which the system of water supply and illumination was improved, and aeration and electrostimulation of the root area were provided.

Different materials were used as soil substitutes, usually natural and artificial zeolites in granulated and tissue form. To study moisture transfer into substrates of various structures, an experimental cuvette was used where data were obtained on zones of excessive moistening in perlite and balkanin substrates (Podolsky and Mashinsky, 1991).

In Svetoblock-M and Malachit plant growth devices, the tissue (fabric) substrates gravilen and vion were employed. Plants were cultivated on balkanine substrate in the Soviet-Bulgarian greenhouse Svet on board Mir station.

The research into the seedlings and tissue culture of plants was made in special temperature controlled units with a temperature range within -4°C - +37°C and also in test-tubes and Petri dishes, for instance, the culture of potato tissue.

To directly retrieve the living tissue aboard the station, the fixatives BK-M and BK-50 were used.

Higher plants at orbital stations were investigated in various morphological forms: seeds, tubers and bulbs, seedlings, vegetative plants, even cacti and a dwarf tree.

The simpliest experiments can be carried out with seeds. However, the results of various authors during short-term flights varied widely; for example, in tests with pea, chromosome aberrations were found in 4 out of 14 experiments, while in wheat in three out of nine. These results were obtained by the pioneer of space studies, prof. Zhukov-Verezhnikov. In Welsh onion, nigella, and maize, chromosome aberrations were not observed, while in the same experiments with winter wheat the effect was obvious. Bur with prolonged seed exposure to space flight (827 days), the result was stunning: almost all arabidopsis seeds were lost (germinating capacity was 0.41% in test and 73.7% in control). Prolonged exposure is associated with a sharp increase in the number of cells with chromosome aberrations (Table 3).

Table 3. Number of Cells with Chromosome Aberrations in *Crepis capillaris L*

Duration (days)	Flight cells with aberrations (%)	Control cells with aberrations (%)
49	0.16 ± 0.12	0.13 ± 0.09
223	1.04 ± 0.10	0.45 ± 0.10
827	13.3 ± 1.26	3.63 ± 0.78

There are other data on impaired structural-functional peculiarities of seeds. Therefore, the stability of some biological parameters of seeds and their free radical state were evaluated. The number of free radicals was determined (by the EPR method) in seeds before and after two (two-month) flights. In 8 out of 15 types, the content of paramagnetic centers was established to increase by 10-30 % as compared with to controls (Table 4). This may lead to a more pronounced mutagenic effect because, as is known, free radicals can impair the functioning of enzymes.

Most of developing mutations are harmful to both the organism and the population as a whole, which makes the protection of genetic apparatus from mutagenic factors an important task. The use of antimutagens is one of the approaches to solving this task. α-Tocopherol, auxin, and kinetin decreased the incidence of chromosome aberrations in-flight in Welsh onion, a species which showed significant adverse effects of space flight exposures (Table 5).

Table 4. Levels of Free Radicals (Relative Units) per Gram of Dry Seed

Count of paramagnetic centers	Experiment 1 (71-day flight) Species	Flight	Control	Experiment 2 (60-day flight) Species	Flight	Control
High	*Allium fistulosum*	1.98	1.06	*Allium montanum*	1.18	1.21
	Linosyris vulgaris	5.03	5.18	*Asparagus officinalis*	1.19	0.83
Median	*Lactuca sativa*	0.87	0.74	*Plantago major*	0.15	0.13
	Picea canadensis	0.65	0.43	*Picea canadensis*	0.14	0.13
	Dianthus barbatus	0.35	0.29	*Papaver orientalis*	0.19	0.19
	Amaranthuspaniculatus	0.67	0.09	*Eredium cicutarium*	0.24	0.24
	Trifolium pratense	0.16	0.09	*Salvia nutans*	0.11	0.10
				Brassica juncea	0.15	0.14
Low	*Pinus sylvestris*	0.10	0.07	*Lotus corniculatus*	0.03	0.03
	Lepidium sativum	0.90	0.10	*Indiganos*	0.03	0.04
	Sinapis arvensis	0.10	0.06			

Table 5. Effects of α-Tocopherol, Auxin, and Kinetin on the Incidence of Chromosomal Aberrations in *Allium fistulosum L.*

Variant	Concentrations of protectors μg/mL)	Percentage of cells with aberrations (mean ±SD)	Student's ratio (control/flight)
Pre-examined		5.71 ± 0.55	
		Experiment 1 (82-day flight)	
Control	-	6.96 ± 0.55	1.61
Control + α-tocopherol	1×10^{-4}	4.00 ± 0.68	3.40
Control + α-tocopherol	1×10^{-2}	3.07 ± 0.59	4.86
Control + auxin	1×10^{-1}	3.42 ± 0.67	4.07
Flight	-	9.72 ± 0.67	3.20
Flight + α-tocopherol	1×10^{-4}	5.13 ± 0.77	3.92
Flight + α-tocopherol	1×10^{-2}	3.57 ± 0.63	5.64
Flight + auxin	1×10^{-1}	5.34 ± 0.89	3.94
		Experiment 2 (522-day flight)	
Control	-	11.36 ± 1.07	4.71
Control + α-tocopherol	1×10^{-2}	5.65 ± 0.77	4.34
Control + kinetin	1×10^{-1}	6.20 ± 0.83	3.82
	-	14.32 ± 0.96	2.06
Flight + α-tocopherol	1×10^{-2}	6.63 ± 0.84	6.85
Flight + kinetin	1×10^{-1}	7.41 ± 0.89	5.27

CONCLUSION

The microgravitational conditions aboard a spacecraft are on the threshold of gravitational sensitivity of such biological objects as higher plants. This threshold is known to lie in the range $1.4 \times 1.0 \times 10^{-3}$ to 1×10^{-4} g_n. On the other hand, the study of the peculiarities of moisture transfer undertaken by us as well as the study of various chemical reactions associated with phase transfer indicate the possibility of distrurbances in chemical reactions under microgravitational conditions.To study different morphological forms of plants under microgavitational conditions, special equipment and other technical facilities have been designed.

Effects related to the accumulation of free radicals in seeds of different plants and an increase in the number of chromosome aberrations in meristemic cells correlating with flight duration have been revealed. To reduce the adverse effect of the space flight factors, approaches based on the use of natural antimutagens have been elaborated.

These studies permit the creation of life-support systems involving the use of higher plants to be regarded as a realizable task.

REFRENCES

1. Laurinanavichyus, R.S., A.V.Yaroshyus *et al.* Methods of experiments in research of the gravity force effect in plant growth and development, in *Biological Studies at Salyut Orbital Stations,* edited by N.P.Dubinin, pp15-20, Nauka, Moscow (1984).
2. Mashinsky, A.L., O.V.Mitichkin and G.M. Grechko in *Organism and Gravity,*edited by A.I. Merkis, 228-237, Vilnyus (1976).
3. Nechitailo, G.S. and A.L.Mashinsky, in *Space Biology. Studies at Orbital Stations,* Mir Publishers, Moscow (1993).
4. Podolsky, I.G., A.L.Mashinsky *et al.* The device for determination of mass exchange properties of capillary-porous systems, *Author's Certificate*[1] 1659790á Inst. of Patent Information, Moscow (1991).
5. Podolsky, I.G. and A.L.Mashinsky, Research of peculiarities of moisture transfer in capillary-porous-soil substitutes in space flight, in *International Symposium Perm-Moscow,* pp. 188, Perm (1991).
6. Sadykov, A.S., V.B.Leontiev *et al.* Systems of chemical processes that depend for biological studies in *Doklady . AN SSSR,* 303(4), pp. 1004-1007 (1988).

NUTRITION AND PRODUCTIVITY FOR BIOREGENERATIVE LIFE SUPPORT

Proceedings of the F4.2 Symposium of COSPAR Scientific Commission F which was held during the Thirty-first COSPAR Scientific Assembly, Birmingham, U.K., 14–21 July 1996

Edited by

S. S. NIELSEN

Department of Food Science, Purdue University, 1160 Smith Hall, West Lafayette IN 47907-1160, U.S.A.

Pergamon

Adv. Space Res. Vol. 20, No. 10, p. 1969, 1997

Printed in Great Britain
0273-1177/97 $17.00 + 0.00

PII: S0273–1177(97)00261–5

PREFACE

In a controlled environment, the hydroponic nutrient solution, photosynthetic photon flux, carbon dioxide level, planting density, and temperature are some of the factors that can be manipulated to affect the productivity and nutrient content of crops for a bioregenerative life-support system. Plan productivity is critical for sustainability of such a system, but these plants must contribute to a safe and nutritious diet for inhabitants of the system.

The papers in this compilation deal with the effect of carbon dioxide level, in controlled environments, on the productivity and nutrient content of the controlled ecological life-support system (CELSS) candidate crops of wheat, potato, tomato, and soybeans. Effects of carbon dioxide vary somewhat by crop and plant part. The papers provide information that will help in the selection of plant species and growing conditions to optimize productivity and to ensure safe and nutritious diets for a CELSS. However, considerable research still is needed to relate plant growing environment in a CELSS to productivity and nutrient composition.

The success of this session is owed to the organizers and to the participants. The editor takes this opportunity to thank P. Chagvrdieff, the Deputy Organizer, who also chaired the session, and to R. Wheeler and F. Salisbury who refereed the papers.

Pergamon

Adv. Space Res. Vol. 20, No. 10, pp. 1971–1974, 1997

Printed in Great Britain
0273-1177/97 $17.00 + 0.00

PII: S0273-1177(97)00262-7

EFFECTS OF MODIFIED ATMOSPHERE ON CROP PRODUCTIVITY AND MINERAL CONTENT

P. Chagvardieff, B. Dimon, A. Souleimanov, D. Massimino, S. Le Bras, M. Péan and D. Louche-Teissandier

CEA, Direction des Sciences du Vivant, Départment d'Ecophysiologie Végétale et de Microbiologie, Centre de Cadarache, F-13108 Saint-Paul-Lez-Durance cédex, FRANCE

ABSTRACT

Wheat, potato, pea and tomato crops were cultivated from seeding to harvest in a controlled and confined growth chamber at elevated CO_2 concentration (3700 $\mu L \cdot L^{-1}$) to examine the effects on biomass production and edible part yields. Different responses to high CO_2 were recorded, ranging from a decline in productivity for wheat, to slight stimulation for potatoes, moderate increase for tomatoes, and very large enhancement for pea. Mineral content in wheat and pea seeds was not greatly modified by the elevated CO_2. Short-term experiments (17 d) were conducted on potato at high (3700 $\mu L \cdot L^{-1}$) and very high (20,000 $\mu L \cdot L^{-1}$) CO_2 concentration and/or low O_2 partial pressure (~ 20,600 $\mu L \cdot L^{-1}$ or 2 kPa). Low O_2 was more effective than high CO_2 in total biomass accumulation, but development was affected: Low O_2 inhibited tuberization, while high CO_2 significantly increased production of tubers.

INTRODUCTION

Controlled and confined environments where humans and plants will be placed during functioning of a Controlled Ecological Life Support System, or CELSS, provide an opportunity to examine the influence of modified atmosphere on plant growth and productivity. Numerous studies have contributed to the characterization of elevating CO_2 concentration in a range compatible with terrestrial global change (Kimball, 1983), fewer data are available on the effects of optimal and supraoptimal CO_2 concentrations on plants. To determine the possibility of maintaining plants and human in the same atmospheric compartment, it is relevant to examine long-term effects of high CO_2 on plants, at levels that may occur in cabin atmospheres (Wheeler et al., 1994). Futhermore, one might wonder whether plant growth could be affected by low O_2 concentration, if separate compartments were designed for optimizing crop productivity. The objective of this study was to compare productivity and mineral content of various crops cultivated together in a sealed chamber at high CO_2 concentration. A preliminary study was also conducted to examine the effect of low O_2 partial pressure and/or high CO_2 concentrations on growth of potato.

MATERIAL AND METHODS

Long-term experiments, from seed to harvest, were conducted in a controlled and confined growth chamber developed at CEA/Cadarache (42 m^3 volume, 12 m^2 usable area). CO_2 concentrations of 350 and 3700 $\mu L \cdot L^{-1}$ were regulated by using an infrared analyser (MAIHAK UNOR 6N). Various crops were cultivated at the same time under similar conditions : PPF was 530 $\mu mol \cdot m^{-2} \cdot s^{-1}$ provided by metal halide lamp (HQI 400W, OSRAM); photoperiod was 16 h light and 8 h dark; temperature was 25 °C/20 °C (day/night); relative humidity was maintained at 65 %. Plants were grown in containers of 80 L vermiculite, irrigated with half-strength Hoagland solution (Hoagland and Arnon, 1950) from 660 mL to 1320 $mL \cdot d^{-1}$, increased as plants grew. The data presented are the replicates of four containers. Seeded crops were wheat cv. Yecora Rojo, durum wheat cv. Ardente, pea cv. Lincoln, tomatoes cvs. Barnauskij Konservnyj and Early Arctic, both parthenocarpic, and potatoes cvs. Haig and Serrana, both of wich were micropropagated *in vitro* before cultivation. Short-term experiments (17 days) were conducted with potato cv. Haig in the C_23A chambers under modified atmosphere (Fabreguettes et al.,1994) in the same conditions as above except PPF was 350 $\mu mol \cdot m^{-2} \cdot s^{-1}$. Modified atmospheres applied to plantlets acclimated under ambient air conditions were: 350 $\mu L.L^{-1}$ CO_2/ 20.6 or 2 kPa O_2; 3700 $\mu L \cdot L^{-1}$ CO_2/ 20.6 kPa O_2; 20,000$\mu L \cdot L^{-1}$ CO_2/ 20.6 or 2 kPa O_2. Gas mixtures were regulated as described previously (Gerbaud and André, 1980). Plant measurements of daily photosynthesis, fresh and dry mass, and mineral content of tissues were determined as previously described (Chagvardieff et al., 1994). Harvest index was defined as the dry-mass ratio of edible part over total biomass (edible part + aerial biomass).

RESULTS AND DISCUSSION

Harvest characteristics of crops grown under 350 and 3700 $\mu L \cdot L^{-1}$ CO_2 are presented in Table 1. For potatoes, high CO_2 slightly increased tuber production by 16 % and 6 % for cvs. Haig and Serrana respectively, with an opposite effect in aerial biomass accumulation. Wheat cv. Yecora Rojo had a marked drop in grain yield (-50 %) at high CO_2, mainly due to a reduction in total biomass accumulation; durum wheat cv. Ardente also decreased aerial biomass, with only a very slight decrease in grain yield (-9 %). Pea biomass production was extremely stimulated by high CO_2, with a particular beneficial effect on seed yield (+245 %). High CO_2 was very beneficial to tomato crop and stimulated both total biomass and fruit production in both cultivars.

Table 1. Harvest characteristics (g $DW \cdot m^{-2}$) of various crops grown under 350 and 3700 $\mu L \cdot L^{-1}$ CO_2

Crop	Cultivar	CO_2 ($\mu L \cdot L^{-1}$) 350 aerial biomass	350 edible part	350 harvest index	3700 aerial biomass	3700 edible part	3700 harvest index	% change in edible part (3700/350)
Potato	*Haig*	317	560	64	504	648	56	+16
	Serrana	421	596	59	372	634	63	+6
Wheat	*Yecora rojo*	899	580	39	654	286	30	-50
	Ardente	1003	370	27	626	340	35	-9
Pea	*Lincoln*	1283	591	32	1892	2040	52	+245
Tomato	*Barnauskij Konservnyj*	248	257	51	294	409	58	+59
	Early Artic	168	123	43	216	180	45	+46

These results are roughly consistent with other data obtained under various CO_2 enrichment studies. Increasing CO_2 concentration on potato crop is favourable to tuber yield in moderate proportion (Cao et al., 1994; Wheeler et al., 1994), whereas supraoptimal CO_2 as been shown to decrease wheat grain yield (Bugbee et al., 1994). In our experiments, wheat cv. Yecora Rojo yield was depressed by high CO_2 but the reduced vegetative growth was unexpected. The large drop observed in this

experiment could be related to an interaction between CO_2 and PPF as mentioned by Wheeler et al. (1991), or effects of the high CO_2 on water uptake (Wheeler et al., 1994). In the same way, the only slight decrease in yield of durum wheat, which is more drought resistant than standard wheat, could be explained by its better water use under high CO_2. A very high yield increase in pea is reported for the first time, the pea crop seems very responsive to high CO_2. Tomato growth and fruit yield were stimulated by high CO_2, and in a consistent manner with the data usually obtained (Yelle et al., 1987).

While numerous attempts have been made to maximize crop productivity in artificial conditions for CELSS, little information is available on mineral composition. Mineral content of edible parts harvested under high CO_2 is presented in Table 2. Mineral characteristics of durum wheat cv. Ardente were unaffected by high CO_2, while wheat cv. Yecora Rojo showed a slightly lower N content at high CO_2, resulting in an increased C/N ratio. High CO_2 decreased pea N, P, and Ca content in comparison with ambient CO_2. There were only slight variations in some mineral contents, showing no dramatic effects of high CO_2. These results are consistent with the hypothesis of mass dilution by carbohydrate synthesis under elevated CO_2 concentration (Porter and Grodzinski, 1989).

Table 2. Mineral content of edible part of plants grown under 350 and 3700 $\mu L \cdot L^{-1}$ CO_2. (% DW)

Crop	CO_2 ($\mu L \cdot L^{-1}$)	C / N	N %	P %	K %	Ca %	Mg %
Wheat	350	12.94	3.46	0.51	0.59	0.033	0.205
(cv Ardente)	3700	12.92	3.47	0.57	0.66	0.034	0.208
Wheat	350	11.68	3.83	0.48	0.45	0.022	0.196
(cv Yecora Rojo)	3700	12.68	3.52	0.52	0.46	0.025	0.202
Pea	350	8.41	5.28	0.64	1.41	0.133	0.191
(cv Lincoln)	3700	8.94	4.97	0.43	1.45	0.101	0.183

A preliminary study on growth under high (3700 $\mu L \cdot L^{-1}$) and very high (20,000 $\mu L \cdot L^{-1}$) CO_2 concentration and/or low O_2 partial pressure (2 kPa) has been conducted on potato cv. Haig. After 17 days under modified atmospheres (Table 3), a significant effect of low O_2 was detected on shoot dry mass, regardless of the CO_2 concentration. At ambient O_2, if no effect was measured on shoot dry mass, high and very high CO_2 induced the production of larger tubers than under ambient CO_2. No tuber formation was detected under low O_2. Total carbon accumulation varied, increasing noticeably from ambient CO_2 treatment to very high CO_2 under ambient O_2, and increasing slightly with CO_2 under low O_2.

Table 3. Biomass distribution and total carbon accumulation in potatoes (cv. *Haig)* after 17 days under modified atmosphere conditions

CO_2 ($\mu L \cdot L^{-1}$) O_2 (kPa)	350 20.6	3700 20.6	20000 20.6	350 2.0	20000 2.0
Shoots (g · plant^{-1} DM)	4.41 ± 0.26^{a}	4.85 ± 0.22^{a}	5.03 ± 0.26^{a}	10.22 ± 0.59^{b}	10.08 ± 0.71^{b}
Tubers (g · plant^{-1} DM)	0.30 ± 0.08	4.16 ± 0.30^{a}	4.56 ± 0.57^{a}	-	-
Total C accumulated (g · plant^{-1})	2.17	3.29	4.05	4.17	4.61

[a, b] Different letters indicate significant differences at $P < 0.05$

Only few data are available for crop cultivation under low O_2 (Parkinson et al.. 1974). At the photosynthetic level, low O_2 could mimic high CO_2 if at an appropriate ratio, but its long term effect is quite unknown. Total biomass accumulation may be largely promoted under low O_2, but the effect on reproductive phase of growth seems somehow deleterious in soybean (Quebedeaux and Hardy, 1975) and wheat (Gerbaud et al., 1988). In potato crop under ambiant O_2 high CO_2 rapidly induced tuber formation (Table 3) but only slightly increased final tuber yield (Table 1), so that the major effect of CO_2 on potato yield seems to be an acceleration of the growth cycle. Under low O_2, no tuberization occurred in the short-term; long-term effects of low O_2 on vegetative phase remain to be investigated.

ACKNOWLEDGMENTS

A.S. was supported by a grant from Ministère de la Recherche et de la Technologie (France). The authors are gratefull for technical assistance of Ecotechnie team (CEA/Cadarache) and thank Dr. Tibbitts for providing potato cv. Haig, Dr. Bugbee for wheat cv. Yecora Rojo and Dr. Philouze for tomato cultivars. This work was partly sponsored by L'Air Liquide.

REFERENCES

Bugbee, B., B. Spanarkel, S. Johnson, O.Monje and G. Koerner, CO_2 crop growth enhancement and toxicity in wheat and rice, *Adv. Space Res.*,14(11), 257-267 (1994)

Cao, W., T.W.Tibbitts and R.M.Wheeler, Carbon dioxide interaction with irradiance and temperature in potatoes, *Adv. Space Res.*,14(11), 243-250 (1994)

Chagvardieff, P., T. d'Aletto and M.André, Specific effects of irradiance and CO_2 concentration doublings on productivity and mineral content in lettuce, *Adv. Space Res.*, 14(11), 269-275 (1994)

Fabreguettes, V., F.Gibiat, J.Pintena, D.Vidal and M.André, The $C_2$3A system: a tool for global control of plant environment and exchange measurements, *24th Int.Conf.Environ.Syst.(ICES), Friedrischaffen, June 20-24, 1994.* Ed.SAE Technical Papers Series, n°941544, Warendale, USA (1994)

Gerbaud, A. and M.André, Effect of CO_2, O_2 and light on photosynthesis and photorespiration in wheat, *Plant Physiol.*,66, 1032-1036 (1980)

Gerbaud, A., M.André, J-P.Gaudillère and A.Daguenet, The influence of reduced photorespiration on long-term growth and development of wheat, *Physiol.Plant.*, 73, 479-485 (1988)

Hoagland, D.A. and D.I. Arnon, The water-culture method for growing plants without soil, Circular 347, California Agricultural Experiment Station, Sacramento, CA, USA (1950)

Kimball, B.A., Carbon dioxide and agricultural yield: An assemblage and analysis of 430 prior observations, *Agronomy Journal*, 75, 779-788 (1983)

Parkinson, K.J., H.L.Penman and E.B.Tregunna, Growth of plant in different oxygen concentrations, *J.Exp.Bot.*,25, 132-145 (1974)

Porter, M.A. and B.Grodzinski, Growth of bean in high CO_2 : effects on shoot mineral composition, *J.Plant Nutr.*,12, 129-144 (1989)

Quebedeaux, B. and R.W.F.Hardy, Reproductive growth and dry matter production of Glycine max (L) Merr, in response to oxygen concentration, *Plant Physiol.*,55, 102-107 (1975)

Wheeler, R.M., T.W.Tibbitts and A.H.Fitzpatrick, Carbon dioxide effects on potato growth under different photoperiods and irradiance, *Crop Sci.*,31, 1209-1213 (1991)

Wheeler, R.M., C.L.Mackowiak, J.C.Sager and W.M.Knott, Growth of soybean and potato at high CO_2 partial pressures, *Adv. Space Res.*,14(11), 251-255 (1994)

Yelle, S., A.Gosselin and M.J.Trudel, Effect of atmospheric CO_2 concentration and root-zone temperature on growth, mineral nutrition and nitrate reductase activity of greenhouse tomato, *J.Amer.Soc.Hort.Sci.*,112(6), 1036-1040 (1987)

Pergamon

Adv. Space Res. Vol. 20, No. 10, pp. 1975–1978, 1997
Published by Elsevier Science Ltd on behalf of COSPAR
Printed in Great Britain
0273-1177/97 \$17.00 + 0.00

PII: S0273-1177(97)00263-9

EFFECT OF ELEVATED CARBON DIOXIDE ON NUTRITIONAL QUALITY OF TOMATO

R. M. Wheeler*, C. L. Mackowiak**, G. W. Stutte**, N. C. Yorio** and W. L. Berry***

**NASA Biomedical Office (Mail Code JJ-G) and*
***Dynamac Corp. (Mail Code DYN-3), Kennedy Space Center, FL 32899, U.S.A.*
****Biomedical and Environmental Sciences, University of California, Los Angeles, CA 90024, U.S.A.*

ABSTRACT

Tomato (*Lycopersicon esculentum* Mill.) cvs. Red Robin (RR) and Reimann Philipp (RP) were grown hydroponically for 105 d with a 12 h photoperiod, 26 °C / 22 °C thermoperiod, and 500 $\mu mol \cdot m^{-2} \cdot s^{-1}$ PPF at either 400, 1200, 5000, or 10,000 $\mu mol \cdot mol^{-1}$ (0.04, 0.12, 0.50, 1.00 kPa) CO_2. Harvested fruits were analyzed for proximate composition, total dietary fiber, nitrate, and elemental composition. No trends were apparent with regard to CO_2 effects on proximate composition, with fruit from all treatments and both cultivars averaging 18.9 % protein, 3.6 % fat, 10.2 % ash, and 67.2 % carbohydrate. In comparison, average values for field-grown fruit are 16.6 % protein, 3.8 % fat, 8.1 % ash, and 71.5 % carbohydrate (Duke and Atchely, 1986). Total dietary fiber was highest at 10,000 $\mu mol \cdot mol^{-1}$ (28.4 % and 22.6 % for RR and RP) and lowest at 1000 $\mu mol \cdot mol^{-1}$ (18.2 % and 15.9 % for RR and RP), but showed no overall trend in response to CO_2 . Nitrate values ranged from 0.19 % to 0.35 % and showed no trend with regard to CO_2 . K, Mg, and P concentrations showed no trend in response to CO_2, but Ca levels increased from 198 and 956 ppm in RR and RP at 400 $\mu mol \cdot mol^{-1}$, to 2537 and 2825 ppm at 10,000 $\mu mol \cdot mol^{-1}$. This increase in Ca caused an increase in fruit Ca/P ratios from 0.07 and 0.37 for RR and RP at 400 $\mu mol \cdot mol^{-1}$ to 0.99 and 1.23 for RR and RP at 10,000 $\mu mol \cdot mol^{-1}$, suggesting that more dietary Ca should be available from high CO_2-grown fruit. Published by Elsevier Science Ltd on behalf of COSPAR.

INTRODUCTION

Information on nutritional quality of plant-produced foods has typically come from field-grown plants exposed to changing meteorological conditions, varying levels of nutrient availability, and nominal ambient CO_2 concentrations of approximately 350 $\mu mol \cdot mol^{-1}$. In addition, crops in the field can experience different degrees of stress from insect herbivory and weed competition. In contrast, food from plants grown in a Controlled Ecological Life-Support System (CELSS) will be maintained under controlled environment conditions free of such stresses. Most likely this would involve the use of electrical lighting, recirculating hydroponic culture (to conserve water and nutrients), and elevated CO_2 concentrations to increase photosynthetic rates and productivity. Indeed, because of the closed nature of a CELSS, CO_2 may even build to levels beyond what is optimal for plants, which typically ranges between 1000 and 2000 $\mu mol \cdot mol^{-1}$ (0.10 to 0.20 kPa) for C_3 plants (Wheeler *et al.*, 1994a; Bugbee *et al.*, 1994).

At NASA's Kennedy Space Center, we have been particularly interested studying the effects of CO_2 concentration on plant growth and composition for a CELSS. The following series of experiments was conducted in plant growth chambers to study to influence of CO_2 concentrations on the nutritional and elemental composition of tomato fruit. Tomatoes have been suggested as a crop for CELSS consideration (Hoff *et al.*, 1982) and have been the focus of extensive CO_2 testing in the past, particularly in greenhouse settings (Madsen, 1974; Hicklenton and

Jolliffe, 1980; Peet and Willits, 1984); however, few data are available on the effects of CO_2 on the nutritional composition of the fruits.

MATERIALS AND METHODS

Tomato (*Lycopersicon esculentum* Mill.) cvs. Red Robin (RR) and Reimann Philipp 75/59 (RP) plants were grown from seed in a walk-in type plant growth chamber. Plants were spaced at 6.7 plants m^{-2} and grown in hydroponic culture trays using a nutrient film technique (Wheeler *et al.*, 1993, 1996). A modified 0.5 x Hoagland nutrient solution (Wheeler *et al.*, 1993, 1996) was circulated through trays at 1.0 to 1.5 $L \cdot min^{-1}$, with solution pH controlled to 5.8 with additions of dilute (0.4 $mol \cdot L^{-1}$) HNO_3, and electrical conductivity adjusted daily to 0.12 $S \cdot m^{-1}$ with additions of a complete stock solution. Lighting was provided by high-pressure sodium lamps and maintained at 500 $\mu mol \cdot m^{-2} \cdot s^{-1}$ PPF with a 12 h photoperiod. Temperatures were cycled between 26 °C (light) and 22 °C (dark), and relative humidity was maintained at 65%. Carbon dioxide levels were controlled to 400, 1200, 5000, 10,000 $\mu mol \cdot mol^{-1}$ (0.04, 0.12, 0.50, and 1.00 kPa) using an infrared gas analyzer and dedicated computer control system. Plants were grown for a total of 105 d, but beginning at 63 d, fruits were harvested weekly as they turned red.

Fruits were macerated in a kitchen blender and then freeze dried for analysis. Dried tissue was analyzed for proximate composition, total dietary fiber (TDF), and nitrate by a commercial laboratory (Nutrition International, Dayton, NJ, USA) following standard AOAC (1990) procedures: moisture by vacuum oven (AOAC method 934-01); ash by muffle furnace (AOAC method 900.02); protein by Kjeldahl nitrogen with 6.25 conversion factor (AOAC method 981.10); fat by acid hydrolysis and ether extraction (AOAC methods 920.39 and 922.06); carbohydrate by difference [100% - (% ash + % fat + % protein)]; nitrate-nitrogen by xylenol method (AOAC method 935.48); and total dietary fiber by enzymatic-gravimetric method (AOAC method 985.29). Tissue elemental composition was analyzed using DC arc inductively-coupled plasma (ICP) spectroscopy (Alexander and McAnulty, 1981).

RESULTS AND DISCUSSION

Analysis of proximate composition and TDF showed no trends with regard to CO_2 concentration (Table 1). (Note, only single samples of pooled tissue from different plants were analyzed in triplicate, hence no statistical analyses were performed). Red Robin (RR) fruit had higher fat and TDF levels than Reimann Philipp (RP) at all CO_2 levels. However, fat levels are low in tomato fruit (Duke and Atchely, 1986) and hence this difference is likely inconsequential for dietary considerations. The higher TDF in RR fruit may be a result of RR having smaller fruit than RP, resulting in proportionately more epidermis (skin) tissue in the samples. In related studies with soybean and potato grown at different CO_2 concentrations, soybean seed showed increased crude fiber with increased CO_2 , but no other difference were noted (Wheeler *et al.*, 1994b). McKeehen *et al.* (1996a) reported decreased ash in lettuce and radish tissue grown at increased CO_2 , but as with soybean and potato, no other trends were apparent.

If proximate composition data for both cultivars and all CO_2 treatments are combined and compared to the average of six entries for field-grown tomato reported by Duke and Atchely (1986), results show that the controlled environment-grown fruits were slightly higher in protein and ash than field-grown fruits (Fig. 1). This trend is consistent with previous analyses of crops grown in NASA's biomass production chamber (Wheeler *et al.*, 1996), yet the distributions of observations (shown as bars on Fig. 1) overlap for all measurements, suggesting that there were no significant differences between fruits from our studies and field-grown tomatoes.

A comparison of elemental analysis showed no trend in response to CO_2 for tissue levels of potassium (K), magnesium (Mg), phosphorus (P), or nitrate-nitrogen (NO_3 -N) (Table 1). However, fruit calcium content showed a clear increase with CO_2 . The increase in fruit Ca levels with increased CO_2 resulted in a similar increase in tissue Ca/P ratio (Fig. 2). The ratio of Ca/P in the diet is known to affect the absorption and retention of calcium in humans (McKeehen *et al.*, 1996a). Except for leafy vegetables, Ca/P ratios are typically low in plant foods, particularly for hydroponically-grown plants that are given plentiful phosphorus in the nutrient solution (McKeehen *et al.*, 1996a). Calcium uptake by plants is known to be closely linked with water movement within the tissue (Wiebe *et al.*, 1977) , and extreme CO_2 levels (*e.g.*, 5000 and 10,000 $\mu mol \cdot mol^{-1}$) have been shown to

Table 1. Effect of CO_2 on Proximate and Elemental Composition of Tomato Fruit

CO_2	cv.	Protein	Fat	Ash	CHO	TDF	K	Ca	Mg	P	NO_3-N
(μmol · mol^{-1})		(%)	(%)	(%)	(%)	(%)	(ppm)	(ppm)	(ppm)	(ppm)	(%)
400	RR	17.7	2.8	10.0	69.5	20.9	33016	198	3056	2752	0.32
	RP	18.2	1.6	5.8	74.4	16.6	39476	956	3201	2558	0.19
1200	RR	17.2	6.2	10.1	66.4	18.2	45491	771	2486	2580	0.14
	RP	17.6	1.4	10.8	70.3	15.9	46127	1714	2813	2518	0.24
5000	RR	21.3	5.7	11.3	61.6	21.5	45575	1662	2072	2328	0.15
	RP	19.1	2.5	12.9	65.5	16.3	55444	2308	2843	2277	0.35
10,000	RR	20.4	5.0	10.4	64.2	28.4	46512	2537	2927	2574	0.25
	RP	20.1	4.0	10.2	65.7	22.6	42540	2825	3087	2294	0.21

cv. = Cultivar; CHO = Carbohydrate; TDF = Total Dietary Fiber; RR = Red Robin; RP = Reimann Philipp.

increase water flow through soybean and potato plants (Wheeler *et al.*, 1993, 1994a). Transpiration and water uptake by tomatoes in these tests were highest at the 5000 and 10,000 treatments as well (Mackowiak *et al.*, unpublished), suggesting that more calcium might have been moved to growing tissues in these treatments. However, related studies by McKeehen *et al.* (1996b) noted no effect of increased CO_2 on Ca content of lettuce leaves. Thus the increase in Ca from CO_2 enrichment may occur only in organs where transpiration rates are typically low (*e.g.*, fruits and tubers).

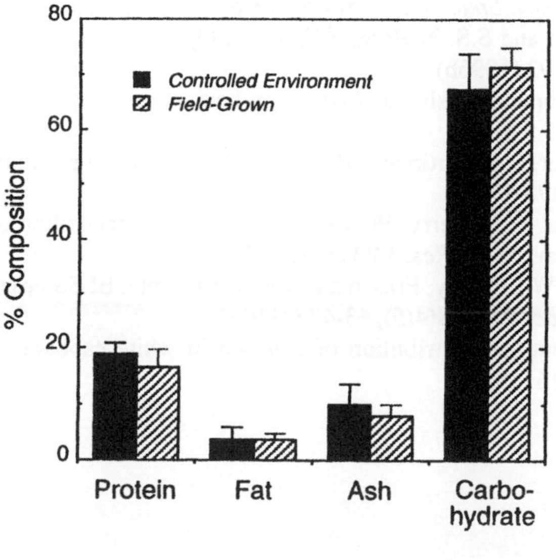

Fig. 1. Proximate composition of controlled environment and field-grown tomato fruit. Bars indicate distribution ranges for observations.

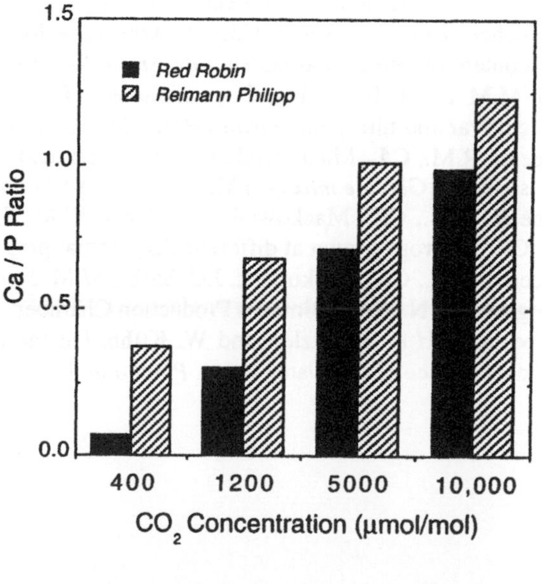

Fig. 2. Carbon dioxide effect on calcium to phosphorus ratio in tomato fruit from cvs. Red Robin and Reimann Philipp 75/59.

CONCLUSION

Growing tomato cultivars Red Robin and Reimann Philipp 75/59 at different CO_2 concentrations had little effect on proximate and elemental composition of their fruit, with the exception of tissue calcium concentration, which increased with increased CO_2. Red Robin tended to be higher than Reimann Philipp in total dietary fiber, but this may be related to the smaller size of the fruits and the higher proportion of epidermis (skin) in the fruit mass. When all treatments were averaged and compared with field-grown tomatoes, ash and protein were slightly higher than field data, but differences were not significant, suggesting that tomatoes should do well in controlled environment settings that might be used for a CELSS.

REFERENCES

Alexander, G. and L. McAnulty, Multielement analysis of plant-related tissues and fluids by optical emission spectrometry. *J. Plant Nutr.*, 3, 55-59 (1981).

AOAC, Official methods of analysis (15th edition). Association of Official Analytical Chemists, Washington, DC. USA (1990).

Bugbee, B., B. Spanarkel, S. Johnson, O. Monje, and G. Koerner, CO_2 crop growth enhancement and toxicity in wheat and rice. *Adv. Space Res.* 14(11), 257-267.

Duke, J.S., and A.A. Atchely, CRC Handbook of Proximate Analysis Tables of Higher Plants. CRC Press, Boca Raton, FL, USA (1986).

Hicklenton, P.R. and P.A. Jolliffe, Alterations in the physiology of CO_2 exchange in tomato plants grown in CO_2-enriched atmospheres. *Can. J. Bot.*, 58, 2181-2189 (1980).

Hoff, J.E., J.M. Howe, and C.A. Mitchell, Nutritional and cultural aspects of plant species selection for a regenerative life support system. NASA Contract Report 166324, Moffett Field, CA, USA (1982).

Madsen, R. Effect of CO_2 concentration on growth and fruit production of tomato plants. *Acta Agricul. Scand*, 24, 242-246 (1974).

Maynard, D.M., A.V. Barker, P.L. Minotti, and N.H. Peck, Nitrate accumulation in vegetables. *Adv. in Agronomy*, 28, 71-118 (1976).

McKeehen, J.D., C.A. Mitchell, R.M. Wheeler, B. Bugbee, and S.S. Nielsen, Excess nutrients in hydroponic solutions alter nutrient content of rice, wheat, and potato. *Adv. Space Res.*, 18(4/5), 73-83 (1996a).

McKeehen, J.D., D.J. Smart, C.L. Mackowiak, R.M. Wheeler, and S.S. Nielsen, Effect of CO_2 levels on nutrient content of lettuce and radish. *Adv. Space Res.* 18(4/5), 85-92 (1996b).

Peet, M.M. and D.H. Willits, CO_2 enrichment of greenhouse tomatoes using a closed-loop heat storage: Effects of cultivar and nitrogen. *Scientia Hort.*, 24, 21-32 (1984).

Wheeler, R.M., C.L. Mackowiak, L.M. Siegriest, and J.C. Sager, Supraoptimal carbon dioxide effects on growth of soybean [*Glycine max* (L.) Merr.]. *J. Plant Physiol.* 142, 173-178.

Wheeler, R.M., C.L. Mackowiak, J.C. Sager, W.M. Knott, and W.L. Berry, Proximate nutritional composition of CELSS crops grown at different CO_2 partial pressures. *Adv. Space Res.* 14(11), 171-176 (1994).

Wheeler, R.M., C.L Mackowiak, J.C. Sager, W.M. Knott, and W.L. Berry, Proximate composition of CELSS crops grown in NASA's Biomass Production Chamber. *Adv. Space Res.* 18(4/5), 43-47 (1996).

Wiebe, H.J., H.P. Schätzler, and W. Kühn, On the movement and distribution of calcium in white cabbage in dependence of the water status. *Plant and Soil*, 48, 409-416 (1977).

Pergamon

Adv. Space Res. Vol. 20, No. 10, pp. 1979–1988, 1997

Printed in Great Britain
0273-1177/97 $17.00 + 0.00

PII: S0273-1177(97)00264-0

CONTROLLED ENVIRONMENTS ALTER NUTRIENT CONTENT OF SOYBEANS

L. J. Jurgonski*, D. J. Smart*, B. Bugbee**, S. S. Nielsen*

**Department of Food Science, Purdue University, West Lafayette, Indiana, 47907-1160, U.S.A.*
***Department of Plants, Soils, and Biometerology, Utah State University, Logan, Utah, 84322-4820, U.S.A.*

ABSTRACT

Information about compositional changes in plants grown in controlled environments is essential for developing a safe, nutritious diet for a Controlled Ecological Life-Support System (CELSS). Information now is available for some CELSS candidate crops, but detailed information has been lacking for soybeans. To determine the effect of environment on macronutrient and mineral composition of soybeans, plants were grown both in the field and in a controlled environment where the hydroponic nutrient solution, photosynthetic flux (PPF), and CO_2 level were manipulated to achieve rapid growth rates. Plants were harvested at seed maturity, separated into discrete parts, and oven dried prior to chemical analysis. Plant material was analyzed for proximate composition (moisture, protein, lipid, ash, and carbohydrate), total nitrogen (N), nonprotein N (NPN), nitrate, minerals, amino acid composition, and total dietary fiber. The effect of environment on composition varied by cultivar and plant part. Chamber-grown plants generally exhibited the following characteristics compared with field-grown plants: 1) increased total N and protein N for all plant parts, 2) increased nitrate in leaves and stems but not in seeds, 3) increased lipids in seeds, and 4) decreased Ca:P ratio for stems, pods, and leaves. These trends are consistent with data for other CELSS crops. Total N, protein N, and amino acid contents for 350 ppm CO_2 and 1000 ppm CO_2 were similar for seeds, but protein N and amino acid contents for leaves were higher at 350 ppm CO_2 than at 1000 ppm CO_2. Total dietary fiber content of soybean leaves was higher with 350 ppm CO_2 than with 1000 ppm CO_2. Such data will help in selecting of crop species, cultivars, and growing conditions to ensure safe, nutritious diets for CELSS.

INTRODUCTION

Soybean is a candidate crop species for a Controlled Ecological Life-Support System (CELSS) because of the relatively high lipid and protein content of its seeds (among the highest of all CELSS candidate crops) and high nutritional quality (Hoff *et al.*, 1982). Soybeans provide a nearly complete protein for vegetarian diets, which typically are low in lipids. Proteins, lipids, and other seed components may change in concentration with growing environment. McKeehen *et al.* (1996a,b) evaluated effects of field vs. controlled environments (CEs) on the composition of edible and inedible plant parts of the CELSS candidate crops rice, wheat, potato, lettuce, and radish. Some information has been available on the composition of soybeans in CEs (Wheeler *et al.*, 1990, 1994), but neither in the detail nor by the methods used to analyze other CELSS candidate crops.

CELSS studies with soybeans have focused on nitrogen (N) nutrition in hydroponic culture (Tolley-Henry and Raper, 1986a,b), as well as irradiance and temperature (Thomas and Raper, 1976; Raper and Thomas, 1978). Most studies of soybean production at various carbon dioxide (CO_2) levels have been done under field conditions and focused on plant development and yield (Rogers *et al.*, 1984, 1986; Jones *et al.*, 1985, Allen *et al.*, 1988, Baker *et al.*, 1989). Noting the relevance and importance of CO_2 levels in a CELSS, Wheeler *et al.* (1990, 1994) studied the proximate composition and caloric value of seeds, pods, leaves, and stems of soybeans grown at four CO_2 levels in the Biomass Production Chamber (BPC) at the Kennedy Space Center. Wheeler *et al.* (1990, 1994) found significant differences between soybean cultivars in proximate composition and some effects of CO_2 levels on protein, lipid, and crude fiber contents. Later, Wheeler *et al.* (1996) reported higher crude fiber contents of soybean seeds from the BPC than from field literature values.

This study builds upon the studies of Wheeler *et al.* (1990, 1994, 1996), using improved methods to determine the effect of environment on chemical composition of various plant parts of soybean cultivars studied by CELSS investigators. The data allow for comparisons with similar data for other CELSS candidate crops (McKeehen *et al.* 1996 a,b), provide information for crop production, and supply information to help develop safe and nutritionally adequate diets for a CELSS.

MATERIALS AND METHODS

Plant Growth, Harvest, and Handling

Soybean (*Glycine max*) cultivars 'Hoyt,' 'PI-494.525,' and 'McCall' were grown under field conditions at Utah State University, and cultivar 'Hoyt' was additionally grown under CE conditions. The CE conditions were as follows: soilless media (1:1, vol/vol, peat:perlite), 12h photoperiod, cool-white fluorescent lamps, 250 $\mu mol \cdot m^{-2} \cdot s^{-1}$ (11 mol $m^2 d^{-1}$) photosynthetic photon flux, 26°C days, 22°C nights, 70% relative humidity, 350 or 1000 ppm CO_2, and harvested at 80d after planting. Field conditions were as follows: silt loam soil, 16h decreasing to 12h photoperiods, 50 decreasing to 30 $mol \cdot m^{-2} \cdot d^{-1}$ photosynthetic photon flux, variable but above normal temperatures, approximately 30% relative humidity, 350 ppm CO_2, harvested 120 days after planting. The concentration of elements used in the nutrient solution for CE studies were as follows for starter and refill solutions, respectively: $Ca_3(NO_3)_2$ - 1, 1 mM; KNO_3 - 1, 5 mM; KH_2PO_4 - 0.5, 0.75 mM; $MgSO_4$ - 0.5, 0.75 mM; K_2SiO_3 - 0.1, 0.1 mM; $Fe(NO_3)_3$ - 5, 1.5 μM; Fe-HEDTA - 15, 5 μM; $MnCl_2$ - 6, 9 μM; $ZnSO_4$ - 6, 2 μM; H_3BO_3 - 40, 40 μM; $CuSO_4$ - 0.6, 0.6 μM; Na_2MoO_4 - 0.09, 0.06 μM. Plant parts (seeds, leaves, stems, pods) were air dried, ground first with a Wiley mill to pass a 3-mm screen, and then with a Udy-Cyclone mill to pass a 1-mm screen.

Analysis of Plant Materials

Ground, triplicate samples were dried using a vacuum oven (AOAC Method 925.09) (AOAC, 1990), so that all data could be corrected for moisture content and expressed on a dry-mass basis (dmb). Samples were analyzed in triplicate for ash content using a muffle furnace (AOAC Method 923.03) (AOAC, 1990), as well as for lipid content by Soxhlet extraction with petroleum ether (AOAC Method 920.39B) (AOAC, 1990).

Because the standard micro-Kjeldahl procedure (AOAC Method 960.52) (AOAC, 1990) measures all protein N and some nitrate N, that method alone is not appropriate to determine the true protein content of plant materials that have a significant fraction of N as nitrate (Goyal and Hafez, 1990; Khanizadeh *et al.*, 1995). Therefore, protein N content was determined in duplicate after precipitating protein with 6% (w/v) trichloroacetic acid (TCA) (Bensadoun and Weinstien, 1976), and then measuring N content of the washed pellet by the micro-Kjeldahl method (AOAC Method 960.52) (AOAC, 1990; McKeehen, 1994). The TCA is thought to precipitate proteins of molecular mass greater than approximately 3,000-5,000 kDa. Total N content was determined in duplicate as described by Goyal and Hafez (1990), using a predigestion procedure to include nitrate N in the Kjeldahl digestion. Total NPN content was calculated as the difference between total N and TCA-N. Nitrate analysis was performed by the HPLC method of Thayer and Huffaker (1980). Nonnitrate NPN content was calculated as the difference between total NPN and nitrate N. All percent N values were multiplied by the factor 6.25 to obtain percent protein. While it is recognized that 5.71 is the common N to protein conversion factor for soybean seeds (Jones, 1931), it is probably inappropriate for inedible parts of plants. Therefore, protein content is reported based both on the 5.71 and the standard 6.25 factors (Table 1).

Total carbohydrate content was calculated by difference, using the TCA-N method to measure protein: % carbohydrate = 100% - (% ash + % lipid + % protein). Total dietary fiber content was determined by the enzymatic - gravimetric AOAC Method 985.29 (AOAC, 1990). Mineral analysis was performed on triplicate

Table 1. Proximate Composition of Soybeans Grown in the Field and Controlled Environment (CEs) at Two Levels of CO_2 Enrichment[1,2]

Plant	Growth Environ-ment	Soybean Cultivar	Proximate Composition (% dwb) Protein[3] x 6.25	x 5.71	Lipid	Ash	Carbo-hydrate[4]
Seed	Field	PI-494	51.0[B]	46.6	14.9[B]	6.0[B]	28.1
		McCall	41.4[D]	37.9	16.9[D]	6.0[B]	35.7
		Hoyt	46.6[a,A]	42.5	18.1[a,A]	5.6[a,A]	29.7
	CE-350[5]	Hoyt	54.1[b]	49.4	20.2[b]	5.9[b]	19.8
	CE-1000	Hoyt	52.6[b]	48.0	19.0[c]	5.9[b]	21.6
Stem	Field	PI-494	6.4[A]	5.9	na[6]	10.5[B]	
		McCall	4.4[B]	4.0	na	6.8[D]	
		Hoyt	5.7[a,A]	5.2	0.7[a]	8.1[a,A]	85.5
	CE-350	Hoyt	20.7[b]	18.9	na	9.2	
	CE-1000	Hoyt	19.1[c]	17.4	0.9[b]	8.2[a]	71.8
Pod	Field	PI-494	5.7[B]	4.7	na	10.5[B]	
		McCall	3.6[A]	3.3	na	11.0[D]	
		Hoyt	4.0[a,A]	3.7	0.8[a]	9.7[a,A]	85.5
	CE-350	Hoyt	7.2[b]	6.6	na	9.8[b]	
	CE-1000	Hoyt	6.7[b]	6.1	1.1[b]	10.5[c]	81.7
Leaf	Field	PI-494	16.2[B]	14.8	3.2[B]	18.5[B]	62.1
		McCall	12.4[D]	11.3	3.3[D]	18.7[D]	65.6
		Hoyt	18.5[a,A]	16.9	2.1[a,A]	14.4[a,A]	65.0
	CE-350	Hoyt	22.5[b]	20.5	2.5[b]	17.9[b]	56.5
	CE-1000	Hoyt	22.2[b]	20.3	2.3[c]	18.1[b]	57.4

[1]Values are the mean of triplicate analyses.
[2]Means within each plant part having a common letter ('Hoyt' in GC, lower case; field-grown cultivars in upper case) are not significantly different by t tests (unequal variances), p>0.05.
[3]Protein content calculated from total N determination.
[4]Calculated as % carbohydrate (dmb) = [100% - (% protein + % fat + % ash)]; % protein in this equation is based on % N x 6.25.
[5]Controlled environment samples grown at 350 ppm or 1000 ppm CO_2.
[6]na = not analyzed due to insufficient sample quantity.

ashed samples by Inductively Coupled Plasma-Atomic Emission Spectroscopy (ICP-AES), as described by McKeehen (1994). Amino acid analysis was performed according to standard hydrolysis and chromatography procedures by a commercial laboratory (AGP Limited, Courtland, Minnesota).

Data from proximate analyses and total N determination were analyzed using multiple t tests of independent samples with unequal variance as described by Steel and Torrie (1980).

RESULTS AND DISCUSSION

Comparison to Literature Values

The proximate composition of field-grown soybean seeds (Table 1) can be compared to USDA Handbook No. 8 values (Haytowitz and Matthews, 1986) of for protein, lipid, ash, and carbohydrate contents (%, dmb) for soybean seeds (43.7, 21.8, 5.3, and 33.0, respectively; protein calculated with factor of 6.25). Protein values are based on a modified Kjeldahl assay to measure total N, whereas handbook values are based on the standard Kjeldahl procedure. Handbook values in Table 1 also were corrected to a dry mass basis to allow for direct comparison. The protein content of field-grown soybeans differed significantly among the three cultivars tested, but values generally were higher than handbook values. High protein contents can occur in well fertilized field environments (Fowler *et al.*, 1990). Lipid contents of field-grown soybeans also differed significantly among cultivars and were lower than handbook values. This inverse relationship between protein and lipid content is typical for field-grown oilseeds (Hanson *et al.*, 1961).

Nitrogen Allocation

Total and protein N. In comparing the field-grown plants, 'PI-494.525' seeds and pods had the highest total N and protein N contents, while 'Hoyt' leaves contained the most total N and protein N. Compared with field-grown plants, 'Hoyt' soybeans grown under high-N, hydroponic culture (i.e. CE) generally had increased total N and protein N for all plant part at both 350 ppm and 1000 ppm CO_2 conditions (Table 2). The only exceptions to this trend were protein N for pods at both CO_2 conditions and protein N for leaves at 1000 ppm CO_2, for which levels were not significantly different. The increases are consistent with data for other CELSS candidate crops (McKeehen *et al.*, 1996a,b), and are probably caused by luxuriant uptake of N by the hydroponically grown plants.

Table 2. Percent Protein (% N x 6.25) as Derived from Total N, Trichloroacetic Acid (TCA) Precipitated N, Nonprotein N, and Nitrate N for Soybeans Grown in the Field and Controlled Environments (CEs) at Two Different Levels of CO_2 Enrichment[1]

Plant Part	Growth Environment	Soybean Cultivar	% Protein from Total N	% Protein from TCA Precipi - tated N	% Protein from Total NPN[2]	% Protein from Nitrate N	% Protein from Nonnitrate NPN[3]
Seed	Field	PI-494	51.0^{B}	39.7^{B}	11.3	0.0	11.30
		McCall	41.4^{D}	33.3^{D}	8.1	0.0	8.40
		Hoyt	$46.6^{a,A}$	$35.1^{a,A}$	11.5	0.02	11.48
	CE-350[4]	Hoyt	54.1^{b}	43.5^{b}	10.6	0.0	10.60
	CE-1000	Hoyt	52.6^{b}	41.4^{c}	11.2	0.0	11.20
Stem	Field	PI-494	6.4^{A}	3.7	2.7	0.02	2.68
		Mc Call	4.4^{B}	2.7	1.7	0.0	1.70
		Hoyt	$5.7^{a,A}$	2.7^{a}	3.0	0.02	2.98
	CE-350	Hoyt	20.7^{b}	7.3^{b}	13.4	.99	12.41
	CE-1000	Hoyt	19.1^{c}	6.6^{c}	12.5	1.04	11.46
Pod	Field	PI-494	5.7^{B}	3.7	2.0	0.02	1.98
		McCall	3.6^{A}	2.4	1.2	0.0	1.20
		Hoyt	$4.0^{a,A}$	2.6^{a}	1.4	0.02	1.38
	CE-350	Hoyt	7.2^{b}	3.1^{a}	4.1	0.09	4.01
	CE-1000	Hoyt	6.7^{b}	3.0^{a}	3.7	0.07	3.63
Leaf	Field	PI-494	16.2^{B}	12.1	4.1	0.05	4.05
		McCall	12.4^{D}	8.3	4.1	0.02	4.08
		Hoyt	$18.5^{a,A}$	13.5^{a}	5.0	0.02	4.99
	CE-350	Hoyt	22.5^{b}	17.2^{b}	5.3	0.95	4.35
	CE-1000	Hoyt	22.2^{b}	14.5^{a}	7.7	2.05	5.65

[1]Means within each plant part having a common letter ('Hoyt' in GC, lower case, field-grown cultivars in upper case) are not significantly different by t tests (unequal variances), p>0.05.
[2]Calculated by difference between % protein from total N and % protein from TCA precipitated N.
[3]Calculated by difference between % protein from total NPN and % protein from nitrate N.
[4]Controlled environment samples grown at 350 ppm and 1000 ppm CO_2.

Seeds, stems, and leaves of plants grown at 350 ppm CO_2 in the CE had significantly higher protein N contents than did those grown at 1000 ppm CO_2. This reduced protein content of leaves at high CO_2 levels is consistent with lower levels of Rubisco at high CO_2.

In testing 'McCall' and 'Pixie' soybeans in the BPC at 500, 1000, 2000, or 5000 ppm CO_2 (Wheeler *et al.*, 1990, 1994), seed protein levels (by standard Kjeldahl procedure) were highest at 1000 ppm (39.3 and 41.9%, respectively) and lowest at 2000 ppm (34.7 and 38.9%, respectively). Protein content of leaves, stems, and pods also varied with CO_2 level (Wheeler *et al.*, 1990, 1994). Wheeler *et al.* (1990, 1994) reported increased rather than decreased levels of leaf protein with increased CO_2 levels, but only for 5000 ppm CO_2.

The levels of total-NPN and NO_3-N varied between cultivars, plant parts, and growth conditions. Field-grown 'McCall' seeds, stems, and leaves generally had lower amounts of total N, protein N, total NPN, and NO_3-N than those plant parts of the other two cultivars. Field-grown plants did not accumulate NO_3-N, whereas plants grown in CEs did. For CE plants, seeds accumulated less NO_3-N than did any other plant part. This is consistent with the findings of McKeehen *et al.* (1996a) who showed that the typical edible portions of rice, wheat, and potato do not accumulate NO_3-N when grown in CEs with excess N. Vegetative material is known to accumulate nitrate and other NPN (Aldrich, 1980). However, soybean leaves, stems, and pods accumulated less NO_3 as a percentage of total N than did the vegetative portions of other CELSS candidate crops (McKeehen *et al.*, 1996a,b).

High NO_3 levels in foods raise human safety concerns (Swann, 1975; Preussmann and Stewart, 1984) but may not be an issue for crops that accumulate nitrate only in inedible parts (McKeehen *et al.* 1996a). For CELSS crops such as lettuce that accumulate NO_3 in edible parts (McKeehen *et al.*, 1996b), nitrate levels may be controlled by limiting NO_3 concentration in the nutrient solution. Plant nonitrate NPN typically includes nucleic acids and low molecular weight organic N compounds such as peptides, amides, and amino acids (Goyal and Huffaker, 1984; Imafidon and Sosulski, 1990). We currently are characterizing the nonnitrate NPN of CE-grown CELSS crops to assess its implications for human safety and nutrition.

Amino acid contents. The seed amino acid contents of field-grown plants differed considerably among cultivars (Table 3), as did the total N and protein N contents (Table 2). In numerous cases for specific amino acids, the range of amino acid contents for the three cultivars did not include the literature value. Considering all the amino acids, the ratios of literature seed values:field 'Hoyt' seed values were 0.84-1.29. For field-grown plants, 'PI-494.525' and 'McCall' seeds had the highest and lowest levels, respectively, of all amino acids, total N, and protein N. As expected, because CE 'Hoyt' seeds had higher levels of total N and protein N than did field seeds, the CE 'Hoyt' seeds likewise had higher amino acid levels. Ratios of mean values for all three cultivars of field seeds:mean value for 350 and 1000 ppm CO_2 CE seeds were 0.81-0.91, while ratios of field 'Hoyt' seed values: mean values for 350 and 1000 ppm CO_2 CE seeds were 0.81 - 0.94. Methionine and cystine were the first limiting amino acids for seeds of all cultivars and growing conditions. The higher total N and protein N contents of seeds and leaves from plants grown at 350 ppm CO_2 compared to 1000 ppm CO_2 translated to increased amino acid levels in 350 ppm CO_2 leaves, butsuch increases were not the case for 350 ppm CO_2 seeds. For each amino acid, the ratios of 350:1000 ppm CO_2 conditions for seeds were 0.98 - 1.08, but for leaves the ratios were 0.72 - 1.20.

Lipid and Ash Contents

The lipid contents of CE seeds and leaves of 'Hoyt' soybeans were significantly higher than those of field-grown seeds and leaves (Table 1). Therefore, both lipid and protein contents of seeds were increased under CE. Ash contents of 'Hoyt' seeds, pods, and leaves also were increased under CEs. Wheeler *et al.* (1996) reported higher seed ash content from the BPC compared to field literature values. Lipid contents of seeds and leaves were seemingly influenced by CO_2 level, but ash contents were not. Both trends are consistent with reports by Wheeler *et al.* (1990, 1994).

Mineral Contents

Seeds of 'Hoyt' soybean plants grown at 350 ppm CO_2 in CE had higher Na, K, P, Mg, Ca, Mn, Fe, and Cu contents than did field-grown or 1000 ppm CO_2 CE Hoyt seeds (Table 4). The other mineral data for Hoyt plant parts varied; however, stems grown at 350 ppm CO_2, pods grown in the field, and leaves grown at 1000 ppm CO_2 tended to have

Table 3. Amino Acid Profiles for Field-Grown Soybeans from the Literature and Soybeans Grown in the Field and Controlled Environments (CEs) at Two Levels of CO_2 Enrichment[1]

Amino Acid	Literature Soybean Seed[2]	Field 'PI-494' Seed[3]	Field 'McCall' Seed[3]	Field 'Hoyt' Seed[3]	CE-350 'Hoyt' Seed[3,4]	CE-1000 'Hoyt' Seed[3,5]	CE-350 'Hoyt' Leaf[3,4]	CE-1000 'Hoyt' Leaf[2,4]	Ratios Literature: Field 'Hoyt' Seeds	Ratios Avg. of 3 Field Seeds: Avg. of 2 CE Seeds	Ratios Field: Avg. of 2 CE 'Hoyt' Seeds	Ratios 'Hoyt' 350:1000 ppm CO_2 Seeds	Ratios 'Hoyt' 350:1000 ppm CO_2 Leaves
Tryptophan	0.52	0.46	0.40	0.44	0.49	0.45	0.25	0.21	1.18	0.91	0.94	1.08	1.2
Lysine	2.39	2.53	2.18	2.50	2.64	2.67	0.86	0.78	0.96	0.90	0.94	0.97	1.11
Histidine	0.90	1.09	0.94	1.05	1.16	1.17	0.34	0.47	0.86	0.88	0.90	0.99	0.72
Arginine	2.74	3.17	2.43	2.95	3.22	3.31	0.90	0.78	0.93	0.87	0.90	0.98	1.15
Aspartic acid	4.60	4.68	3.77	4.30	5.00	5.01	2.05	1.99	1.07	0.85	0.86	1.00	1.03
Threonine	1.49	1.56	1.36	1.53	1.67	1.69	0.71	0.61	0.97	0.88	0.91	0.99	1.16
Serine	2.47	1.96	1.68	1.92	2.18	2.18	0.72	0.65	1.29	0.85	0.88	1.00	1.12
Glutamic Acid	6.95	7.00	5.79	6.45	7.92	7.99	1.59	1.36	1.08	0.81	0.81	0.99	0.74
Cystine[6]	**0.67**	**0.67**	**0.59**	**0.56**	**0.70**	**0.68**	**0.18**	**0.19**	**1.20**	**0.88**	**0.81**	**1.04**	**0.91**
Glycine	1.58	1.71	1.47	1.67	1.86	1.87	0.80	0.71	0.95	0.87	0.89	0.99	1.12
Alanine	1.56	1.85	1.61	1.85	1.98	2.00	0.92	0.77	0.84	0.89	0.93	0.99	1.20
Valine	1.99	1.93	1.69	1.89	2.12	2.13	0.90	0.78	1.05	0.86	0.89	0.99	1.15
Methionine[6]	**0.51**	**0.59**	**0.53**	**0.56**	**0.61**	**0.62**	**0.25**	**0.22**	**0.91**	**0.90**	**0.90**	**0.98**	**1.17**
Isoleucine	2.04	1.79	1.57	1.73	1.99	2.01	0.76	0.68	1.18	0.85	0.87	0.99	1.12
Leucine	2.92	2.92	2.60	2.89	3.33	3.35	1.31	1.14	1.01	0.84	0.87	0.99	0.88
Tyrosine	1.21	1.58	1.37	1.48	1.73	1.71	0.64	0.54	0.82	0.86	0.86	1.01	1.19
Phenylalanine	1.87	1.96	1.68	1.90	2.27	2.18	0.98	0.81	0.98	0.83	0.85	1.04	1.20
Ammonia		0.83	0.76	0.74	0.95	0.96	0.62	0.66					

[1] Amino acids are reported as grams amino acid per 100 g sample (dmb).
[2] Literature values taken from Orr and Watt (1957).
[3] Mean of two determinations.
[4] Controlled environment with 350 ppm CO_2.
[5] Controlled environment with 1000 ppm CO_2.
[6] First limiting amino acids in soybean seed.

Table 4. Mineral Content of Soybean Seeds from the Literature[1] and of Soybean Plants Parts Grown in the Field and Controlled Environments (CEs) at Two Levels of CO_2 Enrichment[2]

Plant Part	Growth Environment	Soybean Cultivar	Na (ppm)	K (%)	P (%)	Mg (%)	Ca (%)	Mo (ppm)	Zn (ppm)	B (ppm)	Mn (ppm)	Fe (ppm)	Cu (ppm)
Seed	Literature		22	.196	.077	.031	.030	-----	53.5	------	27.5	172.0	18.1
	Field	PI-494	31	.169	.082	.026	.034	9.2	60.0	45.8	20.0	106.7	11.7
		McCall	18	.187	.080	.028	.025	5.0	59.2	25.0	23.3	45.0	12.5
		Hoyt	5	.185	.072	.025	.014	7.5	55.0	34.2	28.3	41.7	7.5
	CE-350[3]	Hoyt	41	.227	.079	.034	.026	6.7	41.7	30.8	40.0	51.7	10.0
	CE-1000	Hoyt	0	.200	.074	.032	.025	5.0	26.7	22.5	40.0	42.5	8.3
Stem	Field	PI-494	205	.154	.026	.062	.189	0.0	12.5	20.0	5.0	105.0	3.3
		McCall	188	.168	.023	.039	.104	0.0	15.0	21.7	5.8	52.5	3.3
		Hoyt	118	.131	.021	.063	.126	0.0	12.5	19.2	12.5	29.2	2.5
	CE-350	Hoyt	380	.205	.094	.051	.113	0.8	36.7	13.3	40.8	75.8	12.5
	CE-1000	Hoyt	306	.211	.087	.038	.089	0.8	16.7	10.8	30.0	59.2	10.8
Pod	Field	PI-494	57	.240	.016	.062	.119	0.0	13.3	35.8	10.8	59.2	4.2
		McCall	83	.301	.024	.066	.120	0.0	17.5	28.3	18.3	58.3	5.8
		Hoyt	55	.234	.016	.062	.119	0.0	15.8	39.2	44.2	44.2	5.0
	CE-350	Hoyt	18	.310	.073	.035	.069	0.0	15.8	26.7	29.2	11.7	6.7
	CE-1000	Hoyt	19	.347	.070	.038	.085	0.0	12.5	25.0	30.8	37.5	15.8
Leaf	Field	PI-494	176	.197	.020	.058	.390	0.0	39.2	79.2	24.5	208.3	5.0
		McCall	77	.118	.057	.098	.401	0.0	82.5	85.8	125.8	157.5	6.7
		Hoyt	23	.128	.035	.076	.301	0.0	92.5	64.2	209.2	131.7	5.8
	CE-350	Hoyt	49	.294	.170	.049	.317	0.0	292.5	68.3	162.5	142.5	11.7
	CE-1000	Hoyt	69	.272	.189	.055	.339	0.0	340.0	53.3	166.7	219.2	15.0

[1]Literature values taken from Haytowitz and Matthews (1986).
[2]Analyses on dry mass basis.
[3]Controlled environment samples grown at 350 ppm and 1000 ppm CO_2.

higher mineral contents than did comparable parts grown under the other conditions. Of the field-grown plant parts, 'PI-494.525' seeds and 'McCall' stems had the highest mineral contents of the respective plant parts, whereas field-grown 'Hoyt' plant parts had the smallest amounts of most minerals. Phosphorus and zinc accumulated in the leaves of CE plants, but not in the seeds.

For optimal calcium absorption and retention, a Ca/P ratio of 1 should be maintained in the diet (Stare and McWilliams, 1984). The low Ca levels in the typical edible parts of plant materials other than green, leafy vegetables make it difficult in a strict vegetarian diet to achieve a high enough Ca/P ratio. All soybean seeds had Ca/P ratio significantly less than 1 (0.19-0.41) (Table 5). The Ca/P ratios for stems, pods, and leaves of all field-grown plants were well above 1, with 'PI-494.525' plant parts having the highest overall Ca/P ratios. Due to the increased phosphorus content of all CE soybean parts, the Ca/P ratios for CE plant parts generally were less than the Ca/P ratios for field-grown plant parts.

<u>Total Dietary Fiber (TDF) Content</u>

Of the field-grown leaf samples, 'McCall' leaves had the most TDF (38.6%, dmb), whereas 'Hoyt' leaves had the least (30.1%, dmb). The CE 'Hoyt' leaves had more TDF (39.4%, dmb for 350 ppm CO_2; 35.0%, dmb for 1000 ppm CO_2) than field-grown 'Hoyt' leaves, which differs from the finding that CE lettuce leaves have less TDF than field-grown lettuce leaves (McKeehen *et al.*, 1996b). Increasing CO_2 concentration decreased the TDF of soybean leaves, as was the trend for lettuce leaves (McKeehen *et al.*, 1996b). Wheeler *et al.* (1990, 1994) reported increasing amounts of crude fiber in 'McCall' soybean leaves with increasing CO_2 levels. However, because crude fiber and TDF assays differ in what they measure, values cannot be compared directly.

Table 5. Ratio of Calcium (Ca) to Phosphorus (P) Contents for Soybean Seeds from the Literature and Soybean Plant Parts Grown in the Field and Controlled Environments (CEs) at Two Levels of CO_2 Enrichment

Plant Part	Growth Environment	Soybean Cultivar	Ca/P Ratio
Seed	Literature		0.39
	Field	PI-494	0.41
		McCall	0.32
		Hoyt	0.19
	CE-350[2]	Hoyt	0.33
	CE-1000	Hoyt	0.34
Stem	Field	PI-494	7.35
		McCall	4.49
		Hoyt	6.04
	CE-350	Hoyt	1.20
	CE-1000	Hoyt	1.02
Pod	Field	PI-494	7.64
		McCall	4.90
		Hoyt	7.48
	CE-350	Hoyt	0.95
	CE-1000	Hoyt	1.21
Leaf	Field	PI-494	19.05
		McCall	7.03
		Hoyt	8.56
	CE-350	Hoyt	1.86
	CE-1000	Hoyt	1.79

[1]Literature value taken from Haytowitz and Matthews (1986).
[2]Controlled environment samples grown at 350 ppm and 1000 ppm CO_2.

CONCLUSIONS

Macronutrient and mineral compositon data for soybeans reported here are consistent in numerous ways with such data previously reported for other CELSS candidate crops. Compared to field-grown plants, controlled-environment soybeans had higher total N and protein N for all plant parts, higher nitrate in leaves and stems but not seeds, higher seed lipid content, and reduced Ca: P ratio for stems, pods, and leaves. Carbon dioxide enrichment reduced leaf but not seed protein N and amino acid contents. The macronutrient and mineral composition data now available for numerous CELSS candidate crops point to the need for further studies of N allocation and mineral accumulation in controlled environment plants, as they relate to plant productivity and diet formulation.

REFERENCES

Aldrich, S.R., Protein, Nitrates, and Nitrites in Foods and Feeds, in *Nitrogen in Relation to Food, Environment, and Energy,* Agricultural Experiment Station, pp. 97-110 University of Illinois, Urbana, IL (1980).

Allen, L.H., J.C.V. Vu, R.R. Valle, K.J. Boote, and P.H. Jones, Nonstructural carbohydrates and nitrogen of soybean grown under carbon dioxide enrichment, *Crop Sci.* **28**, 84-94 (1988).

AOAC, *Official Methods of Analysis,* 15th ed. Association of Official Analytical Chemists, Washington, D.C. (1990).

Baker, J.T., L.H. Allen, K.J. Boote, P. Jones, and J.W. Jones, Response of soybean to air temperature and carbon dioxide concentration. *Crop Sci.* **29**, 98-105 (1989).

Bensadoun, A., and D. Weinstein, Assay of proteins in the presence of interfering materials, *Anal.Biochem.* **70**, 241-250 (1976).

Fowler, D.B., J. Brydon, B.A. Darroch, M.H. Entz, and A.M. Johnson, Environment and genotype influence on grain protein concentration of wheat and rye, *Agron. J.* **82**, 655-665 (1990).

Goyal, S.S., and A.A. Hafez, Quantitative reduction and inclusion of plant tissue nitrates in Kjeldahl digestion, *Agron. J.* **82**, 571-576 (1990).

Goyal, S.S., and R.C. Huffaker, Nitrogen Toxicity in Plants, in *Nitrogen in Crop Production,* edited by R.D. Hauck, p. 97, American Society of Agronomy, Madison, WI (1984).

Hanson, W.D., R.C. Leffel, and R.W. Howell, Genetic analysis of energy production in the soybean, *Crop Sci.* **1**, 121-126 (1961).

Haytowitz, D.B., and R.H. Matthews, *Composition of Foods,* USDA Agriculture Handbook No. 8-16, United States Department of Agriculture, Washington, D.C. (1986).

Hoff, J.E., J.M. Howe, and C.A. Mitchell, Nutritional and cultural aspects of plant species selection for a regenerative life support system, NASA Contract Report 166324, Moffet Field, California (1982).

Imafidon G.I., and F.W. Sosulski, Nonprotein nitrogen contents of animal and plant foods, *J. Agric. Food Chem.* **38,** 114-118 (1990).

Jones, D.B., Factors for converting percentages of nitrogen in foods and feeds into percentages of proteins. U.S. Dept. Agric. Circular No. 183. August. USDA, Washington, D.C. (1931).

Jones, P., L.H. Allen, J.W. Jones, and R. Valle, Photosynthesis and transpiration responses of soybean canopies to short- and long-term CO_2 treatments. *Agron. J.* **77**, 119-126 (1985).

Khanizadah, S., D. Busyard, C.G. Zarkardas, Misuse of the Kjeldahl method for estimating protein content in plant tissue, *HortScience* **30**, 1341-1342 (1995).

McKeehen, J.D., Nutrient content of select controlled ecological life-support systems (CELSS) candidate species grown under field and controlled enviornment conditions, M.S. Thesis, Purdue University, West Lafayette, IN (1994).

McKeehen, J.D., C.A. Mitchell, R.M. Wheeler, B. Bugbee, and S.S. Nielsen, Excess nutrients in hydroponic solutions alter nutrient content of rice, wheat, and potato, *Adv. Space Res.* **18,** (4/5), 73-83 (1996).

McKeehen, J.D., D.J. Smart, C.L. Mackowiak, R.M. Wheeler, and S.S. Nielsen, Effect of CO_2 levels on nutrient content of lettuce and radish, *Adv. Space Res.* **18,** (4/5), 85-92 (1996).

Orr, M.L., and B.K. Watt, Amino Acid Content of Foods, Home Economics Research Report No. 4, Household Economics Research Division, Institute of Home Economics, Agricultural Research Service, U.S. Department of Agriculture, Washington D.C. (1957).

Preussmann, R., and B.W. Stewart, *N-Nitroso Carcinogens in Chemical Carcinogens*, 2nd ed., ACS Monograph Ser. No. 182, p. 643-848, Washington, D.C. (1984).

Raper C.D., and J.F. Thomas, 1978, Photoperiodic alteration of dry matter partitioning and seed yield in soybeans, *Crop Sci.* **18,** 654-656 (1978).

Rogers, H.H., N. Sionit, J.D. Cure, J.M. Smith, and G.E. Bingham, Influence of elevated carbon dioxide on water relations of soybeans, *Plant Physiol.* **74,** 233-238 (1984).

Rogers, H.H., J.D. Cure, and J. Smith, Soybean growth and yield response to elevated carbon dioxide. *Agriculture, Ecosystems, and Environment* **16,** 113-128 (1986).

Stare, F.J., and M. McWilliams, *Living Nutrition,* John Wiley and Sons, New York, NY (1984).

Steel, R.G.D., and J.H. Torrie, *Principles and Procedures of Statistics,* 2nd ed., McGaw-Hill Book Co., New York, NY (1980).

Swann, P.F., The toxicology of nitrate, nitrite and N-nitrose compounds, *J. Sci. Food Agric.* **26**, 1761-1770 (1975).

Thayer, J.R., and R.C. Huffaker, Determination of nitrate and nitrite by high-pressure liquid chromatography: Comparison with other methods for nitrate determination, *Anal. Biochem.* **102**, 110-119 (1980).

Thomas, J.F., and C.D Raper, Photoperiodic control of seed filling for soybeans, *Crop Sci.* **16**, 667-672 (1976).

Tolley-Henry, L., and C.D. Raper, Utilization of ammonium as a nitrogen source. Effects of ambient acidity on growth and nitrogen accumulation by soybean, *Plant Physiol.* **82**, 54-60 (1986a).

Tolley-Henry, L., and C.D. Raper, Nitrogen and dry-matter partitioning in soybean plants during onset of and recovery from nitrogen stress. *Bot. Gaz.* **147**, 392-399 (1986b).

Wheeler, R. M., C.L. Mackowiak, J.C. Sager, W. M. Knott, and W.L. Berry, Proximate composition of CELSS crops grown in NASA's biomass production chamber. *Adv. Space Res.* **18** (4/5), 43 - 47 (1996).

Wheeler, R.M., C.L. Mackowiak, J.C. Sager, Proximate composition of seed and biomass from soybean plants grown at different carbon dioxide (CO_2) concentrations. NASA Tech. Memo. 103496. May 1990.

Wheeler, R.M., C.L. Mackowiak, J.C. Sager, W.M. Knott, and W.L. Berry, Proximate nutritional composition of CELSS crops grown at different CO_2 partial pressures. *ADV. Space Res.* **14** 11), 171 - 176 (1994).

Journal Article No 15179 of the Purdue University Agricultural Research Program. Research supported in part by NASA grant NAGW-2329.

INTEGRATION OF BIOREGENERATIVE AND PHYSICAL/CHEMICAL PROCESSES FOR SPACE LIFE SUPPORT SYSTEMS

Proceedings of the F4.9 Symposium of COSPAR Scientific Commission F which was held during the Thirty-first COSPAR Scientific Assembly, Birmingham, U.K., 14–21 July 1996

Edited by

C. A. MITCHELL

NASA Specialized Center of Research and Training (NSCORT) in Bioregenerative Life Support, Purdue University, 1165 Horticulture Building, West Lafayette IN 47907-1165, U.S.A.

Pergamon

Adv. Space Res. Vol. 20, No. 10, p. 1991, 1997

0273-1177/97 $17.00 + 0.00

PII: S0273-1177(97)00930-7

MEMORIAL

SPACE ENGINEERING AND SPACE SCIENCES PIONEER Dr. WILLY Z. SADEH DIES AT AGE 64

Dr. Willy Z. Sadeh was a dedicated and passionate scholar of space engineering and space sciences who believed that any civilization that does not challenge the impossible is doomed to fail...and the impossible for our civilization is the human conquest of the infinite space frontier.

Sadeh was known internationally for his pioneering work in basic and applied space engineering and space sciences research with particular emphasis on the development of a human-tended lunar base. His research concentrated on inflatable structures for a lunar base, waste management systems in a lunar base, physical/chemical and bioregenerative life support systems for a lunar base, fluid management in a closed plant growth chamber in space, conversion of lunar regolith into a friendly soil, radiation protection in space, materials for space structures, and space policy on modeling international cooperation for space exploration.

As a professor of space engineering, Sadeh founded the Center for Engineering Infrastructure and Sciences in Space (CEISS) at Colorado State University in 1987 and served as center director the past ten years. He built the center into one of the premiere graduate level programs in space engineering and space sciences. The center promotes exploration and development of space, and coordinates interdisciplinary research, educational programs, and cooperative projects with academia, industry, and government in the emerging areas of space civil engineering, space life sciences, and space policy. Sadeh developed five novel academic options in Space Civil Engineering, Space Agricultural Sciences, Space Biomedical Sciences, Space Biology, and Space Policy at both undergraduate and graduate levels under the NASA Space Grant College Program.

Sadeh published 244 scientific and technical papers and reports. He received 39 awards in recognition of outstanding achievements and professionalism in education, public service, and research. The more notable awards include 1996 election as a fellow to the American Association for the Advancement of Sciences for pioneering and original research in space engineering and space life sciences, particularly for research in support of the development of a lunar base; 1996 American Astronautical Society Victor A. Prather Award for contributions to the advancement of space engineering and space life sciences; and the 1992 International Astronautical Federation Frank J. Malina Astronautics Medal for demonstrated excellence by an educator in promoting the study of astronautics and related space sciences.

Sadeh was an international advocate for space exploration who helped to establish the engineering and scientific knowledge necessary for the development of a human-tended lunar base and human exploration of Mars in the 21st century. He was a true explorer and husband, father, and grandfather who considered family as important as his work.

Pergamon

Adv. Space Res. Vol. 20, No. 10, p. 1993, 1997

Printed in Great Britain
0273-1177/97 $17.00 + 0.00

PII: S0273-1177(97)00931-9

PREFACE

The stability, sustainability, and robustness of advanced life-support systems for space bases will depend ultimately on the integral functioning of both physico-chemical (P/C) and biological regeneration processes. Bioprocessing of residual and waste biomass from crop production, food preparation, and human and animal excreta has appeal due to low operational power requirements, favorable temperature and pressure requirements, but may be time-limiting with respect to the dynamics of certain recycling requirements. However, bioprocessing may be enhanced substantially by partnership with P/C processes that render residues less recalcitrant to enzymatic and/or microbial attacks in bioreactors, or which greatly spreed rates of bioconversion. In future life-support systems, biological and P/C sub-systems also will back-up each other to ensure the long-term safety and survival of human crews in distant space-based habitats where frequent resupply of consumables from Earth is not an option.

The nine articles included in this compilation deal mainly with ways to treat toxic or recalcitrant by-products of a closed, recycling life-support system A theoretical modeling paper based upon previous results with the BIOS-3 facility in Siberia concluded that catalytic incineration of crop residue in a closed system caused unacceptable reductions in crop performance due to volatile oxidation products. Crops either must be selected for greater resistance to incineration products, "buffers" need to be larger, and/or combustion technologies must be improved. A top-level conceptual design of a Lunar Engineered Closed/Controlled EcoSystem (LECCES) was presented integrating mass flows among human, plant, and animal modules, and a waste-management subsystem combining biological and P/C processes. Another article showed how three bioreactors connected in series could be used to process crop residue in ways that produce an alternate source of human food: anaerobic digestion created volatile fatty acids that subsequently fed on aerobic, edible yeast, and the aerobic effluent was subjected to nitrification. Non-edible solids were incinerated at each step. Potato plants grown hydroponically in effluent from the yeast bioreactor yielded as much total and tuber mass as controls grown on $N0_3^-$ only. Mixed-N controls and plants grown with nutrification bioreactor effluent were repressed in productivity, the latter likely due to nutrient imbalances. For aerobic microbial decomposition of crop residue cellulose, bioreactor retention times of at least ten days will be needed; for inorganic nutrients, $\leq$ 1 day retention times will suffice. Total organic carbon in solution is a problem with short retention times. In a study of the ability of wheat straw ash (from incineration) to provide nutrients for the hydroponic growth of lettuce, nutrient solutions derived from ash suppressed crop growth unless supplemented with reagent grade chemicals that gave the same balance as standard nutrient solutions. One article in the session introduced the Advanced Life Systems for Extreme Environments (ALSEE) project, which proposed to use NASA-developed technologies for reducing waste, recycling and purifying water, as well as production of food in remote communities of Alaska. Another study showed a buildup of trace elements and heavy metals in edible and non-edible crop parts during six months of growth in the BIOS-3 closed system. Mineralization of solid waste by regenerable H_2O_2 at elevated temperature was proposed as a viable waste-processing method in a life-support system with no energy restrictions. A final study correlated the dependence of catabase activity of active sludge microorganisms on substrate concentration in which H_2O_2 is being actively produced in purification tanks.

The organizing committee consisting of Cary Mitchell and Willy Sadeh also chaired the sessions. Nine of the 13 abstracts accepted for presentation were submitted or accepted for final publication. The Main Scientific Organizer wishes to thank the following scientists for critically reviewing manuscripts: B. Finger, J. Garland, M. Kliss, D. Smernoff, R. Strayer, W. Sadeh, R. Wheeler, and K. Wignarajah.

Pergamon

Adv. Space Res. Vol. 20, No. 10, pp. 1995–2000, 1997

Printed in Great Britain
0273-1177/97 $17.00 + 0.00

PII: S0273–1177(97)00932–0

INTERACTION OF PHYSICAL–CHEMICAL AND BIOLOGICAL REGENERATION PROCESSES IN ECOLOGICAL LIFE SUPPORT SYSTEMS

V. Ye. Rygalov, B. G. Kovrov and G. S. Denisov

Open Laboratory-the International Center for Closed Ecological Systems Studies at the Insitiute of Biophysics (Russian Academy of Sciences, Siberian) 660036 Krasnoyarsk-36, Akademgorodok, Russia

ABSTRACT

Catalytic combustion of inedible biomass of plants in ecological Life Support Systems (LSS) gives rise to gaseous oxides (CO_2, NO_2, SO_2, etc.). Some of them are toxic for plants suppressing their photosynthesis and productivity.

Experiments with "Bios-3" experimental LSS demonstrate that a decrease of photosynthetic productivity in a system with straw incineration can jeopardize its steady operation.

Analysis of the situation by a mathematical model taking into account absorption parameters of the system in terms of toxic elements makes it possible to formulate requirements for the structure and operation of LSS to provide for its stability.

Avenues for further investigation of the problem of toxic stability of LSS are proposed.

INTRODUCTION

The fundamental feasibility to create a LSS integrating physical-chemical and biological processes was demonstrated between 1970 and 1980 . During this period the problems relevant to operation stability have been formed. One of them was emergence of toxic oxides in the system atmosphere with combustion of the inedible biomass of plants (Closed System..., 1979; Experimental ecological ..., 1975).

"Bios-3M" LSS experiments described demonstrated that it is not feasible to incinerate at regular intervals the amount of straw required to sustain photosynthetic productivity because of its toxic effect on plants.

Attempts made later were intended to understand the mechanisms of transformation of toxic impurities in Closed Ecological Systems (CES) atmosphere with "Bios-3M" LSS as a case in point. A concept was introduced about the so-called absorption characteristics of closed ecosystems in terms of the toxic elements' inflow which described active and passive mechanisms for removing impurities from CES atmosphere (Rygalov, 1995; Rygalov, 1996). It has been demonstrated that every CES can be specified by a limit admissible flow of toxic gases. When the actual flow of gases is less than the limit admissible the system works. When it is exceeded the processes in the system are disrupted and cease to exist. On these grounds this work makes an attempt to analyze conditions of steady operation of LSS with incineration as a way to regenerate inedible biomass of plants.

MATHEMATICAL MODEL FOR TRANSFORMATION OF TOXIC ATMOSPHERIC IMPURITIES IN LSS

Catalytic incineration is a rapid way to regenerate biogenous elements from deadlock products which are inedible wastes of the plant biomass. Using this method to return carbon into the system atmosphere and some other elements into nutrient solutions, however, gives rise to gaseous oxides toxic for plants. The toxic gases are removed from the system by passive physical-chemical dissolution in nutrient solutions, moisture condensate and binding on the surface of structural elements. The toxic elements get into plants also by dissolving in tissue liquids of leaves and stems. Plants have a certain capacity to detoxify compounds that get inside them by binding them in the tissues (Rygalov, 1995; Rygalov, 1996).

So, conversion of toxic atmospheric impurities in LSS can be described by the following mathematical model:

$$\left.\begin{aligned} \dot{C} &= -\frac{1}{V}\sum_{i=0}^{N} K_{Li}G_i(\alpha_i C - C_i) \\ \dot{C} &= \frac{K_{Li}G_i}{V_i}(\alpha_i C - C_i) - k_i; i = 1,2,..,N \\ \dot{V}_p &= \mu\left(1 - \frac{C_p}{C_m}\right)V_p \end{aligned}\right\} \qquad (1)$$

where C is the toxic gas concentration in the LSS atmosphere;

C_i is the toxic element concentration inside i - absorbent inside the system;
C_p is the toxic element concentration inside the plant biomass;
V_p is the plant mass volume in the system;
V is the atmosphere volume in the system;
G_i is the is the i-absorbent-atmosphere interface area;
V_i is the volume of the i-absorbent;
K_{Li} is the transfer constant dependent on physical-chemical properties of the i-absorbent-atmosphere interface;
α_i is the solubility constant dependent on the physical-chemical properties of the i-absorbent;
k_i is the detoxification constant specifying ability of the i-absorbent to actively transform the toxic agent into nontoxic compounds, other than zero for the plant mass only;
C_m is the constant of plant tolerance to the effect of toxic agent, limit admissible concentration, with $C_p = C_m$ productivity and photosynthesis intensity are equal to zero;
μ is the constant of plant mass specific growth rate in the absence of toxic effect;
i is the number of the i-absorbent;
N is the total number of absorbents in the system;
p is the number of the plant absorbent.

The first two equations of (1) demonstrate the dissolution of the toxic agent in absorbents which is due to the effect of the difference in the concentration at different sides of the interface.

The constant k_i in the second equation implies some absorbents have the capacity to actively transform dissolved toxic agents into nontoxic compounds. In our case this capacity is featured by plants only: $k_p \neq 0$ (Rygalov, 1995; Rygalov, 1996).

The third equation specifies production capacities of the LSS plant cenosis exposed to toxic effect. It involves the assumption that the specific productivity of plants linearly decreases with increasing toxic agent concentration in the plant tissues (Rygalov, 1996).

Equation system (1) describes the case of the so-called pulse combustion when a certain charge of straw is incinerated at regular intervals in LSS. This case is the most realistic in available LSS.

ANALYSIS OF THE MODEL AND ESTIMATES BASED ON "BIOS-3M" EXPERIMENTS-

The processes of dissolution and processing of impurities in the system have been earlier shown (Rygalov, 1995; Rygalov, 1996) to run with characteristic times of about one hour. Meanwhile, the characteristic times associated with plant biomass growth are about one day and more. Therefore, the first two equations of (1) can be analyzed separately from the third one. Multiplying them by respective V_i and summing:

$$\sum_{i=0}^{N} \dot{C}_i V_i = -k_p V_p, \qquad (2)$$

here $i=0$ is relevant to the description of the atmosphere volume.

By virtue of fairly well developed interfaces of absorbents in "Bios-3M" it is safe to assume that the passive redistribution of impurities is rather fast (with characteristic times less than hour (Rygalov, 1995; Rygalov, 1996)) and $C_i \simeq \alpha_i C$. This considered, from (2) after integration yields:

$$C = \frac{M - k_p V_p t}{\sum_{i=0}^{N} \alpha_i V_i} \qquad (3)$$

where M is the amount of the impurity getting into the system atmosphere during a single combustion. By (3) it is possible to evaluate characteristic times of removing the impurities from the atmosphere $t = M / k\,V_p$. For "Bios-3M" LSS $t \sim 1$ hour.

Taking $C_p \simeq \alpha_p C$ into consideration, substitute (3) in the third equation of (1):

$$\dot{V}_p = \mu \left(1 - \frac{\alpha_p}{C_m} \frac{M - kV_p t}{\sum_{i=0}^{N} \alpha_i V_i} \right) V_p \qquad (4)$$

To get an approximate evaluation integrate (4) in the time interval from 0 to $M / (k_p V_p)$. The approximation is to assume:

$$\left. \begin{array}{l} V_p \approx Const \\ \alpha_p V_p << \sum_{i=0}^{N} \alpha_i V_i [Rygalov, 1995; Rygalov, 1996] \end{array} \right\}.$$

Then exactly integrate (4) in the time interval from $\left[T - \frac{M}{(k_p V_p)}\right]$ to T where T is the time interval equal in our case to tens of hours. As a result of the integration:

$$V_{p_T} = V_{p_0} \exp\left(\mu T - \frac{\mu}{2kV_p} \cdot \frac{\alpha_p M^2}{C_m \sum_{i=0}^{N} \alpha_i V_i}\right) \tag{5}$$

From (5) one can easily see that for the plant biomass in LSS not to decrease it is required to hold inequality:

$$\mu T - \frac{\mu}{2kV_p} \cdot \frac{\alpha_p M^2}{C_m \sum_{i=0}^{N} \alpha_i V_i} \geq 0 \tag{6}$$

From (6) after simple transformations:

$$kV_p T \geq \frac{\alpha_p}{2C_m} \cdot \frac{M^2}{\sum_{i=0}^{N} \alpha_i V_i} \tag{7}$$

So, for the LSS with combustion to operate with stability it is necessary to hold the condition or criterion (7), imposing restrictions on the structural and functional characteristics of the system. The better inequality (7) holds the more resistant are the processes in it to the toxic effect. Hence we can see that to enhance the reliability of LSS operation it is necessary:

- to increase values kV_p and C_m which means that it is necessary to have selection work with the plant cenosis and design systems with higher content of the plant mass;
- to decrease the magnitude of M, which primarily means to improve the incineration technology which would release less toxic agents into the atmosphere;
- to increase the value $\sum_{i=0}^{N} \alpha_i V_i$ which means that it is necessary to increase the number and volume of absorbents in the system.

Taking $kV_p = \frac{M}{t}$ ratio (7) can have a fairly simple form. Then, substituting (7) and after simple transformations:

$$\frac{\alpha_p M}{T 2 \sum_{i=0}^{N} \alpha_i V_i} \leq \frac{C_m}{t}, or$$
$$\frac{\alpha_p M}{T \sum_{i=0}^{N} \alpha_i V_i} < \frac{C_m}{t} \tag{8}$$

The latter means that for LSS to work with stability it is necessary for the specific (as per plant mass unit) rate of toxic element supply into the plant to be less than a certain critical value dependent on the physiological parameters of the cenosis.

New one can evaluate the feasibility of (8) for the well known "Bios-3M" LSS (Closed System..., 1979; Experimental ecological ..., 1975). Incineration of straw releases into the system atmosphere a fairly complex mixture of gases. It is assumed that the principal carrier of toxicity in it is NO_2 (Closed System..., 1979). Combustion increased the carbon dioxide concentration in "Bios-3M" atmosphere to 0.6 mg/m^3 (Closed System..., 1979). At the same time the buffer absorption capacity of the system $\sum_{i=0}^{N} \alpha_i V_i$ is able to reduce the initial gas concentration tens of times at a minimum (Rygalov, 1995, Rygalov, 1996). This makes it possible to have the estimate for $\left(M \Big/ \sum_{i=0}^{N} \alpha_i V_i \right) \sim \frac{0.6}{10} = 0.06 mg/m^3$. The water-dissolubility coefficient for NO_2 is $\alpha_p \sim 5.7$ (Concise Chemist's Reference Book, 1964). As indicated earlier, the characteristic times T ~ 10 hours, t ~ 1 hour. C_m can be evaluated by the known values of limit admissible concentrations of NO_2 for plants (Concise Chemist's Reference Book, 1964; Serebryakova, 1980): $C_m \sim (0.01 \div 0.5)$ mg/m^3. Then, inequality (8) takes the form:

$$\frac{5.7 \cdot 0.06}{10} < \frac{0.01 \div 0.5}{1};$$

$$0.034 < 0.01 \div 0.5$$

Approximate nature of the evaluations notwithstanding, it can be seen that inequality (8) is at the edge of feasibility. This indicates that in what concerns toxic contamination of the atmosphere when incinerating plant wastes the operation of "Bios-3M system is not stable. Indeed, experiments demonstrated that incinerating the required amount of the straw in the system at regular intervals was not successful, as during combustion photosynthesis intensity was observed to decrease by more than 30% as compared to the control (Closed System..., 1979). This argues for finding out the limit of admissible concentrations of toxic oxides forming during combustion of inedible biomass of plants for different plant species.

CONCLUSION

1. Removal of plants wastes in LSS by catalytic incineration may give rise to toxic oxides for plants, contamination of the atmosphere and reduce stability of system operation on the whole.

2. It is feasible to formulate a mathematical criterion imposing certain requirements to the ratio of structural and functional parameters of LSS, resistant to catalitic combustion of plants wastes.

3. The criterion formulated makes it possible to evaluate preliminary and to outline possible avenues for enhancing LSS operation reliability: to select plants having enhanced resistantce to gaseous combustion products, to improve combustion technology so that tocsic gases could not enter the atmosphere of the system, to look for alternative ways to process wastes, to increase buffer capacities.

4. For practical applications, to change buffer capacities is the most practicable since the LSS with larger bilks of absorbing substances and a larger contact area between them and atmosphere will be more resistant to toxic gases than other LSS, all other factors being the same.

5. To exactly evaluate LSS operation reliability requires accurate values of the limit admissible for the plant cenosis concentrations of toxic gases and intensities of their detoxification by the plant mass, to be determined in unrelated experiments.

REFERENCE

1.Closed System: Man-Higher Plants (Four-Months Long Experiment). Ed. Prof. Lisovsky, Novosibirsk, "Nauka", 1979, 160 p.

2.Concise Chemist's Reference Book, Compiled by V.I. Perelman, Moscow: "Khimiya", 1964, pp. 313-314.

3.Experimental Ecological Manned Systems. In: *Problems of Space Biology*, v. 28, Ed. ac. V.N. Chernigovsky, Moscow, "Nauka", 1975, 312 p.

4.Rygalov V.Ye. Cultivation of Plants in Space: Their Contribution to Stabilizing Atmospheric Composition in Closed Ecological Systems. Adv. Space Res. Vol. 18, No.4/5, pp. (4/5)165-(4/5)176, 1996.

5.Rygalov V.Ye., Shilenko, M.P., Lisovsky G.M. Minor Components' Composition in Closed Ecological System Atmosphere: Mechanisms of Formation. *46th International Astronautical Congress,* October 2-6, 1995, Oslo, Norway, IAF/IAA-95-G.4.03, 9 p.

6.Serebryakova L.K. Admissible Concentrations of Toxic Substances in Atmospheric Air for Woody Vegetation. In: *Gas Resistance of Plants.*, Novosibirsk, "Nauka", 1980, pp. 184-185.

Pergamon

Adv. Space Res. Vol. 20, No. 10, pp. 2001–2008, 1997

Printed in Great Britain
0273-1177/97 $17.00 + 0.00

PII: S0273-1177(97)00933-2

AN INTEGRATED ENGINEERED CLOSED/CONTROLLED ECOSYSTEM FOR A LUNAR BASE

Willy Z. Sadeh* and Eligar Sadeh**

Center for Engineering Infrastructure and Sciences in Space (CEISS), Colorado State University, U.S.A.

ABSTRACT

Long-term human missions in space, such as the establishment of a human-tended lunar base, require autonomous life support systems. A Lunar Engineered Closed/Controlled EcoSystem (LECCES) can provide autonomy by integrating a human module with support plant and animal modules, and waste treatment subsystems. Integration of physical/chemical (P/C) and biological waste treatment subsystems can lead to viable and operational bioregenerative systems that minimize resupply requirements from Earth.. A top-level diagram for LECCES is developed based on the human module requirements. The proposed diagram is presented and its components are discussed.

INTRODUCTION

Long-range plans for the space program include the establishment of a human-tended lunar base as the stepping stone toward the expansion of humanity into space and the utilization/exploitation of space resources to the benefit of Earth [1,2,3,4,5,6]. Such a mission requires autonomous self-sufficient life support systems since resupply costs from Earth become politically and economically prohibitive to mission implementation. This implies fewer consumables such as H_20 and food that must be shipped, and reductions in the amount of wastes returned to Earth. Currently, US human space missions are of short enough duration that it is more economical to carry all supplies to space and transport back to Earth the wastes produced. For long-term missions this is no longer true. The sheer magnitude of mass flow required between Earth and Moon for consumables drives the design of a Lunar Engineered Closed/Controlled EcoSystem (LECCES).

To be effective, LECCES needs to minimize the resupply of food, H_2O, and O_2 from Earth. Technologies can meet this objective in two ways. First, a minimum scenario, currently used on the space shuttle and planned for the international space station, involves the physical/chemical (P/C) recycling of H_20 and air relying on resupply of food and storage of waste. Second, a maximum scenario, such as LECCES, substantially reduces resupply and storage by materially closing food/waste loops. This operates through bioregenerative cycles integrating humans and plants within a recycling system. Plants utilize radiant energy to synthesize complex organic

[1]Former director and founder of the Center for Engineering Infrastructure and Sciences in Space (CEISS) and professor of space engineering, Colorado State University. Deceased May 12, 1997.
[2]NASA Fellow, CEISS, Colorado State University.

molecules from CO_2 and other nutrients, and provide food and O_2 for human use. The loop is closed by treating human (feces and urine) and plant (inedible biomass) waste products and returning them to an inorganic nutrient pool that is utilized for plant growth. Such a scenario provides for long-term human survivability on a lunar base while minimizing resupply and cost [7].

Lunar base autonomy from Earth requires developing LECCES comprised of human and plant modules, and an optional animal module. The modules must be integrated with a Waste Management System (WMS) to achieve partial and, ultimately, long-term self-sufficiency. P/C and bioregenerative life support systems applicable to LECCES are governed by interfacing its modules with the WMS.

The human module requirements 'drive' the design and operation of LECCES. Definition of the human module requirements is first required to set forth the design criteria for the support plant and animal modules, and WMS. A computer simulation model was developed to predict metabolic massflow rates for each input/output element for the human module with results reported elsewhere [8,9]. P/C characteristics of the human, plant, and animal output metabolic elements (i.e., wastes) provide the basis for identifying and selecting enabling technologies capable of processing and treating wastes for reuse in the modules.

INPUT/OUTPUT METABOLIC ELEMENTS

Schematic diagrams of the input/output metabolic elements in the human, plant, and animal modules are shown in Fig. 1. Input and output metabolic elements are defined as raw materials. Metabolic element inputs to the human module and optional animal module include O_2, dry food (edible biomass), and metabolic/drinking H_20. The input quantities are determined by the energy expenditure rate which is a linear function of body mass with energy activity levels as parameters [8]. Outputs include CO_2, feces (solid and liquid), urine (liquid and solid), insensible H_20 (perspiration/respiration), and animal edible/inedible products.

Metabolic input elements for a plant module include CO_2, radiant energy for photosynthesis, and a hydroponics fed nutrient solution. Outputs are O_2, H_20 vapor through transpiration, edible/inedible biomass, and organic volatiles (e.g., phytotoxic gases). Environmental growth conditions, including photosynthetic photon flux, growth period, planting density, daily photoperiod, CO_2 concentration, relative humidity, dry bulb temperature for germination and harvest, and nutrient solution concentration, determine plant growth rates [9].

Enabling technologies and processes potentially capable of converting raw materials have been recently identified [10,11]. The method used to identify and select enabling technologies serves as the foundation for developing an LECCES top-level system integration diagram (Fig. 2). An important conclusion reached in evaluating viable enabling technologies is that basic and applied research are required even for mature technologies to bring them to a level of applicability and readiness for space applications.

TOP-LEVEL SYSTEM INTEGRATION

An effort was undertaken to develop an LECCES top-level system integration by identifying the interfaces and interactions among the human, plant, and animal modules, and the WMS. To achieve the necessary integration, the WMS consists of the following three subsystems:

1. Raw material sorting/storage;
2. Raw material processing;
3. Processed materials storage/supply.

Support units required for these subsystems include:

1. Environmental monitoring/control (M/C) for modules and WMS subsystems;
2. Backup storage/supply of input metabolic elements (consumables);
3. Nutrient recovery for the plant module;
4. Nutrient backup storage/supply for plant module;
5. Light for the plant module;
6. Residual waste storage/disposal.

LECCES

Steps involved in developing LECCES include:

1. Identification of the category and characteristics of each module input/output metabolic elements;
2. Determination of the metabolic massflow rates of each module input/output elements;
3. Identification of waste treatment processes and associated enabling technologies capable of treating/recycling output metabolic elements for utilitarian reuse.

Completion of these steps provides the basis for estimating the capabilities of LECCES to accommodate a wide range of metabolic massflows. Each output metabolic element is essentially a waste product and, at the same time, represents a raw material requiring treatment prior to its reuse in the modules.

MODULE INTERFACES

Each output metabolic element shown in Fig. 1 must undergo some type of P/C and/or bioregenerative processing prior to its reuse as an input metabolic element [10]. P/C treatment ranges from simple disinfection of humidity condensate for use as potable water to complex separation and processing of human feces to be used as a plant nutrient. Bioregenerative treatment involves converting output metabolic elements to usable input metabolic elements by biological processes.

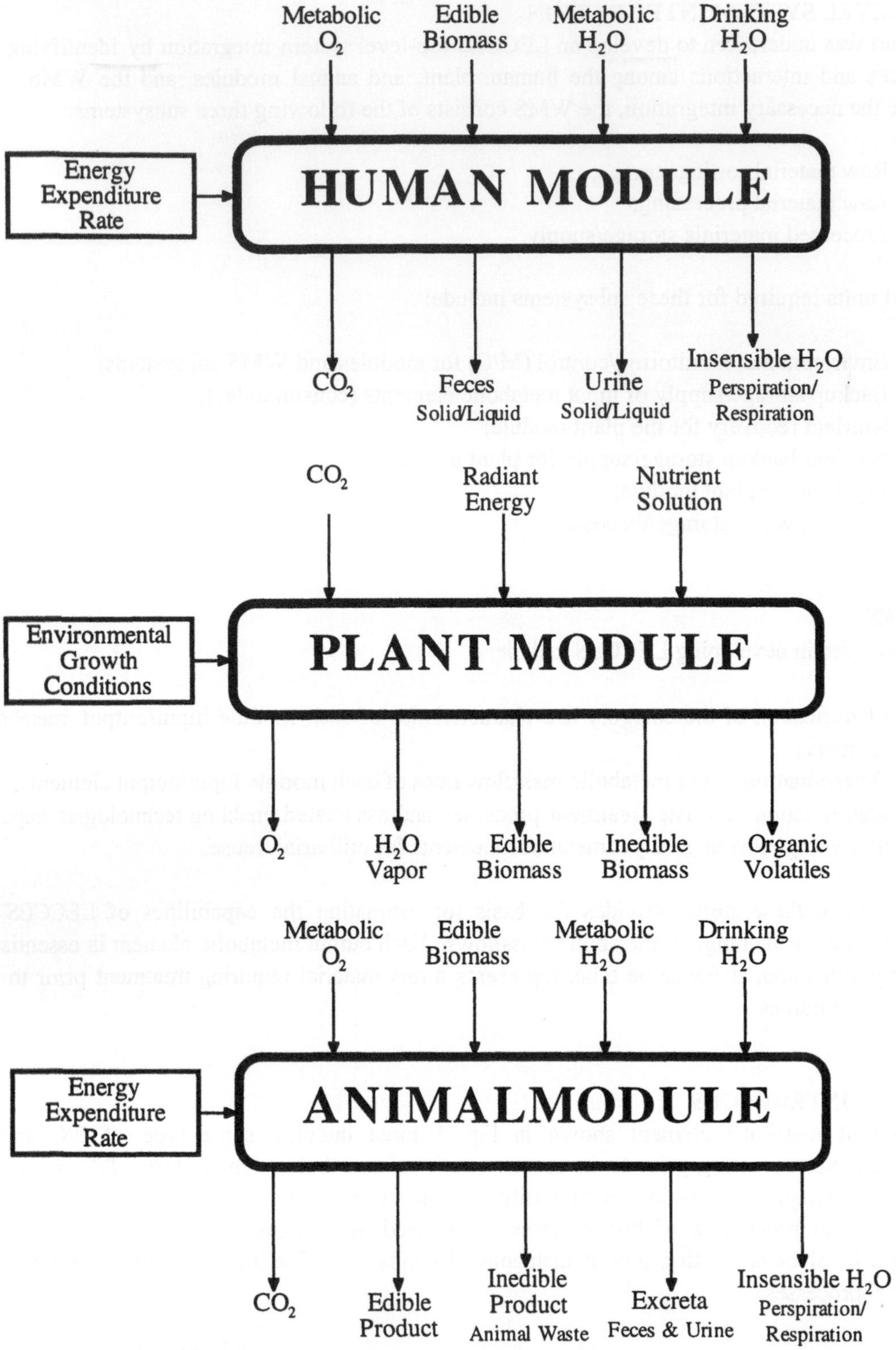

Fig. 1 Input/output metabolic elements for the human, plant, and animal modules.

TOP-LEVEL LECCES INTEGRATION

Integration of human, plant, and animal modules with the WMS is depicted in Fig. 2 LECCES top-level system integration diagram. This diagram provides a conceptual view into collecting, processing, and reusing raw materials as input metabolic elements. The identification of metabolic massflow inputs and outputs for the three modules (Fig. 1), the applicable and associated waste treatment processes, and the required enabling technologies provide the basis for conducting this top-level integration [8,9,10, 11,12,13,14].

Continuous real-time environmental M/C of temperature, pressure, humidity, chemical composition, and metabolic massflow rates is required for each module and WMS subsystem. Support units for each module and WMS subsystem are also shown in Fig. 2 to indicate their input and output roles.

The WMS consists of three basic subsystems: (1) raw material sorting/storage; (2) raw material processing; and (3) processed material storage/supply. Each unit must meet specific functions as briefly outlined below:

1. *Raw material sorting/storage*- Each raw material is sorted prior to processing according to its P/C properties (i.e., gas, liquid, solid) regardless of origin. Sorted raw materials storage ensures continuous massflow rates to the processing subsystem. Gases include O_2, CO_2, N_2, and H_2O vapor. Human feces and urine are consolidated with animal excreta. Animal offal (e.g., blood, feathers, bones, viscera) is collected with inedible plant biomass. Edible biomass from the plant and animal modules are combined.

2. *Raw material processing*- Raw materials is processed by appropriate enabling technologies as required for their reuse. Processed raw materials include: (1) O_2, CO_2, N_2; (2) potable H_2O from H_2O vapor; (3) nutrients from solid and liquid processing. With the completion of the processing, the materials are supplied to the processed materials storage/supply subsystem. Nutrients from the raw material processing subsystem flow to the nutrient recovery support unit for further processing prior to being directed to the plant module. Unusable residual wastes and volatile gases (e.g., organic volatiles) flow to the residual waste storage/disposal support unit.

3. *Processed materials storage/supply*- Required metabolic inputs to each module are furnished from the storage/supply subsystem. Control of the supply of input metabolic elements to the three modules is performed by the storage/supply subsystem. A backup storage/supply support unit serves to supply consumables when demand exceeds supply of processed raw materials. Nutrient backup storage/supply and light support units are required for the plant module to ensure optimum growth conditions.

CONCLUDING REMARKS

The LECCES conceptual design illustrates the feasibility of a regenerative life support system for lunar base human self-sufficiency. Integration of the input/output metabolic massflows for each module and the WMS of LECCES is described using a top-level diagram. The proposed top-level

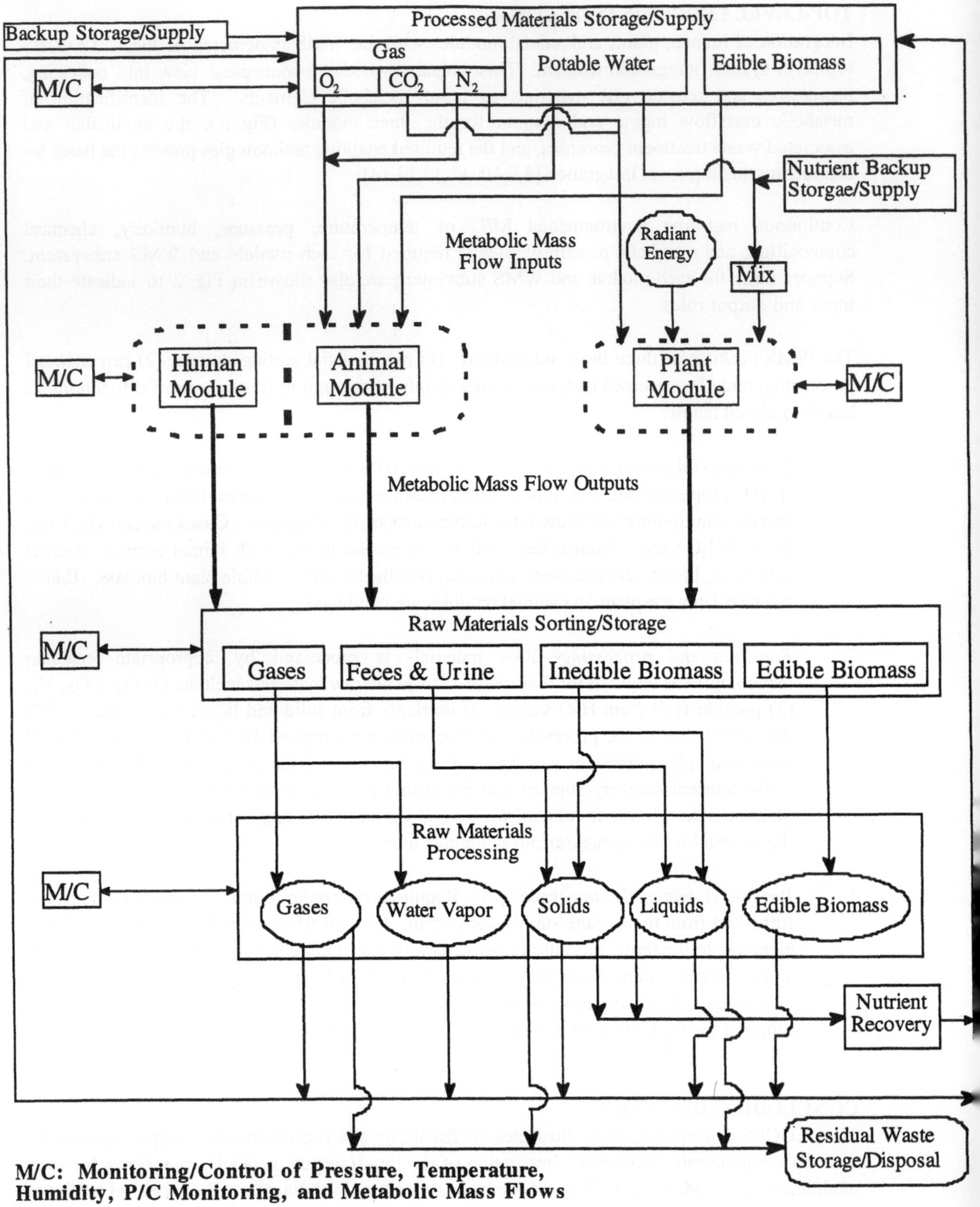

Fig. 2 LECCES top-level system integration diagram.

diagram qualitatively shows the interfaces and interactions among the human, plant, and animal modules, and the raw material sorting/storage, raw material processing, and processed materials storage/supply subsystems of the WMS.

Metabolic massflow quantification has not been yet considered as details of the overall LECCES configuration must first be defined. A range of massflow rates for input/output metabolic elements for each module must be determined prior to hardware selection and development. Future research needs to be directed at developing an expanded version of this top-level diagram including quantification of metabolic massflows and the sizing of WMS subsystems.

From a broad perspective, the development of life support systems evolve toward the emergence of an ecosystem suitable for human settlement on the moon and in space. Attempts to establish LECCES will certainly lead to a much better understanding of the Earth's global environment. As the Earth is a self-sustaining biosphere, the ability of humans to settle on the moon and in space relies on the development of an engineered biosphere outside the Earth's ecosystem.

ACKNOWLEDGMENTS

The support of the NASA Space Grant Program in preparing this paper is gratefully acknowledged. Eligar Sadeh would also like to thank his father, Willy Z. Sadeh, for his invaluable support and professional guidance.

REFERENCES

1. National Commission on Space Report, Pioneering the Space Frontier, Bantam Books, New York, NY, 1986.
2. Ride, S. K., Leadership and America's Future in Space, A Report to the Administrator, NASA, Washington DC, August 1987.
3. NASA, Exploration Studies-Technical Report FY 1988 Status, Vol. I and II, NASA TM 4075, Office of Exploration, NASA, Washington DC, December 1988.
4. NASA, Report of the 90-Day Study on Human Exploration of the Moon and Mars, NASA, Washington DC, November 1989.
5. Augustine Report, Report of the Advisory Committee on the Future of the U.S. Space Program, Superintendent of Documents, U.S. Government Printing Office, Washington DC, December 1990.
6. Stafford Report, America at the Threshold-Report of the Synthesis Group on America's Space Exploration Initiative, Superintendent of Documents, U.S. Government Printing Office, Washington DC, May 1991.
7. MacElroy, R. D., Klein, H. P. and Avener, M. M., The Evolution of CELSS for Lunar Bases, Lunar Bases and Space Activities of the 21st Century, W. W. Mendell, Ed., Lunar and Planetary Institute, 1985, pp. 623-633.
8. Condran, M. J., Sadeh, W. Z., Hendricks, D. W. and Brzeczek M. E., A Model of Human Metabolic Massflow Rates for an Engineered Closed EcoSystem, 19th ICES, San Diego, CA, July 24-26, 1989, SAE Paper 891486.

9. ESA, Life Support and Habitability Manual, Chap. 1, Life Support-Overview of Requirements, Vol. 1, ESA PSS-03-406, 1991.

10. Sadeh, W. Z., Hendricks, D. W. and Condran, M. J., A Review of Enabling Technologies for Treatment of Metabolic Materials, CEISS-TR-4, Colorado State University, Fort Collins, CO, September 1989.

11. Condran, M. J., Preliminary Model for an Engineered Closed EcoSystem for a Lunar Base, CEISS, Civil Engineering Dept., Colorado State University, Fort Collins, CO, M.S. Thesis, Fall 1990.

12. Sadeh, W. Z., Hendricks, D. W. and Condran, M. J., Plant Module in an Engineered Closed EcoSystem (ECES)," CEISS-TR-2, Colorado State University, Fort Collins, CO, July 1989.

13. Sadeh, W. Z., Hendricks, D. W. and Condran, M. J., Animal Module in an Engineered Closed EcoSystem (ECES)," CEISS-TR-3, Colorado State University, Fort Collins, CO, August 1989.

14. Sadeh, Eligar and Sadeh, Willy Z., Bioregenerative Life Support Systems for Long-Term Space Habitation: A Conceptual Approach," Life Support and Biosphere Sciences, Vol. 2, Winter 1996, pp. 161-168.

Pergamon

Adv. Space Res. Vol. 20, No. 10, pp. 2009–2015, 1997

Printed in Great Britain
0273-1177/97 $17.00 + 0.00

PII: S0273-1177(97)00934-4

EVALUATION OF AN ANAEROBIC DIGESTION SYSTEM FOR PROCESSING CELSS CROP RESIDUES FOR RESOURCE RECOVERY

R. F. Strayer, B. W. Finger and M. P. Alazraki

Dynamac Corporation, Mail Code DYN-3, Kennedy Space Center, FL 32780, U.S.A.

ABSTRACT

Three bioreactors, connected in series, were used to process CELSS potato residues for recovery of resources. The first stage was an anaerobic digestor (8 L working volume; cow rumen contents inoculum; fed-batch; 8 day retention time; feed rate 25 gdw day^{-1}) that converted 33% of feed (dry weight loss) to CO_2 and "volatile fatty acids" (vfa, 83:8:8 mmolar ratio acetic:propionic:butyric). High nitrate-N in the potato residue feed was absent in the anaerobic effluent, with a high portion converted to NH_4^+-N and the remainder unaccounted and probably lost to denitrification and NH_4^+ volatilization. Liquid anaerobic effluent was fed to an aerobic, yeast biomass production vessel (2 L volume; *Candida ingens* inoculum; batch [pellicle] growth; 2 day retention time) where the VFAs and some NH_4^+-N were converted into yeast biomass. Yeast yields accounted for up to 8% of potato residue fed into the anaerobic bioreactor. The third bioreactor (0.5 L liquid working volume; commercial nitrifier inoculum; packed-bed biofilm; continuous yeast effluent feed; recirculating; constant volume; 2 day hydraulic retention time) was used to convert successfully the remaining NH_4^+-N into nitrate-N (preferred form of N for CELSS crop production) and to remove the remaining degradable soluble organic carbon. Effluents from the last two stages were used for partial replenishment of minerals for hydroponic potato production.

INTRODUCTION

Integration of resource recovery and biomass production components has been a major focus of research at the Controlled Ecological Life Support System (CELSS) Breadboard Project at the Kennedy Space Center (KSC) since the late 1980's (Mackowiak *et al.*, 1996; Finger and Alazraki, 1995; Strayer and Cook, 1995). Microbiological biodegradative processes have been used to regenerate, from inedible crop residues, the inorganic nutrients that are essential for continuous production of CELSS candidate crops (Garland and Mackowiak, 1990; Garland, 1992a; Garland, 1992b; Garland, Mackowiak, and Sager, 1993; Mackowiak, *et al.*, 1994). Minerals, recovered from crop residues, are incorporated into solutions for replenishment of hydroponic nutrients for growth of CELSS food crops--wheat, potato, and soybeans--in the breadboard-scale Biomass Production Chamber (BPC)(Wheeler, *et al.*, 1996).

Most biological resource recovery research efforts have utilized aerobic, decomposition processes by a mixed community of microorganisms (Shuler, Nafis, and Sze, 1981; Finger and Strayer, 1994; Finger and Alazraki, 1995; Strayer and Cook, 1995). The efforts by Strayer and co-workers at KSC demonstrated successfully the ability of aerobic bioreactors to partially recycle inorganic nutrients from a variety of crop residues and to degrade 80% of soluble organic compounds that leach readily from the crop residues. This latter capability eliminated the negative or inhibitory effects (phytotoxic or allelopathic characteristics, proliferation of surface biofilms, elevated root respiration and, possibly, denitrification) these soluble organic compounds had on hydroponic crop production.

Anaerobic decomposition processes have also been proposed for inclusion in a CELSS (Schwartzkopf, Stroup, and Williams, 1993). In research undertaken at KSC, Schwingel and Sager (1996) investigated the feasibility for utilization of anaerobic degradative processes for mineral recycling and secondary food production from CELSS crop residues. As envisioned, the function of the anaerobic digestor was to produce volatile fatty acids (VFA--acetic, propionic, and butyric acids) that would be utilized for secondary food production in a second aerobic bioreactor stage. This function required manipulation of the anaerobic bioreactor process parameters to prevent methane formation, which is the typical product of anaerobic digestion, and to facilitate conversion of crop residue biomass into VFA. The results of these studies were reported at the last COSPAR (Schwingel and Sager, 1996).

This paper reports results of the integration of the anaerobic bioreactor of Schwingel and Sager (1996) with two other components needed to complete this resource recovery processing scheme: an aerobic, secondary food production component that converts VFA into an edible yeast biomass (single cell protein, SCP); followed by an aerobic nitrification component that also functions to remove biodegradable soluble organic compounds that remain in the liquid output from the yeast production stage. Although Schwingel identified, from the literature (Henry and Thomson, 1979; Anciaux, DeMeyer, Levert, and Vanthournhout, 1989), a purported edible yeast (*Candida ingens*, references) that utilized VFA for biomass production, his tests of this component using anaerobic bioreactor effluent were incomplete. In addition, Schwingel had found that nitrate (from CELSS crop residues) disappeared in the anaerobic digestor, with a majority converted into ammonium. However, a significant amount of the unaccounted nitrate was missing, and presumed converted, via denitrification, into dinitrogen gas. Because nitrate has been the primary nitrogen source for CELSS crops (Stutte, 1996), we added an aerobic, fixed-film bioreactor to process effluent from the yeast production component. This bioreactor functioned to oxidize residual ammonium back into nitrate (i.e., nitrification) and to also biodegrade soluble organic compounds that were not utilized by the yeast.

A major goal of any resource recovery process should be the demonstration of nutrient recycling back to plants. In addition to assessing the feasibility and performance of the three stage system, we also provided effluents from the yeast production and nitrification stages to be tested as replenishment solution for hydroponic production of white potato for an entire crop production cycle. In a companion paper presented here at the 31st *COSPAR* Mackowiak, *et al.* will report these crop production results.

MATERIALS AND METHODS

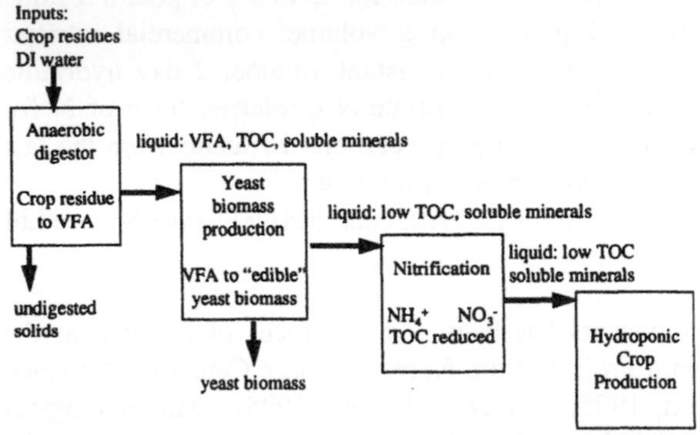

Fig. 1. Schematic diagram of the anaerobic processing scheme

Figure 1 shows the process diagram for the anaerobic processing scheme, consisting of three stages or components: An anaerobic digestor, a yeast biomass production bioreactor and a nitrification bioreactor that also functions to remove soluble total organic carbon (TOC). Material flows serially from one bioreactor to the next. Crop residues and de-ionized water are the inputs, and enter the processing scheme at stage one. The function of the anaerobic digestor is to convert crop residue organic material into volatile fatty acids, mainly acetic, propionic and butyric acids. The effluent from this stage is separated by filtration into solid (retentate) and liquid (filtrate). The liquid contains plant inorganic nutrients and soluble organic compounds (labeled TOC in the diagram) including VFA. The solids are not processed further biologically but would be suitable for incineration or other combustion process.

The function of the second stage is the production of a secondary food source. VFA from stage one is utilized by a yeast, *Candida ingens*, for the production of single cell protein. The yeast grows on top of the solution and is easily harvested with a wire mesh screen. The liquid effluent from this stage contains lower amounts of TOC due to VFA consumption and inorganic crop nutrients. The third stage was designed to perform two functions: to convert ammonium into nitrate, a process termed nitrification, and to further reduce biodegradable soluble organic compounds. These two microbiological processes were incorporated into the process stream in order to make the final liquid effluent suitable for replenishment of hydroponic crop production solutions. Previous experience had shown that biodegradable soluble TOC led to a reduction in crop yields if not removed. Second, the hydroponic growth of crops in a CELSS depends on nitrogen being in the form of nitrate instead of ammonium (Stutte, 1996).

Table 1. List of key operating parameters for each of the three processing components.

Parameter	Anaerobic digestor	Yeast Production Bioreactor	Nitrification Bioreactor
Environmental Variables			
pH	7.3	5.0	7.5
Temperature, (°C)	39	Ambient, ca. 24	ambient, ca. 24
Gas Flow	2 lpm N_2 only at harvest	none	air, 2 lpm
Process Variables			
Inoculum source	Rumen contents	*Candida ingens*	commercial, nitrifiers
Mode of operation	Fed batch	Batch	Continuous
Microorganism distribution	Suspension culture	Pellicle (floats on top)	Fixed-film (biofilm attached to inert media)
Retention time (days)	8	2	2
Substrate addition rate (gdw/day)	25	Effluent equiv. to 20	effluent equiv. to 5
Operating volume (liters)	4-8	2 - 2.5	0.5

Specific operating parameters for each bioreactor are listed in Table 1. The anaerobic bioreactor has been described by Schwingel and Sager (1996) and was adapted for this use from an 8 liter intermediate-scale aerobic bioreactor design (Strayer and Finger, 1994). Modifications were made by us to control pH without the addition of acid. Preliminary tests (not shown) determined that elimination of the continuous N_2 gas flow through the system allowed for adequate pH control without the addition of acid, presumably through production of carbon dioxide which contributed to bicarbonate buffering. During harvest, which required opening ports on top of the bioreactor and a vigorous mixing of bioreactor contents, a nitrogen gas flow of 2 liters min^{-1} was allowed to prevent entry of oxygen. Bioreactor performance was assessed by measuring dry weight losses, VFA production rates, and effluent concentration of soluble TOC and key plant nutrients.

The yeast production bioreactor was a very simple design. The bioreactor operated in a batch growth mode, inoculated with an actively growing, 2 day old agar slant culture of *Candida ingens*. To keep this stage simple and still prevent growth of contaminant bacteria, the bioreactor pH was controlled at 5.0 by computer controlled addition of nitric acid. Henry and Thomas (1979) have shown that this pH favors growth of the yeast while slowing bacterial growth, probably due to bacterial toxicity of VFA at this pH. In addition, bacterial contamination in the input (anaerobic effluent) was removed by filtration through a 0.2 mm pore size membrane filter (Minitan cross-flow filtration unit, Millipore Corp.). Bioreactor performance was assessed by estimating yeast biomass production (dry weight of harvested biomass), VFA consumption rate, and changes in plant nutrient content of the effluent.

The nitrification bioreactor was designed to utilize the ability of microbes to attach to surfaces. Because growth of nitrifying bacteria is extremely slow, fixed-film bioreactors were utilized to retain them in the bioreactor for longer periods than the hydraulic retention time. The available surface area for bacterial attachment was increased by adding inert rings of PVC (1.9 cm diameter by 1.3 cm length) pipe to a cylindrical pipe of clear plastic. The surface area contained in the 2-liter bioreactor was ca. 0.28 m^2. The bioreactor was operated in a continuous mode with input (yeast production bioreactor effluent, diluted 1/4 to lower the input TOC) feed rate equal to effluent output rate of 0.5 liters/day. The retention time of the bioreactor was 2 days, with a liquid volume of 1 liter. Bioreactor performance was assessed by measuring ammonium consumption, nitrate production, and reduction of soluble TOC.

Dry weights of bioreactor inputs and outputs were determined gravimetrically after oven drying at 70°C. Ash free dry weights (volatile solids) were determined gravimetrically after combustion in a muffle furnace at 550°C, Identification and concentration of volatile fatty acids in bioreactor effluents were determined by HPLC (Perkin Elmer HPLC-85B, with an Aminex HPX-87H ion exclusion column, 0.008 N sulfuric acid mobile phase and 210 nm ultraviolet detection). Plant inorganic nutrient cations were determined by ICP and AA spectrometry.

Ammonium, nitrate, and phosphate were determined by Technicon autoanalyzer methods. Soluble TOC was determined by uv assisted persulfate oxidation.

RESULTS AND DISCUSSION

Anaerobic digestion of potato residues through ten weeks of continuous operation was relatively stable. Volatile solid concentration in the effluent, as estimated by ash free dry weight (Figure 2), ranged from 11 to 14 gdw L^{-1} and averaged 12.5 gdw L-1. Fractional dry weight loss, calculated from the ash free dry weight of the feed, averaged 33% +/- 5.2% for the operational period. For comparison, an aerobic digestor fed similar potato residues and also with an 8 day retention time averaged 41% dry weight loss (unpublished results). Figure 2 also shows VFA concentration in the bioreactor effluent at harvest. Acetic acid was the dominant VFA detected (83%, mmoles acetic acid:mmoles total VFA) with minor amounts of propionic and butyric acids also present (ca. 8% each). The low VFA concentration reflects the low conversion of total biomass. VFA yield, in g, is about 72% of the dry weight loss in g, for a reasonable conversion efficiency. Future research should concentrate on ways to improve hydrolysis of crop biomass polysaccharides, as this may be limiting VFA production.

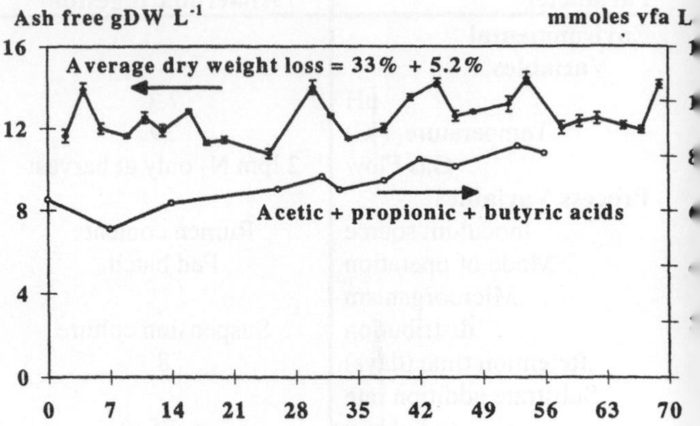

Fig. 2. Temporal variation in key anaerobic digestor response variables during the 10 week operation period. Plotted are volatile solids concentration of the effluent as ash free dry weight, and the VFA concentration of effluent.

Biological performance of the yeast biomass production bioreactor is shown in Figure 3. Two batch runs per week were completed, to match output from the anaerobic digestor and nutrient needs from the integrated crop study. Yeast yields were low over the first ten runs due to poor reproducibility of inoculum preparation. Once a standardized protocol was established with inoculum no older than 2 days, yields were higher and more consistent. Yeast growth also affected the consumption of VFA, as shown on Figure 3. The yeast biomass yield coefficient (g yeast biomass produced per g VFA consumed) averaged 0.4 over the last 10 runs. This coefficient is comparable with others for aerobic microorganisms growing on acetate (0.41 ± 0.07, n = 3; Roels, 1983). VFAs were undetectable in yeast reactor effluents, indicating complete removal of these phytotoxic compounds. Because of the low biomass conversion in the anaerobic digestor, yeast production was only 8% of the volatile solids fed into the anaerobic processing system.

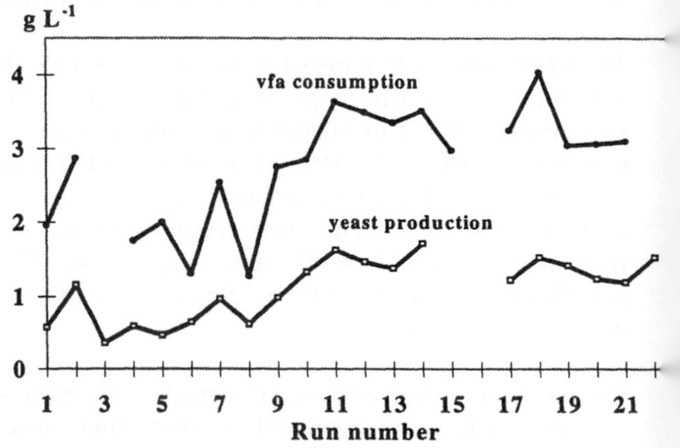

Fig. 3. Variation in yeast production and VFA consumption in the second stage, yeast biomass production bioreactor.

An important goal of biological resource recovery processes in advanced life support is the reduction in soluble TOC levels in bioreactor effluents. Garland and Mackowiak (1990) have stressed that, when untreated crop residue leachate is added to hydroponic crop production systems to replenish inorganic nutrients, the biodegradable soluble TOC that is present contributes to a number of problems. These authors observed that *in situ* microbial degradation of the soluble TOC lead to proliferation of biofilms on hydroponic system root and hardware surfaces, increased root respiration, increased denitrification potential, and lowered crop yield. The fate of soluble TOC after each processing stage of the present study is shown in the first data column of Table 2 and is compared with the soluble TOC of potato residue leachate. Due to VFA production, the soluble TOC in the anaerobic digestor exhibited an apparent increase of 23%. After the next stage, yeast biomass production and concomitant consumption of VFA, soluble TOC were reduced considerably to about half of TOC levels in leachate. The final stage, nitrification, reduced soluble TOC levels further, with only 27% of TOC in leachate remaining. This reduction in soluble TOC is slightly less than a comparable study of aerobic microbial processing of potato residues, which lowered soluble TOC by 80% (unpublished data).

Table 2 also summarizes the mineral composition of effluents from each processing stage, compared with the mineral content of potato residue leachate. Recovery of some important inorganic plant nutrients was poor, notably phosphate (14%) and iron (23%), with recovery of other minor minerals also problematic (Cu, Mn, and Zn). Recovery of magnesium after the anaerobic stage was good, but decreased following the two aerobic stages. Recovery of the major nutrient, potassium, was good when compared with leachate. Recoveries of calcium, a major plant nutrient, and boron were increased by biological processing. Biodegradation of calcium-associated plant organic compounds, such as pectin, may explain the increased recovery of this mineral. We have no explanation for the increase in boron recovery.

These data on nutrient recovery do not provide information on nutrient availability to plants. The companion paper by Mackowiak, et al. (*Adv. Space Res.*, presented at 31st *COSPAR*) presents results of utilization of effluents from the yeast production bioreactor and nitrification fixed-film bioreactor for replenishment of hydroponic solutions for growth of potato from plantlet to harvest--105 day crop cycle.

TABLE 2. Change in effluent mineral composition and soluble TOC with passage through each stage of the processing system. Major and minor minerals (mg L^{-1}, excludes nitrogen, which is covered separately). Recovery after each stage, percentage in parentheses, are compared with the mineral content of unprocessed leachate (aqueous extraction of potato residues, 25 gdw L-1 for 2 hours). Average values for each stage were calculated from assay values for all effluents collected (harvested) and analyzed during the operation of each bioreactor.

Soluble TOC and major minerals

Processing Stage	Soluble TOC-C Avg±95% CI (%)	PO_4-P Avg±95% CI (%)	K Avg±95% CI (%)	Ca Avg±95% CI (%)	Mg Avg±95% CI (%)
Leachate	2427±195 (100)	212±14 (100)	2708±48 (100)	32±1.2 (100)	145±41 (100)
Anaerobe	2991±581 (123)	82±36 (39)	2513±111 (93)	122±20 (383)	149±19 (103)
Yeast	1263±384(52)	69±11 (33)	2323±60 (86)	116±8.9 (363)	131±5 (91)
Nitrification	660±103(27)	30±4 (14)	2539±56 (94)	100±2.1 (312)	108±1 (75)

Minor minerals

Processing Stage	Fe Avg±95% CI (%)	B Avg±95% CI (%)	Cu Avg±95% CI (%)	Mn Avg±95% CI (%)	Zn Avg±95% CI (%)
Leachate	4.7±0.20 (100)	0.52±0.02 (100)	1.00±0.01 (100)	2.22±0.03 (100)	0.70±0.03 (100)
Anaerobe	3.7±0.80 (78)	0.81±0.14 (155)	0.25±0.16 (25)	0.86±0.39 (39)	0.17±0.04 (25)
Yeast	2.2±0.23 (46)	0.80±0.24 (153)	0.92±0.59 (92)	0.58±0.10 (26)	0.37±0.08 (53)
Nitrification	1.1±0.04 (23)	1.11±0.06 (213)	0.37±0.03 (37)	0.14±0.02 (6)	0.23±0.02 (34)

Nitrogen losses during anaerobic processing were substantial (Figure 4). For potato residue leachate, which is the input to the anaerobic bioreactor, all of the inorganic nitrogen is nitrate--1.4 mM NO_3-N. After anaerobic digestion, no nitrate could be found in the effluent, but a fraction of the nitrogen, 57%, was recovered as ammonium. This loss of inorganic nitrogen was probably due to denitrification, as loss by ammonium volatilization would be limited by the bioreactor pH of 7.2 and the lack of gas flow through the system except when the bioreactor was open at harvest. Further nitrogen losses occur in the yeast production bioreactor, but the ammonium is probably incorporated in the yeast biomass and should be recoverable from crew urine and feces after they consume the yeast. Only ammonium was found in yeast bioreactor effluent; nitrate was below detection limits.

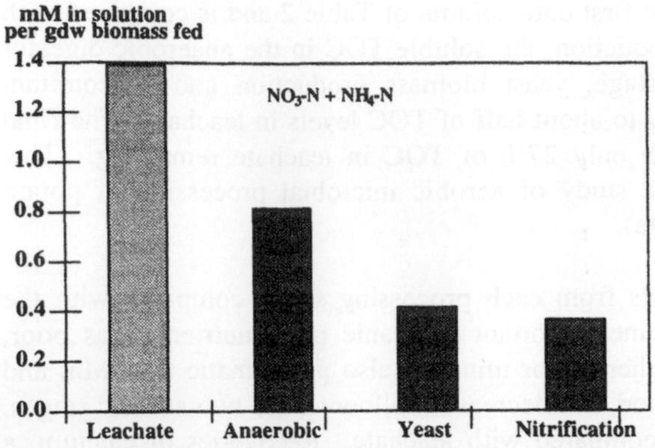

Fig. 4. Effects of anaerobic processing stage on recovery of inorganic nitrogen .

The last stage, nitrification, exhibited only a small nitrogen loss. Recall that one function of this bioreactor was to convert ammonium into nitrate. This was accomplished as all of the nitrogen in the effluent from this bioreactor was nitrate. Nitrate has been the desired N-source for crop production at the KSC ALS Breadboard project (Stutte, 1996) for a variety of reasons. As discussed by Lea-Cox *et al.* (1996), nitrate can accumulate in leaves at high nitrate supply, whereas ammonium, once taken up by roots, must be rapidly assimilated to prevent ammonium accumulation and resulting toxicity. In addition, these authors stress that high ammonium levels in a hydroponic solution have been shown to inhibit uptake of key mineral cations. However, ammonium can be used as part of a mixed N source in hydroponic solutions--5 to 25% of total N (Jones, 1983)--so complete conversion of ammonium to nitrate may not be needed. If a mixed nitrate and ammonium N-source is desired for growth of ALS crops in the future, then the nitrification after an anaerobic digestor or yeast biomass production bioreactor may not be needed.

In summary, we compare anaerobic digestion of potato residues with a typical aerobic potato residue digestion that was conducted at another time in a separate study (i.e., not in a study parallel to this anaerobic processing one) (Table 3). The advantages of an anaerobic digestion system, as presented here, are: additonal food production and better recovery of several major crop nutrients. The major disadvantages are increased system complexity, poor nitrogen recovery, lower waste conversion, and smell. Aerobic digestion, overall, is a simpler system with fewer components. Nitrogen and calcium recoveries and total waste biodegradation are better for aerobic than for anaerobic systems. However, aerobic digestion, as presently configured, does not provide additional food. Emphasis on recovery of minerals with ever simpler components has led to a research focus within ALS that has emphasized aerobic digestion components, including composting, to bioprocess ALS solid wastes. A return to anaerobic digestion as a viable alternative, with secondary food production, will depend on solving the key problems of odor, poor nitrogen recovery, and poor biodegradation.

Table 3. Comparison of aerobic and anaerobic degradation of potato residues

Factor	Aerobic	Anaerobic
Mineral recovery, better for:	N, Ca, B, Cu, Mn	PO_4, K, Mg, Fe
Nitrogen losses	low	Substantial
Degradation	41%	33%
Soluble TOC reduction	75%	73%
Food Production	none	8%, SCP
Odors	low	high (barnyard)
System complexity	1 stage	3 stages
Performance	stable over 10 weeks	stable over 10 weeks

REFERENCES

Anciaux, C.M., M DeMeyer, J.M. Levert, and M. Vanthournhout. Influence of several parameters on the growth of *Candida ingens* in controlled batch culture using volatile fatty acid substrates. *Biol. Wastes* **30**: 21-34 (1989).

Finger, B.W. and R.F. Strayer. Development of an intermediate-scale aerobic bioreactor to regenerate nutrients from inedible crop residues. *SAE Tech. Pap.* 941501 (1994).

Finger, B.W., and M.P. Alazraki. Development and Integration of a Breadboard-scale aerobic bioreactor to regenerate nutrients from inedible crop residues. *SAE Tech. Paper* 951498 (1995).

Garland, J.L. Coupling plant growth and waste recycling systems in a controlled life support system (CELSS). *NASA Tech. Mem.* 107544 (1992a).

Garland, J.L. Characterization of the water soluble component of inedible residue from candidate CELSS crops. *NASA Tech. Mem.* 107557 (1992b).

Garland, J.L. and C.L. Mackowiak. Utilization of the water soluble fraction of wheat straw as a plant nutrient source. *NASA Tech. Mem.* 107544 (1990).

Garland, J.L., C.L. Mackowiak, and J.C. Sager. Hydroponic crop production using recycling nutrients from inedible crop residues. *SAE Tech. Paper* 932173 (1993).

Henry, D.P., and R.H. Thomson. Growth of *Candida ingens* on supernatant from anaerobically fermented pig waste: Effects of temperature and pH. *Appl. Environ. Microbiol.* **37**: 1132-1136. (1979).

Jones, J. B., Jr. A guide for the hydroponic & soilless culture grower. Timber Press, Portland OR. (1983)

Lea-Cox, J.D., G.W. Stutte, W.L. Berry, and R.M. Wheeler. Charge balance – a theoretical basis for modulating pH fluctuations in plant nutrient delivery systems. *Life Support and Biosphere Science*. **3**: 53 – 59.

Mackowiak, C.L., J.L. Garland, R.F. Strayer, B.W. Finger, and R.M. Wheeler. Comparison of aerobically-treated and untreated crop residue as a source of recycled nutrients in a recirculating hydroponic system. *Adv. Space Res.* **18**: 281-287 (1996).

Mackowiak, C.L., R.M. Wheeler, W.L. Berry, and J.L. Garland. Nutrient mass balances and recovery strategies for growing plants in a CELSS. *HortSci.* **29**(5):464 (1994).

Roels, J.A. Energetics and kinetics in biotechnology. Elsevier Biomedical Press, Amsterdam (1983).

Schwartzkopf, S.H., T.L. Stroup, and D. W. Williams. Anaerobically processed waste as a nutrient source for higher plants in a controlled ecological life support system. *SAE Technical Paper* 932248. (1993).

Schwingel, W.R., and J.C. Sager (1996). Anaerobic degradation of inedible crop residues produced in a controlled ecological life support system. *Adv. Space Res.***18**: 293-297 (1996).

Shuler, M.L., D. Nafis, and E. Sze. The potential role of aerobic biological waste treatment in regenerative life support system. *ASME* 81-ENAs-20 (1981).

Strayer, R.F., and K.L. Cook. Recycling plant nutrients at NASA's KSC-CELSS Breadboard Project: Biological performance of the breadboard-scale aerobic bioreactor during two runs. *SAE Tech. Paper* 951708 (1995).

Stutte, G.W. Nitrogen dynamics in the CELSS breadboard facility at Kennedy Space Center. *Life Support and Biosphere Science* **3**: 67 - 74 (1996).

Wheeler, R.M., C.L. Mackowiak, G.W. Stutte, J.C. Sager, N.C. Yorio, L.M. Ruffe, R.E. Fortson, T.W. Dreschell, W.M. Knott, and K.A. Corey. NASA's biomass production chamber: a testbed for bioregenerative life support studies. *Adv. Space Res.* **18**: 215-224 (1996).

Pergamon

Adv. Space Res. Vol. 20, No. 10, pp. 2017–2022, 1997

Printed in Great Britain
0273-1177/97 \$17.00 + 0.00

PII: S0273–1177(97)00935–6

HYDROPONIC POTATO PRODUCTION ON NUTRIENTS DERIVED FROM ANAEROBICALLY-PROCESSED POTATO PLANT RESIDUES

C. L. Mackowiak, G. W. Stutte, J. L. Garland, B. W. Finger and L. M. Ruffe

Dynamac Corporation, Kennedy Space Center, FL, U.S.A.

ABSTRACT

Bioregenerative methods are being developed for recycling plant minerals from harvested inedible biomass as part of NASA's Advanced Life Support (ALS) research. Anaerobic processing produces secondary metabolites, a food source for yeast production, while providing a source of water soluble nutrients for plant growth. Since NH_4-N is the nitrogen product, processing the effluent through a nitrification reactor was used to convert this to NO_3-N, a more acceptable form for plants. Potato (*Solanum tuberosum* L.) cv. Norland plants were used to test the effects of anaerobically-produced effluent after processing through a yeast reactor or nitrification reactor. These treatments were compared to a mixed-N treatment (75:25, NO_3:NH_4) or a NO_3-N control, both containing only reagent-grade salts. Plant growth and tuber yields were greatest in the NO_3-N control and yeast reactor effluent treatments, which is noteworthy, considering the yeast reactor treatment had high organic loading in the nutrient solution and concomitant microbial activity.

INTRODUCTION

Long-duration space habitats containing higher plants will likely require some level of recycling (Barta and Henninger, 1994). Reclaiming water soluble minerals from inedible plant biomass to support hydroponic crop production is an efficient recycling method. It has been predicted that the human food requirement for a space habitat may be 620 g dry mass per person per day and the urine + stool waste component would be approximately 100 g dry mass per person per day (NASA, 1991). If the harvest index for food (g dry mass edible ÷ g dry mass total) is roughly 50% (equal to food dry mass), then approximately 620 g inedible biomass would need to be processed per day. This is six times the amount of the human waste component (excluding trash, gray water, and gaseous wastes). Plants grown hydroponically contain a large amount of nutrients (> 10% ash) in their inedible fraction (McKeehen *et al.*, 1996; Wheeler *et al.*, 1994) or 62 g per person per day, which is more than half the predicted dry mass of the human urine and stool waste stream.

Aerobic bioreactor processing of inedible plant biomass has resulted in nutrient effluents which can support various crop species (Garland *et al.*, 1993; Mackowiak *et al.*, 1994; Mackowiak *et al.*, 1996). Aerobic effluents also supported potato in two different crop production scenarios (batch or continuous culture systems) using recirculating hydroponics for over 400 days (Mackowiak *et al.*, this issue). Through aerobic processing, the organics are microbially degraded and released as CO_2 and H_2O. On the other hand, anaerobic processing produces secondary metabolites that are phytotoxic but can be used as a food source for yeast (yeast reactor), which in turn could be eaten by the crew (Strayer *et al.*, this issue).

The addition of yeast as human food would increase the calculated harvest index of the biomass production component.

Another characteristic of anaerobic processing is that the soluble nitrogen is converted to NH_4^+. Most plants grow well in mixed nitrogen systems (Marshner, 1995) but providing NH_4^+ as the sole source of nitrogen can be detrimental to some crops, even when using pH control (Magalhaes and Huber, 1991). To increase the NO_3:NH_4 ratio, the effluent from the yeast reactor can be treated by a nitrification reactor (Strayer *et al.*, this issue). The use of new and "aged" anaerobic effluents on lettuce seedlings, resulted in growth depression of 80 and 20% and were attributed to the high ammonia levels (Schwartzkopf *et al.*, 1993). We tested anaerobic effluent on hydroponic potato production after it had been treated with a yeast bioreactor or a subsequent nitrification reactor to increase the NO_3:NH_4 ratio.

MATERIALS AND METHODS

Potato (*Solanum tuberosum* L.) cv. Norland were grown from in vitro plantlets in a controlled environment growth chamber using cool-white fluorescent lighting 350 ± 17 μmol m^{-2} s^{-1} photosynthetic photon flux (PPF) at canopy level, with a 12-h photoperiod. Temperature set points were 20/16 °C (light/dark) and humidity held constant at 65%. Carbon dioxide was added to the chamber atmosphere to provide approximately 0.12 kPa partial pressure (1200 ppm). The experimental design consisted of four nutrient tanks, each supporting two trapezoidal-shaped culture trays. Each 0.25 m^2 culture tray contained two plants, however, because of side-lighting, we have calculated that each tray provided the plants 0.3 m^2 (or 6.7 plants m^{-2}). Stainless steel support cages (60-cm height) were placed around the perimeter of each culture tray to restrict plant growth to the allotted area. Fiberglass screening was wrapped around each cage and maintained at plant canopy level to reduce the effect of side lighting. Diurnal photosynthetic rates and conductance were measured 51 days-after-planting (DAP) using a LI-COR 6200 leaf cuvette system. At 105 DAP, all plants were harvested, separated into tissue types and oven-dried at 70 °C for four days to determine dry mass. Subsamples of all tissue were ground through a 2-mm mesh screen in a Wiley mill and analyzed for inorganics using DC arc spectrometry.

The specific protocols for anaerobic processing, yeast bioreactor processing, and nitrification reactor processing are described by Strayer *et al.* (this issue). Briefly, the procedure involved loading the anaerobic bioreactor with oven-dried inedible potato biomass (25 g L^{-1}) with an 8-day retention time. The liquid (effluent) was separated from the solids and the somewhat diluted effluent (equivalent to 20 g L^{-1}) placed into a yeast reactor. The yeast reactor processed the effluent for two days. Effluent from the yeast reactor was used directly for one of the nutrient treatments or it was diluted approximately four times with deionized water and processed for an additional two days in a nitrification reactor. The dilution reduced the total organic carbon (TOC) to levels that would not inhibit microbe metabolism.

Plants were grown hydroponically using recirculating nutrient film technique. Four different nutrient treatments were compared:

A) Nutrient solution comprised solely of reagent-grade chemicals, with N provided as NO_3^- (Table 1).
B) Same as A, with the exception of N being provided as a mixed form (mixed-N) of NO_3^-:NH_4^+ at 75:25.
C) Effluent from yeast bioreactor (ISYB) processing adjusted to provide half the K requirement from effluent.
D) Effluent from nitrification bioreactor (ISNB) processing adjusted to provide half the K requirement from effluent.

Since treatments C and D were not nutritionally complete, reagent-grade salts were used to make up the difference, where appropriate (Table 1). This resulted in the four nutrient solution treatments having comparable nutrients. Since all the systems used recirculating hydroponics, subsequent nutrient replenishments were used (one concentrated stock per treatment) as plants removed nutrients. The ISYB and ISNB replenishments obtained much of their nutrition from the associated bioreactor effluents and the remaining nutrients were amended with reagent-grade salts, similar to the making of the working solutions at the start of the study. The working solution for each treatment was measured for electrical conductivity (EC) on a daily basis and the replenishment stocks added to maintain an EC between 1.2 and 1.3 dS m^{-1}. All but the ISNB replenishment stock should have contained 70 mM N, 56 mM K, 12 mM Ca, 10 mM Mg, 10 mM S, 134 μM Fe, 96 μM Mn, 12.5 μM Zn, 13.5 μM Cu, 124 μM B, and 0.13 μM Mo. Since the effluent going to the nitrification reactor was diluted with water (Strayer *et al.*, this issue), the ISNB replenishment stock had equivalent ratios of nutrients but the overall concentrations were about 30% of the other treatment replenishment stocks (i.e. N= 47 mM, K = 37 mM, etc.). To maintain equivalent EC, more total replenishment volume would be needed for the ISNB treatment.

Table 1. Nutrient concentrations of working solutions prior to adjustments.

Nutrient	Control	Mixed-N	ISYB*	ISNB*
	mM	mM	mM	mM
NO_3-N	7.5	5.6	1.8	1.7
NH_4-N	0.0	1.9	0.2	0.0
PO_4-P	0.5	0.5	< 0.04	< 0.04
K	3.0	3.0	1.5	1.5
Ca	2.5	2.5	0.1	0.1
Mg	1.0	1.0	0.1	0.1
	μM	μM	μM	μM
Fe	50.00	50.00	0.00	0.00
Mn	7.40	7.40	0.10	0.10
Zn	0.96	0.96	0.14	0.13
Cu	1.04	1.04	0.15	0.11
B	9.50	9.50	1.30	1.60
Mo	0.01	0.01	0.00	0.00

* ISYB effluent (0.54 L) and ISNB effluent (2.0 L) were added to the 20 L tank volume to get the listed nutrient levels.

Nutrient solution pH was automatically monitored and controlled at 5.8 with additions of 0.39 M HNO_3 or 0.1 M KOH. Water uptake (from evapotranspiration) was replaced daily by manually adding deionized water to the reservoirs to maintain a constant volume of 20 L. Weekly nutrient solution samples were taken from each reservoir for inorganic analysis using Inductively Coupled Plasma (ICP) spectrometry and Atomic Absorption (AA) spectrophotometry methods. Total microbial cell density was determined for nutrient solutions and rhizospheres using the acridine-orange (AO) method with epiflourescent microscopy (Hobbie *et al.*, 1977).

RESULTS AND DISCUSSION

Total biomass and tuber dry mass yields were greatest for the control and yeast reactor (ISYB) treatments, whereas values for the mixed-N and nitrification reactor (ISNB) treatments were lower (Figure 1).

Table 2. Nutrient composition of solutions and harvested dry biomass.

	Nutrient Solution Composition*				Harvested Shoot Composition				
Treatment	N	P	K	Ca	N	P	K	Ca	K:Ca
	(ppm)	(ppm)	(ppm)	(ppm)	(%)	(%)	(%)	(%)	
Control	62	16.1	76	133	1.6	0.28	9.6	2.0	4.8
Mixed-N	33	6.1	129	121	1.9	0.24	10.5	2.4	4.3
ISYB	50	4.4	62	129	1.8	0.21	7.1	1.8	3.9
ISNB	140	4.2	36	200	1.9	0.28	2.6	3.1	0.8

*Mean values over the entire study.

In regards to tuber dry mass, the control > ISYB > mixed-N > ISNB (1409, 1396, 1089, 892 g m^{-2}, respectively). Treatment ranking for total dry mass was ISYB > control > mixed-N > ISNB (1840, 1835, 1344, 1126 g m^{-2}, respectively). Although there was large variability within the control treatment, production values were similar to the control treatment in other NFT potato studies (Wheeler *et al.*, 1990; Mackowiak *et al.*, 1996a). It was surprising that the mixed-N and ISNB treatments did relatively poorly. Based upon weekly canopy height data, it appears that these treatments were adversely affected by the fifth week (Figure 2). Midday leaf photosynthetic rates at 51 DAP were not exceptionally different among treatments, where rates averaged 13.7 ± 3.0 mol m^{-2} s^{-1}. However, midday leaf conductance was highly correlated with final yields, where $r^2 = 0.99$.

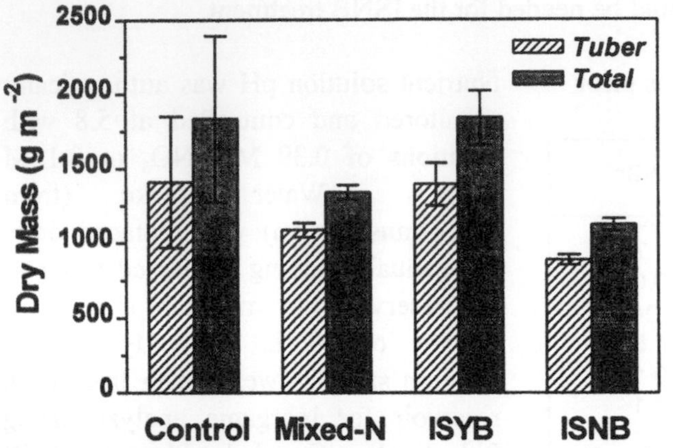

Fig. 1. Nutrient treatment effects on dry mass. Vertical lines = standard deviations (n =2).

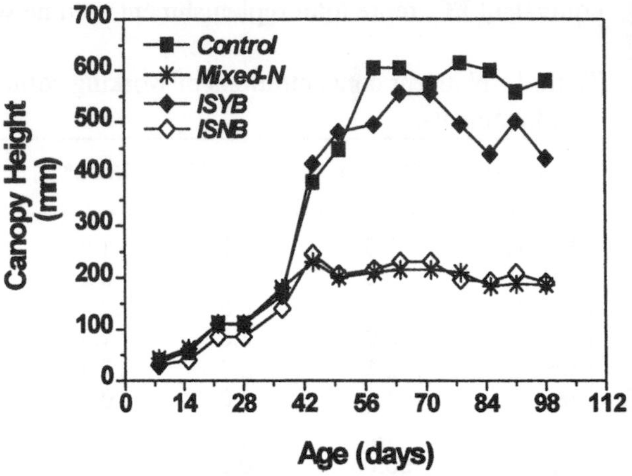

Fig. 2. Nutrient treatment effect on canopy height.

Based on nutrient solution and shoot tissue analyses results, it is likely that nutritional imbalances were the cause for the poor yields for the mixed-N and ISNB treatments (Table 2). In the case of the mixed-N treatment, total N in solution was rather low compared to the other treatments, although shoot tissue levels were similar among treatments at harvest (Table 2). The plants from the mixed-N treatment were light green in color, suggesting nitrogen deficiency (Ulrich, 1993). Plant growth is directly related to nitrogen supply and suboptimal levels cause an increase in leaf starch (Marschner, 1995). In the mixed-N treatment starch content of leaves at 51 DAP was nearly three times greater (353 μg mg^{-1} leaf dry mass) than leaves from the control treatment (127 μg mg^{-1} leaf dry mass). Nitrogen volatilization may had occured in the stored replenishment concentrate resulting in less N being added back to the working solution for the mixed-N treatment. In addition, since very little HNO_3 was being added for pH adjustment, the plants were not receiving that additional source of N, which can be a significant share of the total N supplied to the system (Mackowiak *et al.*, 1996b).

The ISNB treatment was probably K deficient, where solution and tissue levels were much lower than the other treatments (Table 2), and tissue levels were lower than the 6% suggested for potato leaves (Mills and Jones, 1996). In addition, the K tissue values were much lower than other cations, such as Ca, resulting in nutrient imbalances (Table 2). Plants on this treatment were very dark green with some leaf puckering, similar to K deficiency symptoms shown by Ulrich (1993). This treatment used a more dilute stock replenishment (60% of the control) because effluent going to the nitrification reactor needed to be diluted with water (Strayer, this issue). Although nutrient ratios were equal among the replenishment solutions of the four treatments, greater amounts of liquid would have to be taken up by the plants in the ISNB treatment to acquire the same amount of total nutrients. A slight miscalculation in formulation would be exaggerated in the replenishment additions.

Unlike the mixed-N and ISNB treatments, the ISYB treatment resulted in dry mass yields equal to the control treatment (Figure 1). Values for all parameters were quite similar, except for acid use efficiency (AUE), which was at least 20% greater for the ISYB treatment (Table 3). Similar trends have been seen when using aerobic effluents for hydroponic crop production (Mackowiak et al., 1996b). These effluents contain recalcitrant organics which have been shown to improve pH buffering (MacCarthy *et al.*, 1994). Although NH_4-N is noted for reducing pH and thus acid requirements in plant growth systems (Marschner, 1995), it is unlikely that ammonium alone played a role in lowering the acid requirements in the ISYB treatment since the solution levels were < 3% of the total N replenishment (Table 1) (Jones, 1983). It appears that much of the ammonium that should have been in the ISYB effluent was lost (volatilization and nitrification) during processing and storage.

Table 3. Recycling effects on productivity, water use, and acid use for pH control.

Treatment	Total Biomass	Edible Biomass	Harvest Index	WUE*	AUE*
	(g mol^{-1} PAR)*	(g mol^{-1} PAR)*	(%)	(g kg^{-1} H_2O)	(g mol N^{-1})
Control	0.83	0.64	77	5.86	1.70
ISYB	0.83	0.63	76	5.58	2.18

*WUE = water use efficiency, AUE = acid use efficiency, PAR = photosynthetically active radiation.

Although microbial densities were greater in the nutrient solution of the ISYB treatment, rhizosphere (were microbial activity is greatest) densities were similar (Figure 3). This suggests that the higher organic load from the ISYB treatment increased activity in the nutrient solution but it did not affect the rhizosphere or plant productivity.

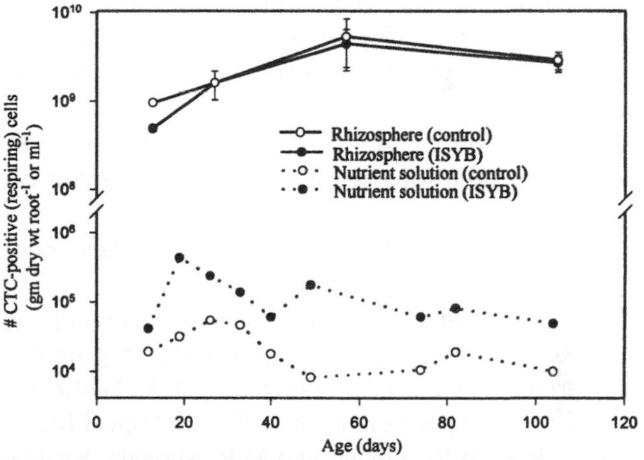

Fig. 3. Nutrient treatment effect on microbial densities of the nutrient solution and rhizosphere.

CONCLUSIONS

Anaerobic effluent which had been treated for two days in a yeast reactor was an acceptable source of nutrients for potato in recirculating hydroponics, and yeast reactor processing eliminated the need for processing through a nitrification reactor. This treatment may have been less successful if more of the ammonium remained in the ISYB reactor effluent. Although effluent from the yeast reactor treatment had negligible levels of ammonium, it had greater acid use efficiency, which may reflect in improved buffering capacity of effluent-based nutrient solutions. Plant growth was depressed in the mixed-N and nitrification reactor effluent treatments, which likely was due to nutrient imbalances, that could be easily corrected in future studies. The increased organic load in the yeast reactor effluent treatment had no apparent effect on microbial activity in the rhizosphere, although activity was greater in the nutrient solution. This suggests that as long as there is a healthy and active microbial community in the rhizosphere, the plants can tolerate some organic loading of the nutrient solution.

REFERENCES

Barta, D.J. and D.L. Henninger, Regenerative life support systems: Why do we need them?, Adv. Space Res. 14(11), 403-410 (1994).

Garland, J.L., C.L., Mackowiak, and J.C. Sager, Hydroponic crop production using recycled nutrients from inedible crop residues, SAE Technical Paper 93173, 1993.

Hobbie, J.E., R.J. Daley, and S. Jasper, Use of nucleopore filters for counting bacteria for fluorescent microscopy, Appl. Environ. Microb. 33, 1225-1228 (1977).

Jones, J.B., The nutrient solution, in J.B. Jones ed. A Guide for the Hydroponic and Soilless Culture Grower, pp 48-50, Timber Press, Portland OR, USA (1983).

MacCarthy, P., R.L. Malcolm, C.E. Clapp, P.R. Bloom, An introduction to soil humic subastances, , in P. MacCarthy, C.E. Clapp, R.L. Malcolm, and P.R. Bloom eds. Humic Substances in Soil and Crop Sciences: Selected Readings, pp. 1-12, ASA, Inc. and SSSA, Inc., Madison WI, USA (1990).

Mackowiak, C.L., J.L. Garland, and G.W. Stutte, Growth regulator effects of water soluble materials from crop residues for use in plant hydroponic culture, Proceeding 21st Annual PGRSA (1994).

Mackowiak, C.L., J.L. Garland, R.F. Strayer, B.W. Finger, and R.M. Wheeler, Comparison of aerobically-treated and untreated crop residue as a source of recycled nutrients in a recirculating hydroponic system, Adv. Space Res., 18(1/2) 281-287 (1996a).

Mackowiak, C.L., J.L. Garland, and J.C. Sager, Recycling crop residues for use in recirculating hydroponic crop production, Acta Horticulturae, 440:19-24 (1996b).

Mackowiak, C.L., R.M. Wheeler, G.W. Stutte, N.C. Yorio, and J.C. Sager, Use of biologically reclaimed minerals for continuous hydroponic potato production in a CELSS, Adv. Space Res., (this issue).

Magalhaes, J.R. and D.M. Huber, Response of ammonium assimilation enzymes to nitrogen form treatments in different plant species, J. Plant Nutri., 14(2), 175-185 (1991).

Marschner, H., Mineral Nutrition of Higher Plants, H. Marschner ed. Academic Press, New York, USA (1995).

McKeehen, J.D., C.A. Mitchell, R.M. Wheeler, B. Bugbee, and S.S. Nielsen, Excess nutrients in hydroponic solutions alter nutrient content of rice, wheat, and potato, Adv. Space Res., 18(4/5), 73-83 (1996).

Mills, H.A. and J.B. Jones, Plant Analysis Handbook II, MicroMacro Publishing, Inc., Athens GA, USA (1996).

NASA, Environmental control and life support system architectural control document, SSP 30262 Revision D, Space Station Freedom Program Office, NASA, Reston VA, USA (1991).

Schwartzkopf, S.H., T.L. Stroup, and D.W. Williams, Anaerobically-processed waste as a nutrient source for higher plants in a controlled ecological life support system, SAE Technical Paper 932248.

Strayer, R.F., B.W. Finger, and M.P. Alazraki, Evaluation of an anaerobic digestion system for processing CELSS crop residues for resource recovery, Adv. Space Res. (this issue).

Ulrich, A., Potato, in W.F. Bennett, ed., Nutrient Deficiencies and Toxicities in Crop Plants, APS Press, St. Paul, MN, USA (1993).

Wheeler, R.M., C.L. Mackowiak, J.C. Sager, W.M. Knott, and W.L. Berry, Proximate composition of CELSS crops grown in NASA's Biomass Production Chamber, Adv. Space Res., 18(4/5), 43-47 (1994).

Wheeler, R.M., C.L. Mackowiak, J.C. Sager, W. M. Knott, and C.R. Hinkle, Potato growth and yield using nutrient film technique (NFT), Amer. Pot. J. 67:177-187 (1990).

Pergamon

Adv. Space Res. Vol. 20, No. 10, pp. 2023–2028, 1997

Printed in Great Britain
0273-1177/97 $17.00 + 0.00

PII: S0273-1177(97)00936-8

EFFECTS OF BIOREACTOR RETENTION TIME ON AEROBIC MICROBIAL DECOMPOSITION OF CELSS CROP RESIDUES

R. F. Strayer, B. W. Finger and M. P. Alazraki

Dynamac Corporation, Mail Code DYN-3, Kennedy Space Center, FL 32780, U.S.A.

ABSTRACT

The focus of resource recovery research at the KSC-CELSS Breadboard Project has been the evaluation of microbiologically mediated biodegradation of crop residues by manipulation of bioreactor process and environmental variables. We will present results from over 3 years of studies that used laboratory- and breadboard-scale (8 and 120 L working volumes, respectively) aerobic, fed-batch, continuous stirred tank reactors (CSTR) for recovery of carbon and minerals from breadboard grown wheat and white potato residues. The paper will focus on the effects of a key process variable--bioreactor retention time--on response variables indicative of bioreactor performance. The goal is to determine the shortest retention time that is feasible for processing CELSS crop residues, thereby reducing bioreactor volume and weight requirements. Pushing the lower limits of bioreactor retention times will provide useful data for engineers who need to compare biological and physicochemical components. Bioreactor retention times were manipulated to range between 0.25 and 48 days. Results indicate that increases in retention time lead to a 4-fold increase in crop residue biodegradation, as measured by both dry weight losses and CO_2 production. A similar overall trend was also observed for crop residue fiber (cellulose and hemicellulose), with a noticeable jump in cellulose degradation between the 5.3 day and 10.7 day retention times. Water-soluble organic compounds (measured as soluble TOC) were appreciably reduced by more than 4-fold at all retention times tested. Results from a study of even shorter retention times (down to 0.25 days), in progress, will also be presented.

INTRODUCTION

Recent research at the Controlled Ecological Life Support System (CELSS) Breadboard Project at the Kennedy Space Center has focused on the integration of resource recovery and biomass production components (Mackowiak, *et al.*, 1996; Finger and Alazraki, 1995; Strayer and Cook, 1995). A primary objective is to regenerate, from inedible crop residues, inorganic nutrients that are essential for continuous production of CELSS candidate crops. Recovered minerals are incorporated into a solution for replenishment of hydroponic nutrients for growth of CELSS food crops--wheat, potato, and soybeans--in the breadboard-scale Biomass Production Chamber (BPC) (Wheeler, *et al.*, 1996).

Initial CELSS resource recovery research at KSC discovered that most nutrients could be recovered readily by a simple aqueous extraction of inedible crop residues--stems, leaves, and roots (extraction conditions were 50 gdw crop residues L^{-1} at ambient temperature for 2 hours) (Garland and Mackowiak, 1990; Garland, 1992a; Garland, 1992b; Garland, Mackowiak, and Sager, 1993; Mackowiak, *et al.*, 1994). However, direct utilization of these crop residue "leachates" was found to inhibit plant growth, probably by direct or indirect phytotoxic effects of soluble organic material that was also extracted from the residues along with the plant nutrients. Subsequent studies determined that microbiological treatment of the crop residues could reduce soluble organic content and eliminate plant growth inhibition. Thus, in one simple bioreactor step, three simultaneous biological and chemical processes were accomplished: nutrient extraction/release from inedible biomass, biodegradation of soluble crop residue organics, and (partial) biodegradation of insoluble crop residue organics such as cellulose and hemicellulose.

Bioreactor studies at both a laboratory (8 L bioreactor working volume) (Finger and Strayer, 1994) and breadboard (120 L working volume) (Finger and Alazraki, 1995; Strayer and Cook, 1995) scale at KSC have explored a limited range of key bioreactor process variables, including retention time (combined hydraulic and solid). Minimizing retention time for bioreactors would be desired for an Advanced Life Support system to minimize key costs of resource recovery biological processes (Drysdale, 1995). Basic bioreactor engineering literature defines retention time (also called mean residence time or nominal holding time) as bioreactor volume divided by the volumetric flow rate through the bioreactor (with inflow equal to outflow)(Bailey and Ollis, 1986). By minimizing retention time, volume and mass (ALS "costs") will also be minimized.

Because of limited biodegradation success, including longer retention times of 48 days, a concerted research effort was undertaken to determine, in greater detail, the effects of this important variable to microbiological resource recovery components. Two laboratory, intermediate scale bioreactor studies were designed to address specifically the effects of retention time on crop residue biodegradation. This paper reports the results of these two studies after summarizing results from previously published breadboard-scale studies that used a more limited range of retention times (8, 24 and 48 days).

MATERIALS AND METHODS

KSC Breadboard Project bioreactors for resource recovery have been designed and fabricated at two scales: laboratory scale (8 liters working volume) and breadboard scale (120 liters working volume). These sizes were selected so bioreactor effluents (from biodegradation of crop residues) could be used to replenish hydroponic solutions for crop growth studies. Crop studies were conducted at a "laboratory" scale of about 0.5 to 1 m^2 plant growing area and a breadboard scale (KSC Biomass Production Chamber or BPC) with 20 m^2 plant growing area. A detailed description of these two bioreactor designs and operations are presented elsewhere (Finger and Strayer, 1994; Strayer and Cook, 1995; Finger and Alazraki, 1995).

Specific studies of bioreactor retention time were conducted at the laboratory scale with two identical, replicate bioreactors (termed intermediate-scale aerobic bioreactors, I-SABs, in Finger and Strayer, 1994). In one study, bioreactors were run at different retention times: 1.3, 1.8, 2.7, 5.3, 10.7, and 21.3 days. The shortest time we could test, 1.3 days, was dictated by bioreactor design/geometry--only 3/4 of the bioreactor could be harvested daily without making drastic modifications to harvest hardware and established operational protocols. Separate bioreactor runs were done for each retention time. The duration of each run lasted for a minimum of three retention times after stabilization of bioreactor response variables (i.e., CO2 production, dry weight losses). At the end of each run, entire bioreactor contents were harvested and the bioreactor dismantled and cleaned for the next run.

As will be shown later, results from this study suggested that faster retention times should be attempted. Because feeding and harvest of bioreactors is a manual operation, faster retention times would require 24 hour care. The study was designed to run continuously, with retention times decreasing successively after response variables had stabilized and samples collected (duration of at least 5 retention times). Retention times tested were: 2, 1, 0.5 and 0.25 days, run sequentially without bioreactor disassembly and cleaning between retention time "runs".

Environmental variables and process variables, other than retention time, were kept constant. For both retention time studies, substrate concentration was 20 gdw crop residues $liter^{-1}$. Wheat residues (cultivar Yecora rojo, grown in a controlled greenhouse environment) were used for the first study and were obtained from Dr. Bruce Bugbee, Utah State University , then oven-dried and milled to 2 mm diameter. For the second study, potato residues from the KSC Biomass Production Chamber (BPC) were used after freeze-drying and milling (2 mm diameter). Bioreactor inocula were from a commercial source (Gempler's) as well as microflora native to the crop residues and de-ionized water source. Bioreactor pH was controlled at 6.5 by addition of 1 N nitric acid, temperature was controlled at 35°C, and dissolved oxygen was kept above 2.0 mg L^{-1} by an air flow of 7.5 liters min^{-1} and a stirring rate of 360 rpm. Dissolved oxygen, pH, temperature, gas flow rate, and offgas CO_2 concentration were monitored continuously as described by Strayer and Finger (1994). Dry weight (70°C overnight) and ash-free dry weight (i.e., volatile solids, 550°C overnight) of bioreactor contents were determined in triplicate at each harvest.

RESULTS AND DISCUSSION

In initial shake flask studies on crop residue biodegradation at room temperature, process variables were screened to determine levels that gave the maximum dry weight loss. Retention times tested were 2, 4, 8, and 16 days. Dry weight losses for the 8 and 16 day retention time were best, with little difference observed between the two (Finger and Strayer, 1994).

Results from these studies were used design intermediate scale aerobic bioreactors (I-SABs) of 8 liter working volume. A series of eight short I-SAB runs (15 to 24 days duration) were conducted to compare degradation rates at 4 and 8 day retention times with two solids loading rates, 20 and 40 gdw wheat residue addition day^{-1}. Results showed that the best degradation occurred at a retention time of 8 days at the 20 gdw day^{-1} addition rate (Finger and Strayer, 1994). This retention time was used for subsequent, longer duration ISAB runs with both wheat and potato residues, with the bioreactor effluent often used for nutrient recycling studies.

The eight day retention time was also used as one of the design criteria for our 120 liter breadboard-scale aerobic bioreactor (B-SAB). The main function of this bioreactor is to process inedible crop residues (with graywater, urine, and human solid waste processing to be added later) with the goal of supplying inorganic nutrients for replenishment of hydroponic solution for up to 2 growing levels of the KSC BPC (5 m^2 growing area per level). Operation of this bioreactor has supported two breadboard studies: a batch growout of wheat (70 days duration, bioreactor run for over 40 days) and long-duration potato study (418 days) (see Finger and Alazraki, 1995; Strayer and Cook, 1995). For the wheat study, bioreactor retention time was 8 days. For the potato study, bioreactor retention time varied between 24 and 48 days. These unrealistically long retention times resulted from two factors. First, the bioreactor was designed for a "worst case" crop--wheat--with a low harvest index and high amount of inedible biomass to be processed. Second, the effluent output had to match the nutrient demands of the potato, which were variable over the longer, 4 crop-cycle study.

Bioreactor biological performance during the wheat growout is shown in Figure 1 and for the longer potato growout in Figure 2. Even though direct comparisons cannot be made due to residue compositional differences between the two crops, the 30 - 35% biodegradation of wheat residues at an 8 day retention time is significantly lower than the 50 to 55% biodegradation of potato at retention times 3 to 6 times greater. In separate laboratory studies that used intermediate-scale bioreactors and 8 day retention times, 40 to 45% biodegradation of both wheat and potato residues could be obtained (results not shown).

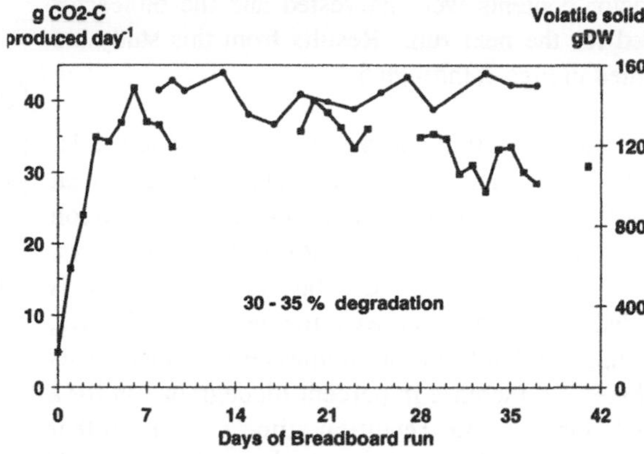

Fig. 1. Biological performance of B-SAB operated at an 8 day retention time for 41 days. Bioreactor response variables: carbon mineralization--closed squares with units on the left Y-axis, and volatile solids content--closed circles with units on the right Y-axis. Feeding rate: 120 g C day^{-1} . The bioreactor would have contained 2150 gdw volatile solids (average) if degradation had not occurred.

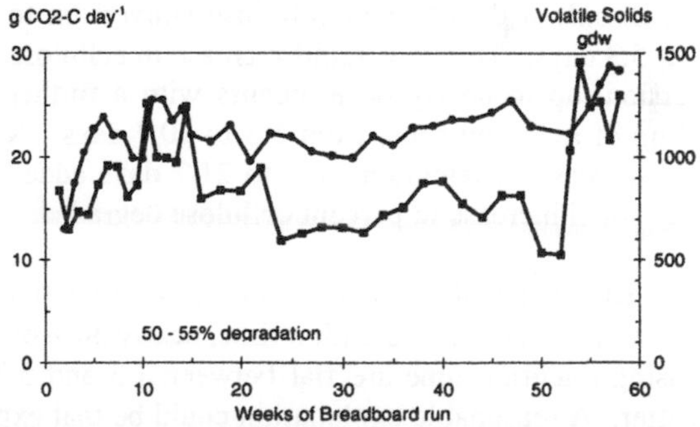

Fig. 2. Biological performance of B-SAB operated at 24 and 48 day (weeks 24 to 48) retention times for 418 days. Plotted are carbon mineralization--closed squares and volatile solids content--closed circles. Feeding rate: 23.8 g C day^{-1} for 48 day retention time and 47.5 g C day^{-1} for 24 day retention time. The bioreactor would have contained 2300 gdw volatile solids (average) if degradation had not occurred.

As mentioned in the Materials and Methods section, intermediate scale studies were designed to better explain the effects of retention time on biodegradation. In the first study, duplicate I-SABs were run at 1.3, 1.8, 2.7, 5.3, 10.7, and 21.3 day retention times, with wheat residues (20 gdw L^{-1}) as the feedstock. Duplicate bioreactor runs were done for each retention time. The duration of each run lasted for a minimum of three retention times after bioreactor response variables (i.e., CO2 production, dry weight losses) had stabilized. At the end of each run, entire bioreactor contents were harvested and the bioreactor cleaned for the next run. Results from this study are presented in Figs. 3 through 5.

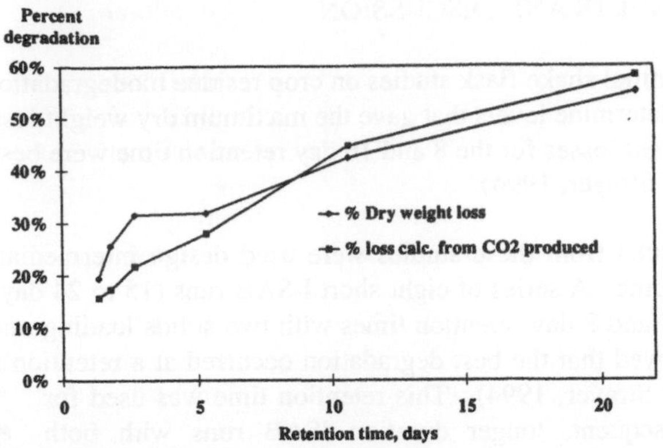

Fig. 3. Effects of retention time on biodegradation of wheat residues. Two separate estimates of degradation are shown, dry weight (volatile solids) loss and carbon mineralization (as CO_2 production)

Carbon mineralization and dry weight (volatile solids) loss both indicate that decomposition of wheat residues appears to be related directly to retention time (Figure 3). At fast retention times of 1.3 days, between 16 and 20 % of the added biomass is degraded. At the other extreme tested, 21.3 days, between 54 and 58% of the biomass is degraded. This is a three-fold increase in percent biodegradation for a 16-fold increase in retention time. For further comparison, in the past we had conducted ISAB studies using 8 day retention times and wheat residues to obtain 40 - 45% degradation. Extrapolation from Figure 3. indicates an 8 day retention time would have been in the vicinity of 38% degradation.

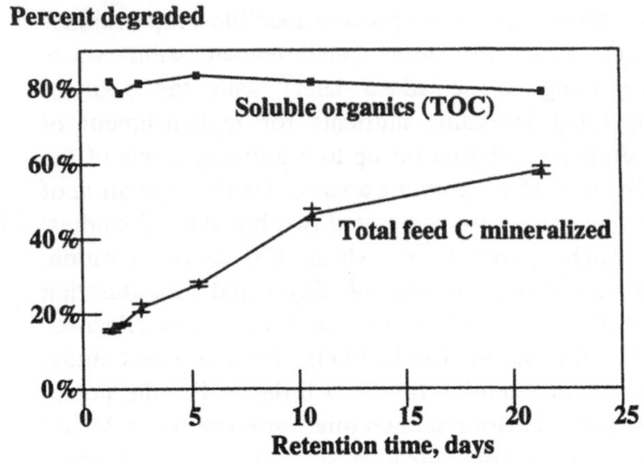

Fig. 4. Effects of retention time on degradation of soluble organic compounds (measured as soluble TOC) from wheat residues. Compared with total wheat residue biodegradation (as C mineralized).

The effects of retention time on degradation of wheat residue fibers is shown in Figure 4. Cellulose degradation at fast retention times is very low, 12% at 1.3 days. A four-fold increase in retention time results in only a doubling in cellulose conversion to 25% at 5.3 days. The most rapid increase in cellulose conversion, up to nearly 60%, occurs with a further doubling of retention time, from 5.3 to 10.7 days. A further increase in retention time to 21.7 days added another small increase in percent cellulose degraded.

The shape of the hemicellulose vs. retention time curve in Figure 4 shows a different response from cellulose. Even at a fast retention time, nearly 30% of hemicellulose is biodegraded. Increases in degradation with increasing retention time are flat between 1.3 and 5.3 days retention time, with a significant increase occurring thereafter. A reasonable explanation could be that exposed hemicellulose is easily biodegraded, then, as cellulose is degraded, more hemicellulose is exposed and, thus, also biodegraded at retention times exceeding 10.7 days.

Lignin, an important plant fiber component in field grown crops, is extremely low in crops grown in controlled environments (1 to 5%, data not shown). Thus, the effects of retention time on lignin degradation are not shown.

Figure 5 shows the effects of retention time on degradation of the soluble organic compounds, measured as soluble TOC, that quickly leach from wheat residues. At all retention times studied, from 1.3 to 21.3 days, soluble TOC degradation was near 80%. These soluble organics are so easily biodegraded that we did not approach limiting degradation at the fastest retention time we tested. Also replotted in Figure 5 is data from Figure 3 on total Carbon mineralization for comparison. At the fastest retention time, only 16% of total carbon was mineralized compared with the 80+% of soluble carbon.

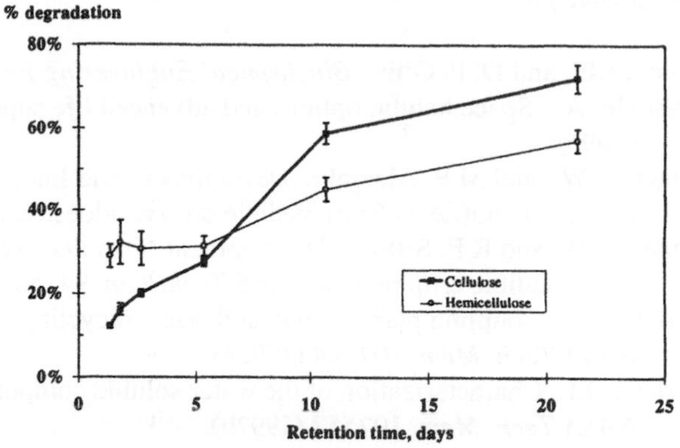

Fig. 5. Effects of retention time on percent biodegradation of wheat residue cell wall constituents--cellulose and hemicellulose.

The results from Figure 5 suggested retention time limits to degradation of soluble organic carbon had not yet been reached at 1.3 days. A second study was designed to try to find these limits. Unfortunately, this study could not be conducted until almost a year later and wheat residue biomass was no longer available. The study used potato residues from a KSC BPC harvest. Figure 6 shows that biodegradation of total potato residue was adversely affected at retention times faster than 1 day. Degradation of soluble TOC was greater than 80% , with more variability than for wheat residues (Figure 5). A high value for one of the ISABs at 0.5 days retention time was caused by a pH control problem one day before the end of this run. TOC data for this retention time were excluded from calculations shown in Figure 6. These data indicate taht the effects of fast retention times on soluble TOC degradation may still have not reached a limit at 0.25 days.

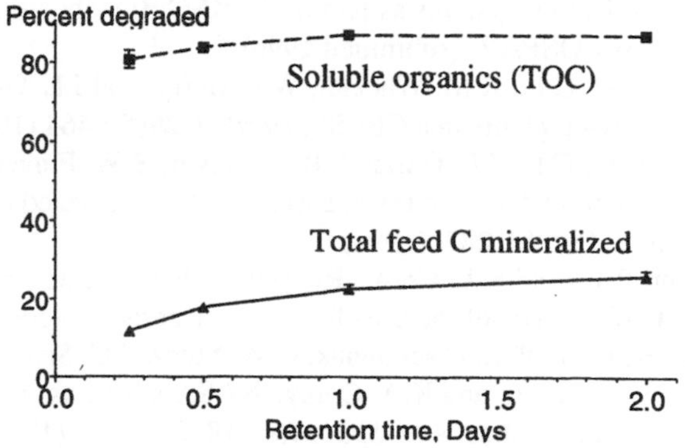

Fig. 6. Effects of fast retention time on degradation of soluble organic compounds (measured as soluble TOC) from potato residues. Compared with total wheat residue biodegradation (as C mineralized).

CONCLUSIONS and FUTURE DIRECTIONS

If the goal of biological resource recovery is to maximize degradation of crop residue organic carbon, including cellulose, then bioreactor retention times of 10 days or longer will be needed. This implies higher costs in the form of volume and mass, because retention time is related directly to bioreactor volume. At 10 days or longer retention times, soluble TOC degradation is maximal. Although not shown in this paper, mineral recovery is adequate at these retention times (Mackowiak, et al. 1996).

If the goal of biological resource recovery is to provide inorganic nutrients with the smallest volume bioreactor possible, then a retention time of 1 day, or less, will suffice. Soluble TOC, which contributes to problems in hydroponic systems if not removed, is biodegraded at all retention times tested. Total biomass degradation and decomposition of cellulose and other polysaccharides will be minimal at fast retention times.

At the KSC-Breadboard Project, near-term research will focus on fast retention times, probably no faster than 1 day until harvest procedures can be automated. We will be adding graywater to bioreactor processing for integration into the breadboard by March 1997 and urine processing by March 1998. Both of these liquid waste streams contain easily biodegradable soluble organic compounds that, like soluble TOC from crop residues, should be eliminated by passage through a 1 day retention time bioreactor . Some of the specifics for this future research are the topic of another COSPAR paper (Garland *et al.*, submitted).

REFERENCES

Bailey, J.E., and D. F. Ollis. *Biochemical Engineering Fundamentals.* McGraw-HIll, New York (1986).

Drysdale, A. Space habitat options and advanced life support design constraints. *SAE Tech. Paper* 951690 (1995).

Finger, B.W., and M.P. Alazraki. Development and Integration of a Breadboard-scale aerobic bioreactor to regenerate nutrients from inedible crop residues. *SAE Tech. Paper* 951498 (1995).

Finger, B.W. and R.F. Strayer. Development of an intermediate-scale aerobic bioreactor to regenerate nutrients from inedible crop residues. *SAE Tech. Pap.* 941501 (1994).

Garland, J.L. Coupling plant growth and waste recycling systems in a controlled life support system (CELSS). *NASA Tech. Mem.* 107544 (1992a).

Garland, J.L. Characterization of the water soluble component of inedible residue from candidate CELSS crops. *NASA Tech. Mem.* 107557 (1992b).

Garland, J.L. and C.L. Mackowiak. Utilization of the water soluble fraction of wheat straw as a plant nutrient source. *NASA Tech. Mem.* 107544 (1990).

Garland, J.L., C.L. Mackowiak, and J.C. Sager. Hydroponic crop production using recycling nutrients from inedible crop residues. *SAE Tech. Paper* 932173 (1993).

Garland, J.L., C.L. Mackowiak, R.F. Strayer, and B.W. Finger. Integration of waste processing and biomass production systems as part of the KSC Breadboard Project. Submitted to *Adv. Space Research* as part of the 31st COSPAR. (submitted 1996).

Mackowiak, C.L., R.M. Wheeler, W.L. Berry, and J.L. Garland. Nutrient mass balances and recovery strategies for growing plants in a CELSS. *HortSci.* **29**(5): 464 (1994).

Mackowiak, C.L., J.L. Garland, R.F. Strayer, B.W. Finger, and R.M. Wheeler. Comparison of aerobically-treated and untreated crop residue as a source of recycled nutrients in a recirculating hydroponic system. *Adv. Space Res.* **18** :281-287 (1996).

Strayer, R.F., and K.L. Cook. Recycling plant nutrients at NASA's KSC-CELSS Breadboard Project: Biological performance of the breadboard-scale aerobic bioreactor during two runs. *SAE Tech. Paper* 951708 (1995).

Wheeler, R.M., C.L. Mackowiak, G.W. Stutte, J.C. Sager, N.C. Yorio, L.M. Ruffe, R.E. Fortson, T.W. Dreschell, W.M. Knott, and K.A. Corey. NASA's biomass production chamber: a testbed for bioregenerative life support studies. *Adv. Space Res.* **18**: 215-224. (1996).

Pergamon

Adv. Space Res. Vol. 20, No. 10, pp. 2029–2035, 1997
Published by Elsevier Science Ltd on behalf of COSPAR
Printed in Great Britain
0273-1177/97 $17.00 + 0.00

PII: S0273–1177(97)00937–X

RECYCLING OF INORGANIC NUTRIENTS FOR HYDROPONIC CROP PRODUCTION FOLLOWING INCINERATION OF INEDIBLE BIOMASS

D. L. Bubenheim* and K. Wignarajah**

**NASA Ames Research Center, Space Technology Division, Regenerative Life Support Branch, Moffett Field CA 94035*
***Lockheed Martin Engineering and Science Corporation, Ames Research Center, Space Technology Division, Regenerative Life Support Branch, Moffett Field CA 94035*

ABSTRACT

The goal of resource recovery in a regenerative life support system is maintenance of product quality to insure support of reliable and predictable levels of life support function performance by the crop plant component. Further, these systems must be maintained over extended periods of time, requiring maintenance of nutrient solutions to avoid toxicity and deficiencies. The focus of this study was to determine the suitability of the ash product following incineration of inedible biomass as a source of inorganic nutrients for hydroponic crop production. Inedible wheat biomass was incinerated and ash quality characterized. The incinerator ash was dissolved in adequate nitric acid to establish a consistent nitrogen concentration is all nutrient solution treatments. Four experimental nutrient treatments were included: control, ash only, ash supplemented to match the control treatment, and ash only quality formulated with reagent grade chemicals. When nutrient solutions were formulated using only ash following incineration of inedible biomass, a balance in solution is established representing elemental retention following incineration and nutrient proportions present in the original biomass. The resulting solution is not identical to the control. This imbalance resulted in a suppression of crop growth. When the ash is supplemented with reagent grade chemicals to establish the same balance as in the control - growth is identical to the control. The ash appears to carry no phytotoxic materials. Growth in solution formulated with reagent grade chemicals but matching the quality of the ash only treatment resulted in similar growth to that of the ash only treatment. The ash product resulting from incineration of inedible biomass appears to be a suitable form for recycle of inorganic nutrients to crop production.

INTRODUCTION

Recovery of resources from waste streams in a space habitat is essential to minimize the resupply burden and achieve self sufficiency. In a Controlled Ecological Life Support System (CELSS) human wastes and inedible biomass will represent significant sources of secondary raw materials supplying carbon, water, and inorganic nutrients necessary to support the plant-based life support functions. Future implementation and reliance on such a regenerative life support system requires that waste treatment technologies provide a form of recovered resource of appropriate quality to insure support of the crop plant component. Further, these systems must be maintained over extended periods of time, requiring maintenance of nutrient solutions to avoid toxicity and deficiencies.

The major waste streams of concern are from human activities and plant wastes (1,4,12). Many physical, chemical and biological technologies have been considered for use as the primary processor for resource recovery in a regenerative life support system (1,3,4,5,9,12). Resource recovery from solid wastes such as inedible biomass has concentrated on physical and chemical processes because of the speed with which the process occurs and die to the implied purity and separation of the product streams. The three primary physical/chemical technologies currently being investigated are incineration, super critical water oxidation (SCWO), and steam reformation (2,4). The suitability of the products resulting from processing of wastes utilizing these technologies is, however, not well defined.

Incineration has been identified as a high priority candidate technology for resource recovery from solid wastes and as a result we have been systematically defining the quality of solid and gaseous products (3,9,11) as well as the suitability of these products for input to crop production (6,7,8). Previously we had characterized the influence of incineration conditions on elemental retention during incineration of inedible biomass (3). In this paper we quantify the suitability of the ash product resulting from incineration of inedible biomass to provide the inorganic source of nutrients required to support the life support functions of plants in a regenerative life support system.

MATERIALS AND METHODS

Incineration of Inedible Biomass. Wheat straw, from a crop grown under a standard set of environmental conditions to maintain elemental quality, served as the model inedible wheat biomass. The straw was incinerated at 800°C under a set of conditions previously described (3). Biomass and resulting ash quality were determined using inductive coupled plasma (ICP), ion chromatography (IC), and thermal arc spectroscopy.

Recovered Incinerator Ash Study Treatments. Four experimental nutrient treatments were included in the study:

1) Control - standard nutrient solution for lettuce formulated with reagent grade chemicals.
2) Ash Only - incinerator ash, dissolved in nitric acid, as the only source of nutrients in solution.
3) Supplemented Ash - incinerator ash, dissolved in nitric acid, supplemented with reagent grade chemicals to match the control solution.

4) Simulated Ash "Reagent Grade Ash" - formulated with reagent grade chemicals to match the quality of the Ash Only treatment.

The incinerator ash was dissolved in adequate nitric acid to establish a consistent nitrogen concentration is all nutrient solution treatments.

Crop Culture. Four hydroponic systems were arranged on two benches in a greenhouse. Each hydroponic system included six growing containers on each of the two greenhouse benches. A randomized, complete-block, experimental design was utilized with each greenhouse bench representing a block. The recirculating hydroponic system consisted of four identical, physically separated units, to provide individual treatments. Each unit consisted of three major components:

1. A 100 L capacity reservoir for nutrient solution and a magnetic drive pump with polypropylene coated impeller and pump body for solution distribution.
2. Twelve 6.4-L, plant growing containers with a growing surface of 28 x 18 cm. Each container had six 1.5 cm diameter holes to accommodate the lettuce plants.
3. A nutrient solution recirculation system with a manifold to distribute nutrient solution from the reservoir to each container and a drain from each container for solution return to the reservoir.

Lettuce seeds were germinated under fluorescent lamps in a germination tray containing one-fifth strength nutrient solution (7). On day three, nutrient levels were increased to one-third strength. On day five, uniform seedlings were transferred to the greenhouse production system containing the nutrient treatments. The pH of the nutrient solutions was maintained between 5.7-6.3. Electrical conductivity was monitored and nutrient stocks added daily as required to maintain conductivity at 980 $\mu S\ cm^{-1}$. Water lost from the systems due to transpiration was replaced with deionized water. Temperature in the greenhouse was maintained between 21° and 24 °C. Naturally occurring photoperiod was not altered. Sequential, destructive, whole-plant harvests were made at 3, 5, 7, 9, 12, 16, and 22 days after transfer to the greenhouse hydroponic production system and exposure to the nutrient treatments. At each harvest six plants were selected from each treatment and separated into leaves, stem and roots. Fresh mass for all parts and leaf area were determined immediately; dry mass for each plant part was determined by drying all plant material in an oven until a stable mass reading was reached.

RESULTS AND DISCUSSION

The ash quality following incineration of inedible biomass is determined by the nutrient proportions present in the original biomass and elemental retention during incineration (3). When the product ash resulting from incineration of wheat straw is solubilized in nitric acid it can be diluted to formulate a hydroponic solution. The nutrient solution formulated using only incinerator ash had a quality different from that of the control solution (table 1). Based on providing as near to 100% of the potassium requirement as possible, approximately 100% of the manganese is supplied while most of the other essential nutrients are present in lesser proportions. Molybdenum is the obvious exception, being supplied at over 4 times the level typical in this nutrient solution. A clear phytoxicity threshold has not been defined for exposure to Molybdenum and thus toxicity effects were not expected.

Table 1. Model of Nutrient Supply for Growing Lettuce Utilizing Incinerator Ash.

Nutrient	Lettuce Hydroponic Solution (mM)	Elemental Composition of Ash (mg/g)	Proportion Supplied by Ash Alone (%)
K	5	792	101
Mg	1.5	17.3	24
Ca	2.5	133.3	67
B	0.0466	0.8942	44
Cu	0.00015	0.0006	3
Fe	0.096	0.4667	7
Mn	0.0045	0.5333	108
Mo	0.0005	0.1333	41.6
Zn	0.0038	0.004	8
P	1	28	60

There appears to be no toxicity associated with the product ash from incineration of inedible biomass. The growth of lettuce plants is almost identical in treatments having equivalent nutrient composition whether derived from ash, in whole or in part, or from reagent grade chemicals (Fig. 1). The imbalance of the nutrient composition for both the ash only and simulated ash treatments resulted in suppression of growth in both cases. While it is not possible to quantify the individual contribution of each deficient inorganic in these treatments, there was a clear visual observation of classic zinc deficiency symptoms in both the ash only and simulated ash treatments. All observed plant parameters including fresh and dry mass of plant parts and total plant mass as well as leaf area displayed similar patterns of growth (Figs. 1 & 2)

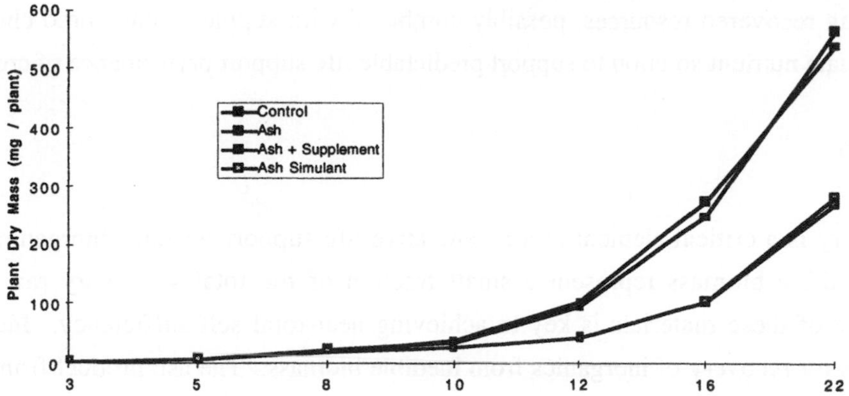

Figure 1. Growth of Lettuce as Influenced by Source of Nutrients.

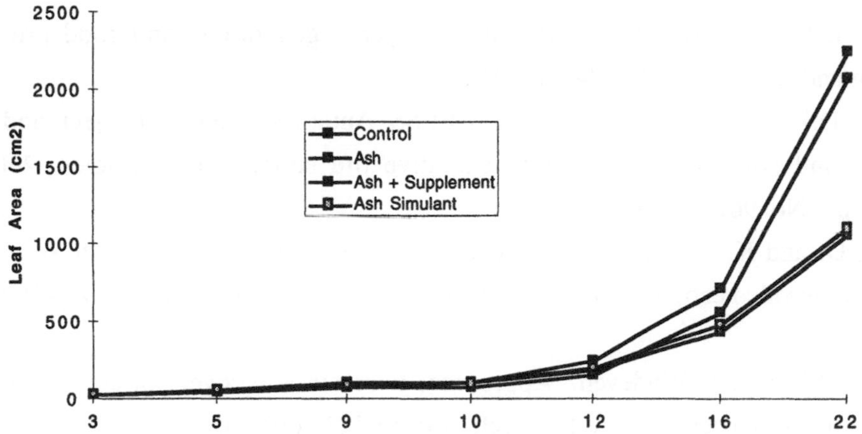

Figure 2. Leaf area (cm^2 $plant^{-1}$) of Lettuce as Influenced by Source of Nutrient.

Incineration is an attractive approach to processing of solid wastes for resource recovery. Incineration products are well segregated and this work demonstrates that the resulting ash is of sufficient quality to serve as a nutrient source for hydroponic crop production. The imbalance in nutrient composition of the ash derived nutrient solution compared with the standard hydroponic solution does result in a suppression of growth when used as the sole source of nutrients in hydroponic production of lettuce. Consideration of the inorganic quality of combined waste streams flowing from a human habitat such as planetary and lunar bases will provide a different final recovered nutrient profile (10). Careful formulation of nutrient solutions utilizing recovered resources, possibly combined with supplemental stored chemicals, should provide an adequate nutrient solution to support predictable life support performance of crop plants.

CONCLUSIONS

Resource recovery is a critical element in a regenerative life support system. Inorganic plant nutrients contained in inedible biomass represent a small fraction of the total secondary raw resource pool, however, recycle of these materials is key to achieving near-total self sufficiency. Incineration is an effective process for recovery of inorganics from inedible biomass. The ash product from incineration is suitable for direct utilization as a crop production input. Maintaining near optimal growth will, however, require supplement of the ash derived nutrient solution. Incineration has potential application in the resource recovery system of a CELSS for the purpose of inorganic recovery.

REFERENCES

1 Bubenheim, D.L. Plant for Water recycling, oxygen regeneration, and food production. Waste Management and Research, 9:435-443. (1991).
2. Bubenheim, D.L. and M.T. Flynn. The CELSS Antarctic Analog Project and validation of assumptions and solutions regarding regenerative life support technologies. 26th ICES, SAE Technical Paper No. 961589. (1996).
3. Bubenheim, D. and K. Wignarajah. Incineration as a method for resource recovery from inedible biomass in a controlled ecological life support system. Journal of Life Support and Biosphere Science 1:129-140 (1995).
4. Bubenheim , D.L. and T. Wydeven. Approaches to resource recovery in controlled ecological life support systems. Advances in Space Research. 14:113-123. (1994).
5. Bubenheim, D.L., M. Bates and M. Flynn. An approach for development of regenerative life support systems for human habitats in space. 25th ICES, SAE Technical Paper: no. 951730. (1995),
6. Bubenheim, D., M. Flynn, M. Patterson, and K. Wignarajah. Incineration of biomass and utilization of product gas as a CO_2 source for crop production in closed systems: Gas quality and phytotoxicity. (In this COSPAR - Advances in Space Research issue) (1996).
7. Bubenheim, D., K.Wignarajah, W. Berry, and T. Wydeven. Phytotoxic effects of gray water due to surfactants. J. Amer. Hort. Sci. 121 (In Press) (1997).
8. Patterson, M.T, K. Wignarajah and D.L. Bubenheim. Biomass incineration as a source of CO_2 for plant gas exchange: phytotoxicity of incinerator-derived gas and analyses of recovered evapotranspired water. Journal of Life Support and Biosphere Science. (In Press) (1996).
9. Pisharody, S., B. Borchers and G. Schlick. Solid waste processing in a CELSS: Nitrogen recovery. Journal of Life Support and Biosphere Science. 3:61-65 (1996).

10. Wignarajah, K. and D.L. Bubenheim. The integration of waste processing with crop production in regenerative life support. (In this COSPAR - Advances in Space Research issue) (1996).

11. Wignarajah, K., S. Pisharody, M. Flynn, D.L.Bubenheim, B. Potter and C. Klein. Nitrogen sources of NO_x generated during incineration of inedible plant biomass. Journal of Life Support and Biosphere Science (InPress). (1996).

12. Wydeven, T. A survey of some regenerative physico-chemical life support technologies. NASA Tech. Memo. # 101004, NASA-Ames research Center, Moffett Field, CA (1988).

Pergamon

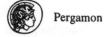

Adv. Space Res. Vol. 20, No. 10, pp. 2037–2044, 1997
Published by Elsevier Science Ltd on behalf of COSPAR
Printed in Great Britain
0273-1177/97 $17.00 + 0.00

PII: S0273–1177(97)00938–1

APPLICATION OF NASA'S ADVANCED LIFE SUPPORT TECHNOLOGIES IN POLAR REGIONS

D. L. Bubenheim* and C. Lewis**

**NASA Ames Research Center, Mail Stop 239-15, Moffett Field, California 94035-1000 U.S.A.*
****Center for Applied Research and Agricultural and Forestry Experiment Station,*
Department of Natural Resources, University of Alaska Fairbanks, Alaska, U.S.A.

ABSTRACT

NASA's advanced life support technologies are being combined with Arctic science and engineering knowledge in the Advanced Life Systems for Extreme Environments (ALSEE) project. This project addresses treatment and reduction of waste, purification and recycling of water, and production of food in remote communities of Alaska. The project focus is a major issue in the state of Alaska and other areas of the Circumpolar North; the health and welfare of people, their lives and the subsistence lifestyle in remote communities, care for the environment, and economic opportunity through technology transfer. The challenge is to implement the technologies in a manner compatible with the social and economic structures of native communities, the state, and the commercial sector. NASA goals are technology selection, system design and methods development of regenerative life support systems for planetary and Lunar bases and other space exploration missions. The ALSEE project will provide similar advanced technologies to address the multiple problems facing the remote communities of Alaska and provide an extreme environment testbed for future space applications. These technologies have never been assembled for this purpose. They offer an integrated approach to solving pressing problems in remote communities.

Published by Elsevier Science Ltd on behalf of COSPAR.

INTRODUCTION

Development of regenerative life support systems is essential for NASA to realize the fundamental goal of future exploration of the Moon and Mars. The regenerative life support system referred to as a Controlled Ecological Life Support System (CELSS) has the goal to achieve the greatest practical level of mass recycle and provide self sufficiency and safety for humans (Straight, Bubenheim, Bates, and Flynn, 1994; Bubenheim and Flynn, 1996). A research and development program is providing the suite of technologies required to recover resources from the waste stream and utilize plants to produce food, purify water, and revitalize the atmosphere of space bases human habitats.

NASA's advanced life support technologies are being combined with Arctic science and engineering knowledge to address the unique needs of the remote communities of Alaska through the Advanced Life Systems for Extreme Environments (ALSEE) project. ALSEE is a collaborative effort involving the NASA, the State of Alaska, the University of Alaska (UA), the North Slope Borough of Alaska, and the National Science Foundation (NSF). The focus is a major issue in the State of Alaska and other areas of the Circumpolar North; the health and welfare of its people, the subsistence lifestyle in remote communities, economic opportunity, and care for the environment. The project provides efficient treatment and reduction of waste, purification and recycling of water, and production of food and energy. (Bubenheim and Lewis, 1995; Lewis, Bubenheim and Stanford, 1995).

The scientific and technological capability enabling the transfer of advanced life support technologies to the Circumpolar North is provided by the CELSS Antarctic Analog Project (CAAP). CAAP is a joint NSF and NASA project for the development, deployment and operation of CELSS technologies at the Amundsen-Scott South Pole Station (Roberts, Chiang, Lynch, and Smith, 1992). NASA goals are technology selection, system design and methods development of regenerative life support systems for planetary and Lunar bases and other space exploration missions. Food production, water purification, and waste recycle and reduction provided by CAAP will improve the quality of life for the South Pole inhabitants, reduce logistics dependence, and minimize environmental impacts associated with human presence on the polar plateau.

CELSS ANTARCTIC ANALOG PROJECT

The general objectives of the Antarctic Analog Program are first to take advantage of similarities in mission fidelity among future NASA missions and current missions to the Antarctic continent. These efforts combine resources expended by several organizations and provide a return of understanding and increased capability not previously available. In the specific case of the CAAP, the high degree of functional equivalency among the application at the South Pole Station and analogous operations in future space missions help understand the complex nature of regenerative system operations and interrelated elements with the human habitat.

The opportunity for continuous operation of regenerative life support technologies and integrated systems supporting humans in an extreme and life threatening situation provides a real and controlled method for establishing the abilities of these systems. The extended period of isolation for the crew at the South Pole Station can provide a good measure of operational transparency and implementation and validation of system management improvements. Further, accomplishing these technical and system management objectives under the real and analogous constraints such as power, volume, and logistics and at an appropriate scale will provide mission planners with a path for fitting regenerative life support systems within the constraints of future space exploration missions.

IMPLICATIONS OF CAAP

One of the fundamental purposes of CAAP is to transfer space technologies to the NSF. Additionally, CAAP offers potential for direct transfer of space technologies to the private sector, many of which hold high potential for commercialization. The CAAP project has led to consideration of using CELSS technologies in other remote communities to address health, social, economic, and environmental issues. The CAAP provides the enabling scientific and engineering base for this transfer.

ALASKA OPPORTUNITIES

The Arctic, specifically the ALSEE Project, provides an opportunity for NASA to fulfill another of its goals - the transfer of technologies to remote communities and Native people. This goal is parallel to that of the UA, the land grant university of the State of Alaska: provide service to the state by conducting research and education programs appropriate for and transferring knowledge and technology to the peoples of Alaska for the betterment of their social and economic well being.

The problems of obtaining adequate pure drinking water and disposing of liquid and solid waste in the Arctic have led to unsanitary and socially unacceptable conditions (Committee on Arctic Social Sciences, 1993). Past solutions such as honey buckets and open lagoons are damaging to human health and the environment (Office of Technology Assessment, 1994). ALSEE provides for identification and development of solutions using applications of technologies developed through NASA research and by teaming with academic, industrial, and Native experts in remote community support. The partnership of UA scientists, educators, NASA researchers and community leaders and experts can provide a holistic approach to the water and sanitation problems in the circumpolar north.

The ALSEE Project will use a technology suite, including hydroponics and aquaculture (Bubenheim, 1988, 1991; Lewis, Norton, and Thomas, 1980) in the waste and water treatment system and provide fresh food products to remote areas. Researchers and educators can offer secondary and post-secondary educational opportunities in the areas of science, business and management, and provide training in the operation of ALSEE systems. Spin-off technologies and food product sales offer business opportunities that would not otherwise have existed in the far north, a region where there are few employment possibilities.

APPLICATION IN REMOTE COMMUNITIES OF ALASKA

Alaska offers many of the desirable features of space analogs but a much greater potential benefit exists for transfer and utilization of NASA technologies and systems in remote communities. The presence of a community of people in the remote regions of the State sharing culture and ancestry presents a challenge not found in the Antarctic. Effective transfer of technologies requires careful consideration and understanding of the needs of a community and social and economic constraints. An education and training component will also be necessary.

Background: The Problem

Sanitation and a safe water supply, environmental protection, food supplies, and psychological respite from harsh environments have been and still are concerns for those living in the Arctic (Berman and Leask, 1994). The literature abounds with research information, descriptions and applications of various technologies, and discussions of ways in which hardships of life in the Arctic can be mitigated (Committee on Arctic Social Sciences, 1993; Office of Technology Assessment, 1994). The indigenous peoples of the Arctic embrace the climate as a part of their culture. Those who have immigrated to the Arctic do not. However, Native peoples are absorbing advanced technology into their lifeways. Western culture and economies are becoming more available to them and are impacting village economies and subsistence practices (Geier, Lewis, and Greenberg, 1996).

Solid and liquid waste management is a problem in the Arctic. The U.S. Arctic is referred to as the forgotten America because of the voluminous problems associated with sanitation. When oil moneys became available to the state of Alaska and some Native corporations in the early 1980s, attempts were made to adapt sewage collection technology (utilidors and buried pipes) used in urban Alaska (Brown, 1994; Olofsson and Iwamoto, 1994; Shiltec Engineering, 1993). Problems of adaptation including remoteness, small size, lack of infrastructure, and absence of a trained labor pool were identified by the Committee on Arctic Social Sciences (1993). Capital costs are high as are costs of operating and maintaining the systems.

Environmental resources are assets in Arctic lifeways supplying fish, game, and plants. Early tribes were nomadic and populations were small. Human impact on the environment was slight and tribes moved when resources became scarce for any number of reasons not all related to human impact. The advent of permanent structures such as missions and schools lent a permanence to the nomadic lifestyle. Permanent settlements increased human impact on the environment A recent statement by one of the elders from the community of Grayling on the lower Yukon River is an excellent summation of the problem: We must save something [referring to people and the environment]. We cannot continue along the paths of the recent past. But we do not know what to do (Deacon, 1994).

The cultural legacy of Native Alaskans is hunting and gathering. However, these foodways have been acculturated and western foodways have increasingly been incorporated in individual consumption patterns (Swanson and Lewis, 1992). Domesticated plants and animals form the base of western foods. Native peoples in Alaska do not have an agrarian component in their culture. Furthermore, the climate in Arctic regions does not lend itself well to food production unless it is accomplished in some type of controlled environment.

Thus, there is little agricultural production in the Arctic villages. Barges ship bulk items and non-perishable foods during the approximately 120 days the rivers are free of ice. Perishable food products are transported by air; prices are high (Stetson and Assoc., 1996) and multiple transfers adversely affect food quality.

Hundreds of detailed descriptions of indigenous cultures have been published since the 1880s (International Indigenous Commission, 1991). A common thread emphasizes the pressures of cultural change and the effects of harsh climates on the inhabitants. Throughout the circumpolar Arctic, the Native and non-Native populations alike experience the impact of a constellation of stresses. Cultural change is one. Villagers strive to find a path to maintaining the ways of the elders and remaining in their villages while at the same time realizing that there is little in the way of a cash economy that will allow them to stay in the traditional home and still be a part of the technologically advancing world. Feelings of loneliness and isolation, inability to control self-destiny, and indecisiveness may lead to high suicide rates and alcohol and drug consumption. Community contact, recreation opportunities, and relief from darkness and cold can alleviate stress to a degree.

A Possible Solution to the Problem

ALSEE combines NASA derived advanced life support technologies with university, industry and Native science, engineering, and tradition. CAAP provides the enabling scientific and engineering base for the Alaska demonstration. The challenge is to implement the technological capabilities provided by CAAP in a manner compatible with the social and economic structures of Alaska communities, the state, and the commercial sector.

The ALSEE project is community oriented. It is dedicated to enhanced community well-being through improvement in sanitation and availability of safe water, and enhancement of education and training programs. It addresses improvement in the quality of life for individuals in rural Alaska by providing a facility that offers respite from the harsh environment, recreation opportunities and a gathering place, and improved quality and variety of fresh foods. Finally, the ALSEE project offers job possibilities in and outside the community by offering advanced training in water and waste treatment, and the opportunities for business development of products associated with ALSEE.

The mission of ALSEE reflects its intentions: transfer new and existing technologies to communities, offer opportunities for applied and basic research to continuously improve extreme environment waste treatment systems, improve the quality of life for people living in extreme environments while remaining cognizant of traditional Native lifeways, reduce environmental impact of water and waste disposal, and incorporate education and training at all levels of ALSEE project development. The six-step process adopted by the ALSEE partners contributes to accomplishing the mission.

1. Define the problem through research and by asking questions.
2. Define the problem factors and their relationships.
3. Understand the problem factors and the requirements for solutions.
4. Formulate appropriate objectives.

5. Identify and/or develop appropriate technology and education/training programs.
6. Put the technology and educational suites in place.

ALSEE IN ALASKA

The ALSEE project in Alaska will be conducted in phases (Table 1). The first phase will emphasize team building and transfer of personnel, technologies and knowledge among the partners. In the second phase, the research and demonstration testbed will be constructed and operated at a site in Alaska. This phase includes initial demonstration of the suite of technologies and integrated systems and should evolve to include regular service to the host community. The third phase will transfer the ALSEE Project to the host village(s) which will retain responsibility for commercial development. UA and NASA will provide technical and educational support.

Table 1. Benchmarks in Phase 1 and Phase 2 of ALSEE in Alaska.

PHASE 1	PHASE 2
Technology transfer and pilot testing at NASA site	Construction of facility
Testbed site selection	Job creation
Alaska testbed site	Demonstration
Technology development in Alaska	Commercial involvement
Educational relationships	Implementation and dissemination of results
Economic studies	
Alliance formation	
Social impact/integration studies	

Development of educational and training materials and appropriate workshops and courses is concurrent in all phases of the ALSEE project. Although these programs will focus on training people to operate ALSEE facilities, the courses and materials will be applicable to 2-year degree and certificate programs in water and waste treatment. These same courses will be applicable to 4-year degrees in general science, education, and more specialized science degrees. There is also a potential to offer courses in management and marketing using the ALSEE facility as a training center. K through 12 programs will be targeted with teacher training.

The city of Barrow, Alaska, was selected as the first ALSEE site using criteria developed by the partners (Table 2). It is the largest community on the north slope of Alaska, is in the North Slope Borough, the largest borough in area in the state. Barrow is the largest community in the Borough with a population of 3,800. The total population of the Borough is approximately 5,000. Barrow has a strong commitment to improvement of sanitation facilities. Residents have experience with water and waste treatment. The ALSEE facility would not service the entire population. Rather, it would be sized to service a population of approximately 250. Technology could be transferred to smaller villages. Barrow is also the site of Arctic Suvunmun Ilisagvik College, which offers 4-year degree programs, 2-year associate degrees and certificate programs. One of these programs is in water and waste technology which awards both a 2-year degree and a certificate.

Table 2. Criteria for Site Selection

CRITERIA
A formally organized community having demonstrated past ability to collaborate with outside business
Appropriate experience with relevant technologies
Educational infrastructure including college level courses
Appropriate work force to support installation and operation
Accessible via scheduled transportation
Existing infrastructure which could accommodate the testbed without major construction and changes
A hub for smaller, remote communities

IMPLICATIONS OF ALSEE

One of the most important products of ALSEE is education A cadre of research and education professionals and industrialists is associated with the ALSEE project. Their expertise is available to the community

ALSEE projects produce products. Potable water and high-quality fresh foods and flowers from the Farm, as in CAAP are two which are immediate and obvious. Others may include innovative methods to produce energy, system controls, building design, and waste recovery units. All offer potential manufacturing and marketing opportunities for entrepreneurs.

The ALSEE facility will have an area that can be used as a gathering place. The atmosphere of this open space will reflect the mild, temperate climate necessary for food production. Although harsh climates and long periods of darkness are a part of the lifeways of those native to the Arctic, a change from them may be welcome. Arctic residents may appreciate the opportunity to visit a warmer and sunnier space whenever they desire.

REMARKS

Conventional technological solutions which work well elsewhere have often been tried in rural Alaska. Unfortunately, these approaches attempted to solve only one segment of an inter-related problem set, or conversely, a single technology was applied to the complete set. This lack of a systems approach has led to only partial solutions or immediate or short-term failure in practically every case.

The unique features of the ALSEE project are the transfer of advanced technologies and their integration to address the multiple problem set facing the remote communities of Alaska; sanitation and safe water, environmental concerns, food supplies, and psychological factors., The ALSEE suite of technologies offers an integrated approach to solving problems. This holistic approach would not be possible if the technologies were offered individually without consideration of impacts on culture and technical capabilities in the community.

The ALSEE project offers the potential for development of new industries in Alaska as advanced technologies are demonstrated and transferred to the commercial sector. The self-sufficiency aspects of the ALSEE approach may be uniquely suitable to the self-sufficient history of Native communities and at the same time may provide for some reduction of governmental subsidies associated with supporting remote communities. This would hopefully aid rural community residents to pursue their lifeways of choice and have an enhanced quality of life.

While the scope of the ALSEE Project seems idealistic, the outcome of such research will yield improvements in the quality of life for Arctic communities. For over 30 years, NASA improvements in materials and development technologies have helped create many products that have enhanced our quality of life. University of Alaska researchers and educators have provided opportunities for Alaskans since 1817. It is fittingly appropriate that Arctic rural communities will receive an opportunity to become direct recipients through the ALSEE Project.

REFERENCES

Amber, M. 1993. What can indigenous economics tell us about todays society? Business Alert. First Nations Development Institute. 8(2):1-7.

Berman, M. and L. Leask. 1994. Violent death in Alaska: Who is most likely to die? Alaska Review of Social and Economic Conditions 29(1):1-12.

Brown, F.I. 1994. Anaktuvuk Pass water and sewer supplemental work book. Capital Improvement Program Management, North Slope Borough, Barrow, AK.

Bubenheim, D.L. (1988, 1991). Plants for water recycling, regeneration, and food production. *Waste Mgt. and Research* (1988) 9:435-443, Waste Mgt. in Space: A NASA Symposium, Oct. 1991. *J. Int. Solid Wastes and Public Cleansing Assoc. ISWA.*

Bubenheim, D.L. and M.T. Flynn. 1996. The CELSS Antarctic Analog Project and Validation of Assumptions and Solutions Regarding Regenerative Life Support Technologies. 26th ICES, SAE Technical Paper No. 961589. (1996).

Bubenheim, D.L. and C. Lewis. 1995. Application of NASAs advanced life support technologies for waste treatement, water purification and recycle, and food production in polar regions. Presentation to the 2nd Circumpolar Agric. Conf., Tromso, Norway, Sept. 4-7.

Committee on Arctic Social Sciences. 1993. Arctic contributions to social science and public policy. National Academy Press, Washington, D.C.

Deacon, H. 1994. Speech at the Holy Cross Subregion of the Doyon Region meetings, Anvik, AK, June 13, 14.

Geier, H., C.E. Lewis, and J.A. Greenberg. 1996. An economic profile for Grayling, Anvik, Shagaluk, and Holy Cross. Agric. and For. Exp. Sta., Univ. of Ak Fairbanks, Bull. (in press).

International Indigenous Commission. 1991. Indigenous peoples traditional knowledge and management practices. A Report to the United Nations Conference on Environment and Development, United Nations, NY.

Lewis, C.E., R.A. Norton , and W.C. Thomas.1980. Controlled-environment agriculture: a pilot project. Bulletin No.55. Alaska Agricultural Experiment Station, Fairbanks, AK.

Lewis, C.E., D.L. Bubenheim, and K.L. Stanford. 1995. Advanced Life Systems for Extreme Environments: An Arctic application. Presentation to the 2nd Circumpolar Agric. Conf., Tromso, Norway, Sept. 4-7.

Office of Technology Assessment. 1994. An Alaskan challenge: Native village sanitation. Congress of the United States, Washington, D.C.

Olofsson, J.A. and L.A. Iwamoto. 1994. Selection of sanitation alternatives: a strategy for remote Alaskan communities. 7th International Cold Regions Engineering Specialty Conference, Edmonton, Alberta, Canada, March 7-9.

Roberts, C.A., E. Chiang, J.T. Lynch, and P.D. Smith. 1992. Challenges of the U.S. Antarctic program in the decade of the 90s. SAE Tech. Paper #921128, 22nd Int. Conf. on Env. Systems, Warrendale, PA., pp.7-8.

Shiltec Engineering. 1993. Village water and sewer project: preliminary engineering report, North Slope Borough. Shiltec Northern Engineering, Environmental Engineering, Anchorage, AK.

Stetson and Assoc. 1996. Alaskan food costs. Stetson and Assoc., Quarterly Rpt., March 1996, Fairbanks, AK.

Straight, C.L., D.L. Bubenheim, M.E. Bates, and M.T. Flynn. 1994. The CELSS Antarctic analog project: an advanced life support testbed at the Amundsen-Scott South Pole Station, Antarctica. Life Supp. and Biosph. Sci. Cognizant Comm. Corp. 1:52-60.

Pergamon

Adv. Space Res. Vol. 20, No. 10, pp. 2045–2048, 1997

Printed in Great Britain
0273-1177/97 $17.00 + 0.00

PII: S0273–1177(97)00939–3

ELEMENT EXCHANGE IN A WATER-AND GAS-CLOSED BIOLOGICAL LIFE SUPPORT SYSTEM

I. V. Gribovskaya, Yu. A. Kudenko, J. I. Gitelson

Institute of Biophysics (Russian Academy of Sciences, Siberian Branch)

ABSTRACT

Liquid human wastes and household water used for nutrition of wheat made possible to realize 24% closure for the mineral exchange in an experiment with a 2-component version of "Bios-3" life support system (LSS) Input-output balances of revealed ,that elements (primarily trace elements) within the system. The structural materials (steel, titanium), expanded clay aggregate, and catalytic furnace catalysts. By the end of experiment, the permanent nutrient solution, plants, and the human diet gradually built up Ni, Cr, Al, Fe, V, Zn, Cu, and Mo. Thorough selection and pretreatment of materials can substantially reduce this accumulation.
To enhance closure of the mineral exchange involves processing of human- metabolic wastes and inedible biomes inside LSS. An efficient method to oxidize wastes by hydrogen peroxide in a quartz reactor at the temperature of 80 ^{O} C controlled electromagnetic field is proposed.

INTRODUCTION

The last 5 month-long "man-higher plants" experiment conducted in "Bios-3"at the Institute of Biophysics (Russian Academy of Sciences, Siberian Branch) in Krasnoyarsk in 1985 achieved the highest (more them 90%) closure of the system on gas and water. 70% needs in nutrition calories of the 2-men crew was achieved due to growing 14 species of plants (Lisovsky, 1985).

Mineral elements required for the plants and crew were mostly supplied into the system from prestored stock with hygiene means (soap, tooth-powder, table salt, lyophilized food, mineral elements as salts and acids for plants nutrition). On the other hand, some elements we re removed from the system with solid human wastes, kitchen wastes, inedible biomes of plants, and analysis samples. The substance regenerated wihin the sistem supplied a part of plants demand for mineral elements. Filtered household water and liquid wastes of the crew were fed from the living quarters into the plant growing module.

Wheat was particularly useful in the system because its edible part is out of contact wıth the nutrient solution, while its inedible part (straw, leaves) absorbs sodium from the nutrient solution. Our investigations have shown, that percentage (of the total supply) of required amount of mineral elements from human liquid wastes was: nitrogen 59, sulfur - 28, potassium - 19, phosphorus - 17, magnesium - 13. Total closeness of the mineral exchange in "Bios-3" did not exceed 24%.

RESULTS AND DISCUSSION

Element-by-element balance measured at the inputs and outputs of the system - K, Na, Ca, Mg, P, S, N, Fe, Mn, Cu, Zn, Cr, Mo, V, Ti, Al, Cd, Sn, Pb, As - revealed that elements (mostly trace elements) are supplied from within the system. Amounts of Ni, Cr, Al, Pb recorded at the output was 10-20 times that at the input; Sn, Ti, Zn, V, Fe - 2-4 times . The source of the elements indicated were structural materials. New steel that had not undergone preliminary treatment was the sourse of Ni, Cr. The catalytic furnace (430 ^{o}K) for straw incineration enriched the ash with Zn, Cr, V, Ti, Fe from the corroding catalysts, since these elements are components of the catalysis chemical composition. Some elements - Al, Pb, Ni, Ti, Zn, V, Cu came from the expanded clay contacting with the nutrient solution.

Amount of elements in the permanent solution increased with their concentration in plants (Figure 1). On the figure showes, that among the plants under investigation (wheat, cabbage, dill, radish, onion, chufa, beet, salad) accumulation to the end of expeiment of Ti, Al, Fe, Cr was the highest in cabbage and chufa leaves- 5-10 times. Cucumber and sorrel changed the mineral composition the least.By the end of experiment the amaunt of elements supplied to the crew with food also increased. Nickel and chromium were recorded to exceed daily average by a factor of 3 guota for human. This had no negative impact on the health of the crew; yet in a longer experiment this might have negative effect on the performance of the system.

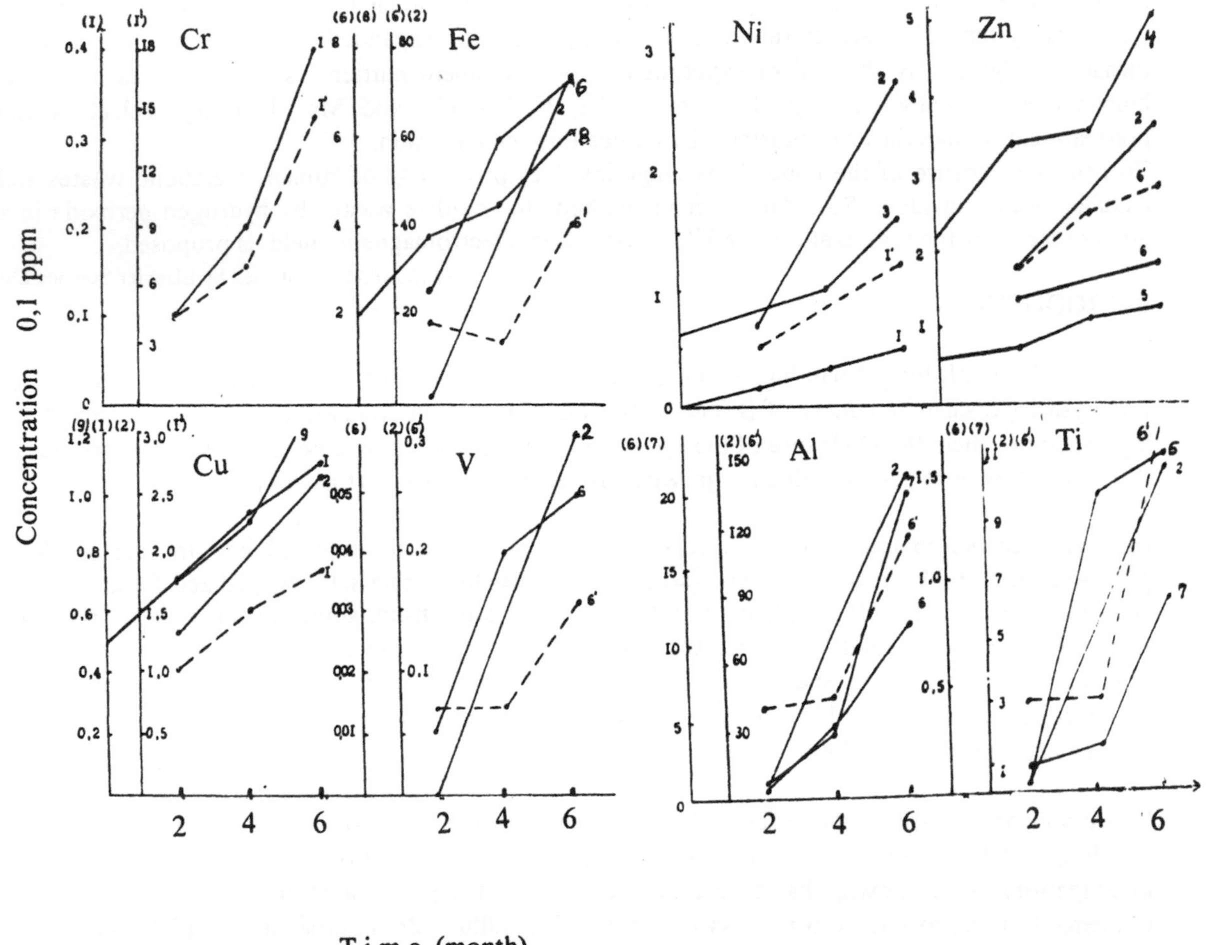

Fig.1.Dynamics of microelements content in edible and inedible biomes of plants in the course of 6-month experiment: 1-wehat grain, 1 -wehat straw, 2- cabbage, 3 -sorrel, 4 -radish roots, 5 -onion, 6-chufa storage root, 6 -chufa leavs, 7 -dill, 8 -beet roots, 9 -potato tubers.

In the 5 month-long experiment to follow which used only tested structural materials (titanium, steel, acrylic glass, silicon rubber and new catalysts) the additional supply of elements to the plants and the crew was substantially smaller and kept within the norm.

CONCLUSION

On Enhancing the Closeness of Mineral Exchange

Enhancement of closeness in terms of mineral elements involves intrasystem processing of the wastes of the crew and the plant component to returning elements in cycle fornutrition of plants. Available methods for mineralizing wastes of LSS vital activity are divided into two types: dry mineralization in air flow at high temperatures 1270 ^{O}K (Labak et al., 1973), 430 ^{O}K with catalysts (Lisovsky et al.,1979), and wet combustion (Jonson, 1985) at the temperature of 470-670 ^{O} K and the pressure of 220-270 atm in the pure oxygen atmosphere (Jonson, 1985), The trouble with the former dry mineralisation is formation of dioxins, CO, SO_2 , NO_2, application of acid extraction (Gribovskaya, 1995) to dissolve caking ash. The other incineration method produce smaller amount of noxious emissions, but is complicated in the engineering aspect, by troubles with removal and dissolving of ash, which may line the surface of the reactor and conduits as a deposit, besies structural materials are sourse of toxic concentration of Cr^{+3} ,Cr^{+6} by high temperature and presure (Bramlette 1990).

The wet mineralization method proposed by us for processing of wastes (faces, urine, unable part of plants, household water) is free from these shortcomings. Running at 80-90 ^{O} C under normal atmospheric pressure no water-insoluble residue are produced.Gases noxious to humans and plants do not form, the mineral substances are produced in the form of solution which can be employed to grow plants. Power reguirement to maintain the reaction was 15-25 W, frequency 50 qz .

The mineralization method under development is based on utilization of nascent oxygen, its active forms and OH-radicals. In an experimental quartz glass reactor a mixture of wastes with hydrogen peroxide (30%) was exposed to controlled electromagnetic field to release the amount of oxidizers reguired to decompose the organic matter .

Hydrogen peroxide can be prestored or produced inside LSS from water with electric energy without additional chemical agents. Energy expenditure is 7.5 kWh per 1 kg of peroxide. To process daily amount of solid and liquid wastes and inedible biomes requires 4-5 l of peroxide, i.e. 10 kWh/man daily. Given an LSS with no energy restrictions (nuclear powered) mineralization by the proposed method is advantageous over other mineralization techniques by its engineering simplicity, reliability, direct utilization for plants. The solution coming out of the reactor may have salt sediment, but it is suspended and does not clog the reactor and is easily dissolved by distilled water.

REFERENCES:

1.Bramlette,T.T., Mills,B.E., Hencken,K.R., Brynildson,M.E., Johston,S.C., Hruby,J.M., et.al. Distraction of DOE/DP Surogatte Wastes with Supercritical Water Oxidation Technology. Report SAND 90-8229. Sandia National laboratiries, Albuguerque,N.M.,1990.

2.Gribovskaya, I.V., Gladchenko, I.A., Zinenko, G.K., Extraction of Mineral Elements from Inedible Wastes of Biological components of a Life Support System and their Utilization for Plant NutritionWydeven, T., We, Adv.Space-Res. Vol. 18, 4/5, pp. (4/5)93-(4/5)97, 1996.
3.Jonson, C. C., Wudeyen, T., Wet Oxidation of a Spacecraft Model Waste. Paper 851372. *15th Intersociety Conference on Environmental Systems*, San Francisco, California. SAE International Warrendale PA:1985.
4.Labak, L. J., Remus, G.A., Shapira, J., Dry Incineration of Wastes for Aerospace Waste Management Systems, Paper 72-ENAV-2, *Environmental Control and Life Supporort System Conference*, San Francisco, CA. Warrendale PA: SAE International: 1972.
5.Lisovsky, G.M., Closed System: Man-Higher Plants, Nauka, Novosibirsk, 1979.

Pergamon

Adv. Space Res. Vol. 20, No. 10, pp. 2049–2052, 1997

Printed in Great Britain
0273-1177/97 $17.00 + 0.00

PII: S0273-1177(97)00940-X

A NEW ENZYMATIC TECHNIQUE TO ESTIMATE THE EFFICIENCY OF MICROBIAL DEGRADATION OF POLLUTANTS

A. B. Sarangova and L. A. Somova

Institute of Biophysics (Russian Academy of Sciences, Siberian Branch) Krasnoyarsk 660036, Russia

ABSTRACT

Dynamics of active sludge microorganism activity in aerotanks under chemostat conditions has been studied. Dependence of microorganism catalase activity has been found to depend on residual substrate concentration in proportion to the biomass of microorganisms. Experimental data and field observations has formed the basis to develop a technique to evaluate in relative units the amount of the substrate consumed by biocenosis of the active sludge in the air tanks of purification facilities.

INTRODUCTION

Microbiological processes degrading organic substrates are in their nature a totality of enzymatic reactions. To control a man-made ecosystem involves knowledge of classical laws of growth and enzymatic kinetics of microorganisms (Varfolomeyev and Kalyuzhny, 1990).

Quality or intensity of microbial degradation of substrates is determined by measuring of soluble organic substrates in the system. BOD_5 (biochemical oxygen demand) technique takes 5 days to yield a result while the COD (chemical oxygen demand) technique is associated with certain difficulties (Lurie, 1974).

All this calls for development of new enzymatic techniques to rapidly estimate substrate concentration in systems with microbial degradation of organic compounds. A method we have developed for estimating catalase activity may be such a technique.

Induction of the catalase enzyme in a cell is connected with hydrogen peroxide forming in reduction of oxidized flavoproteins in the respiration chain (Mikhlin D.M., 1960). Hydrogen peroxide is an extremely toxic intermediate product of cell metabolism in the substrate oxidation cycle and catalase disrupts it. The catalase activity is in proportion to the amount of hydrogen peroxide, because catalase is the only enzyme having no saturation stage (Cornish-Bowden, 1979).

The aim of this work is to study dependence of catalase activity of active sludge microorganisms developing under chemostat conditions on residual substrate concentration in an aerotank, and to develop an integral method to estimate efficiency of microbial degradation of soluble organic compounds.

MATERIALS AND METHODS

The object under study was biocenosis of active sludge microorganisms in aerotanks of municipal treatment facilities. The sewage water arrives continuously, the rate being is 2890 m^3/h. The mean values of BOD_5 and COD of sewage water are 100-200 mg/l and 200-300 mg/l respectively. The microorganism biomass was counted according to the standard technique (dry weight, g/l) (Lurie, 1974). Residual substrate concentration in the air tanks was evaluated by BOD_5 and COD techniques (Lurie, 1974). Catalase activity of microorganisms was evaluated by a modified permanganometry method (Sarangova and Somova, 1994). To eliminated metabolite effect the active sludge was washed in phosphate buffer with pH 6.8. The error of the method is 5%.

DYNAMICS OF CATALASE ACTIVITY OF ACTIVE SLUDGE MICROORGANISMS

According to the technological requirements of sewage water treatment (Fig. 1) the oxidizing cycle takes 385 minutes. In the first 110 minutes the sewage water does not arrive and the microorganisms are starving. In this period the active sludge biomass is 8 g/l owing to the sludge coming back from secondary sedimentation tanks. The sewage water arrives for the next 110 minutes, since that moment the active sludge biomass reduces to 2.7 g/l to remain further constant (Fig. 2, curve 1).

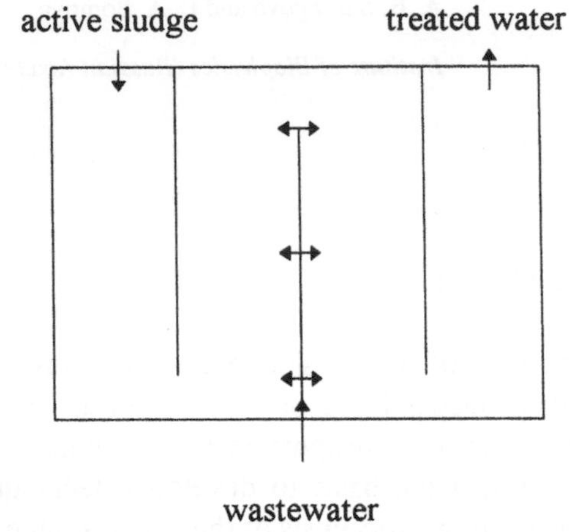

Fig. 1. Sampling procedure in treatment facility air tanks.

From BOD_5 and COD dynamics it is possible to see that in this period the major part of contaminants in the sewage water oxidizes (Fig. 2, curves 2 and 3). In the last 80 minutes the sewage water does not arrive.

Our observations show that the changes in the active sludge biomass are similar to microorganisms' catalase activity dynamics after the 55th minute, or the induction of the enzyme (Fig. 2, curves 1, 4).

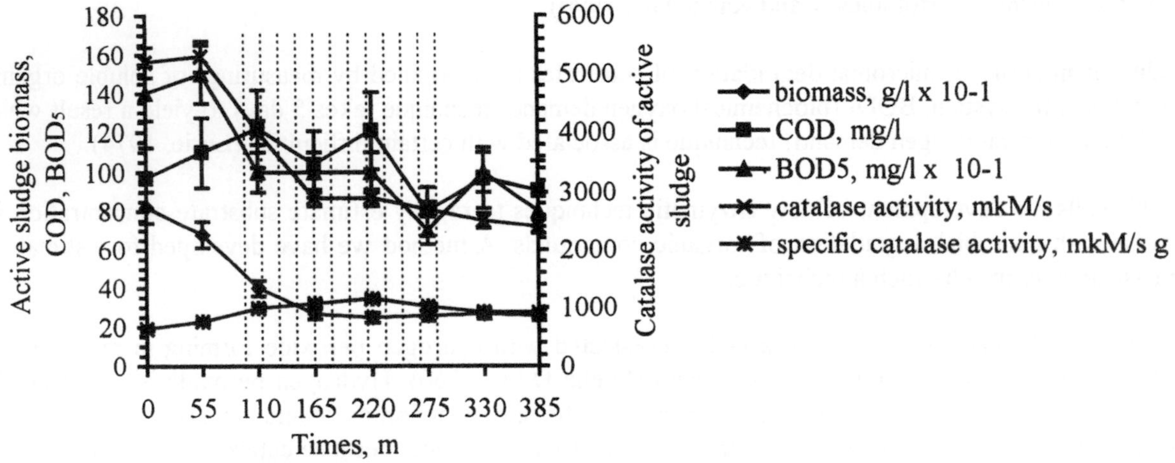

Fig. 2. Purification of sewage waters by active sludge microorganisms in aerotank.
Shadowed areas - arrival of sewage water.

Conversion of absolute values of catalase activity into specific ones revealed similarity between dynamics of specific catalase activity and that of COD (Fig. 2, curves 2, 5). During the first 110 minutes of "starvation" organic substrate concentration in the air tank increases probably due to intensive lysis processes in the active sludge and activity of catalase also increases (Fig. 2, curves 2, 5). H_2O_2 production is metabolism intensive, the increasing concentration of organic contaminants is likely to increase the turnover rate of the enzyme. After the substrate is oxidized (Fig. 2, curve 2) and supply of the sewage water is terminated, the enzyme turnover rate decreases and specific activity of the catalase smoothly decreases (Fig. 2, curve 5).

METABOLIC EFFECT ON CATALASE ACTIVITY

During "starvation" microorganisms grow on lysis products, even though their biomass decreases dissolved with the sewage water (Fig. 2, curve 1). Specific activity of the catalase in microorganisms increases and at the final stage of purification (at 220 min) is twice as large as the specific activity of the catalase of "starving" microorganisms (Fig. 2, curve 5).

Specific activity of the catalase of microorganisms in the presence of lysis products and without them during the "starvation" period and at the final stage of purification practically do not differ. Substantial difference was observed during arrival of the sewage water into the aerotank, the specific catalase of the active sludge in the absence of lysis products being considerably less than in the active sludge in the air tank proper (Fig. 3, curves 1, 2).

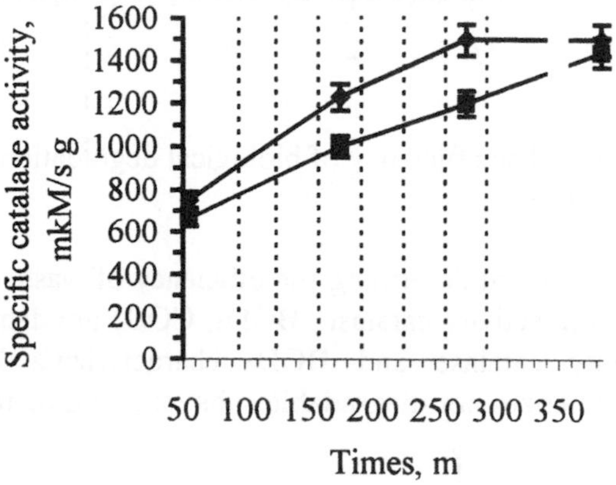

Fig. 3. Effect of lysis products on specific catalase activity of active sludge microorganisms in aerotanks.
1 - catalase activity of microorganisms in the presence of lysis products, mkM·s/g,
2 - catalase activity of microorganisms in the absence of lysis products, mkM·s/g
Shadowed areas - arrival of sewage water

Under chemostat conditions the biomass was growing in accordance with the supplied substrate (Varfolomeyev S.D. and Kalyuzhny S.V., 1990), hence the catalase is initiated in proportion to the biomass of microorganisms (Fig. 2, curves 1, 2). Dynamics of specific activity of the microorganism catalase in the absence of exometabolites specifies induction of this enzyme in connection with increasing biomass of the active sludge (Fig. 3, curve 1). Arrival of organic substrate in the air tank makes increase the rate of catalase turnover according to the amount of H_2O_2 forming in the cell (Fig. 3, curve 2).

The material obtained formed the basis to deduce an equation to evaluate the amount of the substrate consumed by active sludge microorganisms with account of specific activity of catalase in relative units:

$$A_c/A_p = (S_o - S)/S_o$$

where A_c is specific activity of the active sludge catalase in absence of exometabolites,
A_p is specific activity of the active sludge catalase in their presence,
S_o is concentration of the arriving substrate,
S is residual concentration of the substrate at the air tank output.

Field observation over the process of microbial degradation of substrates in the aerotank lend support to the validity of this statement. The correlation coefficient of the method proposed with BOD_5 is 95%.

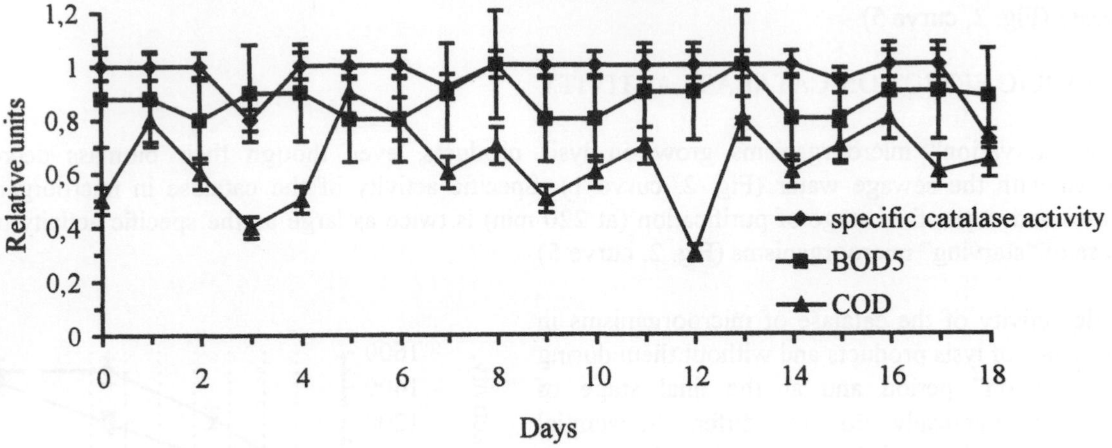

Fig. 4. The efficiency of biological degradation of substrates by active sludge of municipal purification facilities

The results of determing the efficiency of waste water biological treatment with the help of three methods (specific activity catalase, BOD_5, COD) are demonstrates (Fig. 4.). There is correlation between specific activity catalase and BOD_5 characterizing biological degradation of organic compounds. COD characterizes non-degradable substrates and therefore has no correlation with specific activity catalase.

CONCLUSION

Experimental data and field observations made possible to establish relationship between the activity of catalase of microorganisms and the organic substrate concentration. Feasibility to evaluate intensity of microbial degradation of contaminants by catalase activity of microorganisms in air tank active sludge has been shown. Application of the method proposed provides rapid evaluation (15-20 min) of the process of biological purification of sewage waters and offers wide prospects for monitoring operation of biological purification facilities. Besides, the method allows to estimate the reducer component functioning within small closed ecosystems (Sarangova A.B., Somova L.A., Pis'man T.I., Advances in Space Research (in press)).

REFERENCES:

Cornish-Bowden, E., Fundamentals of Enzymatic Kinetics, Moscow: Mir, 1979, 280 p.
Lurie Yu. Yu., Chemical Analysis of Industrial Waste Waters, Moscow: 1974. (in Russian)
Mikhlin D.M., Biochemistry of Cell Respiration, Moscow: USSR Academy of Science, 1960, 414 p. (in Russian)
Sarangova A.B., Somova L.A., Catalase Activity of Microorganisms in Small Ecosystems. *AMSE Transactions "Scientific Siberian"*, A, Vol. 14, AMSE Press, 1994, pp. 125-139.
Varfolomeyev S.D., Kalyuzhny S.V., Biotechnology: Kinetic Fundamentals of Microbiological Processes, Moscow: Vysshaya Shkola, 1990, 296 p. (in Russian)

AUTHOR INDEX